THE CONFERENCE ON
L-FUNCTIONS

THE CONFERENCE ON
L-FUNCTIONS

Fukuoka, Japan 18 – 23 February 2006

Editors

Lin Weng

Masanobu Kaneko

Kyushu University, Fukuoka, Japan

World Scientific

NEW JERSEY • LONDON • SINGAPORE • BEIJING • SHANGHAI • HONG KONG • TAIPEI • CHENNAI

Published by

World Scientific Publishing Co. Pte. Ltd.
5 Toh Tuck Link, Singapore 596224
USA office: 27 Warren Street, Suite 401-402, Hackensack, NJ 07601
UK office: 57 Shelton Street, Covent Garden, London WC2H 9HE

British Library Cataloguing-in-Publication Data
A catalogue record for this book is available from the British Library.

THE CONFERENCE ON *L*-FUNCTIONS

ISBN 13 978-981-270-504-4
ISBN 10 981-270-504-X

Printed in Singapore by World Scientific Printers (S) Pte Ltd

Preface

From February 18 to 23, 2006, with the support from our distinguished speakers and participants, the Conference on L-Functions was successfully held at Nishijin Plaza, Fukuoka, Japan. This volume is an invaluable limited collection of the works related to the conference.

The aim of the conference is to provide a common platform for mathematicians working on L-functions to communicate with each other more effectively. To a certain degree, this has been fulfilled: More than 130 participants attended the conference. We are also quite encouraged by the fact that many many young researchers and students were present. In fact, much more has been achieved – Not only our lectures cover numerous topics related to L-functions, they also reflect the current, most advanced and extremely important aspects of L-functions.

Each volume should send its clear message to the readers. Ours is as follows: On one hand, the theory of L-functions has already become very much diverse and is naturally embedded into various branches of mathematics; on the other hand, the theory is also pretty much integrated. Accordingly, in this volume we collect the papers of Bruggeman, Ibukiyama and Kim related to automorphic forms; the algebraic theory of Hida on $\mathcal{L}$-invariant of p-adic L-functions; the analytic results of Jutila, Motohashi and Suzuki on the circle method, Sieve method and the Riemann Hypothesis of rank 3 zeta; the conjecture of Kaneko on multiple zeta values; the articles of Matsumoto *et al.* on zeta functions of root systems and of Wakayama *et al.*'s on spectral zeta functions; the paper of Murty on a special class of L-functions, and the geometric approach of Weng to L-functions.

The conference was sponsored by JSPS and a COE program at the Graduate School of Mathematics, Kyushu University, Japan in relation with Akito Futaki, Ryoichi Kobayashi, Iku Nakamura, and Mitsuhiro T. Nakao, among others. We express our gratitude to them. Clearly, without the help from a large group of students, the conference would not run so smoothly. Thank them all, our dear young fellows.

The Editors

List of Participants

S. Akiyama, Niigata University
H. Aoki, Tokyo University of Science
N. Aoki, Rikkyo University
R. Bruggeman, Universiteit Utrecht
D. Burns, King's College London
J. Choy, Korea Institute for Advanced Study
N. Dan, The Romanian Academy
S. Djoko, Kyushu University
S. Egami, Toyama University
I. Fesenko, Nottingham University
M. Furusawa, Osaka City University
H. Gangl, Durham University
Y. Gon, Kyushu University
T. Goto, Tokyo University of Science
K. Gunji, Tokyo University
Y. Hamahata, Tokyo University of Science
M. Hanamura, Tohoku University
S. Harada, Kyushu University
S. Haran, Israel Institute of Technology
Y. Hashimoto, Kyushu University
T. Hayashi, Kyushu University
T. Hibino, Toua University
H. Hida, UCLA
K. Hiraga, Kyoto University
M. Hirano, Ehime University
T. Hiranouchi, Kyushu University
S. Hirayama, Kyushu University
T. Ibukiyama, Osaka University
Y. Ichihara, Hiroshima University
A. Ichino, Osaka City University
K. Ihara, Nagoya University
T. Ikeda, Kyoto University
N. Ishii, Osaka Pref. University
Y. Ishikawa, Okayama University
K. Itakura, Tokyo University
T. Ito, Kyoto University
D. Jeon, Korea Institute for Advanced Study
D. Jiang, University of Minnesota
B. Jun, Korea Institute for Advanced Study
M. Jutila, University of Turku
J. Kajikawa, Kyushu University
M. Kaneko, Kyushu University
S. Kang, Korea Institute for Advanced Study
T. Kashio, Osaka University
Y. Kato, Tohoku University
Y. Kawamoto, Kyushu University
H. Kawamura, Hokkaido University
H. Kim, University of Toronto
K. Kimoto, University of the Ryukyus
M. Koike, Kyushu University
T. Komatsu, Kyushu University
K. Konno, Fukuoka University of Education
T. Konno, Kyushu University
R. Kozuma, Kyushu University
M. Kurihara, Keio University
J. Lagarias, University of Michigan
E. Lapid, Hebrew University
J. Lee, Hongik University
M. Lee, Korea Institute for Advanced Study
Y. Lee, Seoul National University
T. Machide, Hokkaido University
Y. Maeda, Hokkaido University
K. Matsumoto, Nagoya University
S. Matsumoto, Kyushu University
K. Matsuno, Tokyo Metropolitan University
M. Mera, Kyushu University
T. Miyazaki, Keio University
Y. Mizuno, Osaka University
Y. Mizusawa, Sophia University
S. Mochizuki, Tokyo University
D. Morishita, Nagoya University
M. Morishita, Kyushu University
Y. Motohashi, Nihon University
S. Muneta, Kyushu University
K. Murty, University of Toronto
H. Naito, Kagawa University
T. Nakahara, Saga University
S. Nakajima, Gakushuin University
D. Nakajo, Kyushu University
T. Nakamura, Nagoya University
Y. Nanbara, Kyushu University
H. Ochiai, Nagoya University

T. Oda, University of Tokyo
T. Okazaki, Osaka University
K. Onodera, Keio University
K. Ota, Kyushu University
M. Ozaki, Kinki University
Y. Ozeki, Kyushu University
K. Prasanna, UCLA
N. Ramachandran, University of Maryland
Y. Sakai, Kyushu University
F. Sato, Rikkyo University
M. Shimura, Chuo University
M. Spiess, University of Bielefeld
K. Suzuki, Kyushu University
M. Suzuki, Nagoya University
Y. Taguchi, Kyushu University
Y. Takakura, Kyushu University
A. Tanaka, Kyushu University
T. Tanaka, Kyushu University
H. Taya, Tohoku University
A. Teshima, Kyushu University
M. Tien, Tsing Hua University
H. Tokitsu, Kyushu University
L. Truelsen, University of Aarhus
T. Tsuji, Tokai University
H. Tsumura, Tokyo Metropolitan University
H. Tsutsumi, Osaka Univ. of H. & S. Science
S. Uchiyama, NTT
A. Umegaki, Sophia University
T. Umeno, Kyushu University
E. Urban, Columbia University
R. Vidunas, Kyushu University
C. Virdol, Nagoya University
I. Wakabayashi, Seikei University
S. Wakatsuki, Kyoto University
M. Wakayama, Kyushu University
S. Wakiyama, Kyushu University
L. Weng, Kyushu University
A. Yamagami, Kyoto University
M. Yamagishi, Nagoya Inst. of Technology
K. Yamanohira, Kyushu University
Y. Yamasaki, Kyushu University
T. Yamazaki, Tsukuba University
S. Yasuda, Kyoto University
T. Yasuda, Kyushu University
H. Yoshida, Kyoto University
M. Yoshida, Kyushu University
J. Yu, Tsing Hua University
D. Zagier, MPIM Bonn

Contents

Quantum Maass Forms

R. BRUGGEMAN

In this paper, we propose a definition of the concept of quantum Maass forms. On the basis of this definition, we discuss the relation with Maass forms and with cohomology groups.

In [9], D.Zagier defined the function $\varphi : \mathbb{Q} \to \mathbb{C}$ by

$$\varphi(\xi) = e^{\pi i \xi/12} \sum_{n=0}^{\infty} \prod_{j=1}^{n} \left(1 - e^{2\pi i j \xi}\right) . \tag{0.1}$$

For $\gamma = \pm \begin{pmatrix} a & b \\ c & d \end{pmatrix} \in \mathrm{PSL}_2(\mathbb{Z})$, $c > 0$, this function satisfies

$$\varphi(\xi) - v(\gamma)(c\xi + d)^{-3/2}\varphi(\xi) = g_\gamma(\xi) , \tag{0.2}$$

with a multiplier system v of weight $-\frac{3}{2}$. The function g_γ is the restriction to $\mathbb{Q}$ of an element of $C^\infty(\mathbb{R})$.

The name *quantum modular forms* is used for functions on $\mathbb{Q}$ that satisfy a modular transformation property modulo simpler functions on $\mathbb{R}$, in Zagier's example the smooth functions g_γ.

The word "quantum" has no physical meaning here. Quantum modular forms are not modular forms, but have some relation to holomorphic modular forms (to the Dedekind eta function in Zagier's example).

The subject of this paper is the definition of *quantum Maass forms*, related to $\mathrm{PSL}_2(\mathbb{Z})$-invariant real analytic eigenfunction of the Laplace operator on the upper half plane that are not holomorphic. A definition is proposed in §1.1.5, and extended in §1.2.2. Section 2 gives results that suggest that these definitions are sensible.

Acknowledgements. Don Zagier introduced me to quantum modular forms and quantum Maass forms. The results in this paper are closely connected to results on the relation between Maass forms and cohomology, in a project of John Lewis, Don Zagier and me. I have profited from Tobias Mühlenbruch's comments on a previous version of this paper.

It was a pleasure to attend the conference on *L*-functions in Fukuoka, where I presented these results. I am grateful for the invitation.

1 Quantum Maass forms associated to Maass cusp forms and Eisenstein series

We work with the discrete subgroup $\Gamma = \mathrm{PSL}_2(\mathbb{Z})$ of $G = \mathrm{PSL}_2(\mathbb{R})$. By $\left[\begin{smallmatrix} a & b \\ c & d \end{smallmatrix}\right]$ we denote $\pm\left(\begin{smallmatrix} a & b \\ c & d \end{smallmatrix}\right) \in G$.

The space $\mathcal{E}_s^\Gamma$ consists of the complex valued functions u on the upper half plane $\mathfrak{H}$ that satisfy

(i) Γ-*invariance:* $u(\gamma z) = u(z)$ for all $\gamma \in \Gamma$.

(ii) *Eigenfunction of Laplace operator:* $-y^2\left(\partial_x^2 + \partial_y^2\right)u = s(1-s)u$ for some $s \in \mathbb{C}$, the *spectral parameter*.

Our main interested is in the case $0 < \operatorname{Re} s < 1$. For convenience, we assume $s \neq \frac{1}{2}$.

The space $\mathcal{E}_s^\Gamma$ has infinite dimension. The finite dimensional subspace space Mf_s of *Maass forms* is defined by the condition

(iii) *Polynomial growth:* $u(z) = O\left(y^A\right)$ as $y \to \infty$ for some $A > 0$.

The space $\mathsf{Mf}_s^0 \subset \mathsf{Mf}_s$ of *Maass cusp forms* consists of u satisfying

(iii′) *Quick decay:* $u(z) = O\left(y^{-A}\right)$ as $y \to \infty$ for all $A > 0$.

1.1 Quantum Maass forms associated to Maass cusp forms

The space Mf_s^0 is non-zero only for a discrete set of $s \in \frac{1}{2} + i\mathbb{R}$. Each $u \in \mathsf{Mf}_s^0$ has a Fourier expansion

$$(1.1) \qquad u(z) = \sum_{\neq 0} a_n \sqrt{y} K_{s-1/2}(2\pi|n|y) e^{2\pi i n x}.$$

A Maass cusp form is *even* if $a_{-n} = a_n$ for all n, and *odd* if $a_{-n} = -a_n$. This gives a decomposition $\mathsf{Mf}^0_s = \mathsf{Mf}^{0,+}_s \oplus \mathsf{Mf}^{0,-}_s$, where $\mathsf{Mf}^{0,+}_s$, respectively $\mathsf{Mf}^{0,-}_s$, is the subspace of even, respectively odd, Maass cusp forms.

The main theorem in [3] associates to each $u \in \mathsf{Mf}^{0,\pm}_s$ objects of three other types:

(a) (See Maass, Chap. IV & V of [5].) With a_n as in (1.1), the *L-function* is $L(\rho) = 2\sum_{n=1}^{\infty} a_n n^{-\rho}$. Let $\gamma_s(\rho) = \frac{1}{4}\pi^{-\rho}\Gamma\left(\frac{\rho-s+\frac{1}{2}}{2}\right)\Gamma\left(\frac{\rho+s-\frac{1}{2}}{2}\right)$. The completed L-function, $L^*(\rho) = \gamma_s\left(\rho + \frac{1\mp 1}{2}\right)L(\rho)$, has a holomorphic continuation to $\rho \in \mathbb{C}$, and satisfies $L^*(1-\rho) = \pm L^*(\rho)$.

(b) The *periodic function* on $\mathbb{C} \smallsetminus \mathbb{R}$

$$(1.2)\qquad f(\tau) = \frac{1}{2}\pi^s\Gamma(1-s)\cdot\begin{cases}\sum_{n=1}^{\infty} n^{s-\frac{1}{2}} a_n e^{2\pi i n\tau} & \text{if } \tau \in \mathfrak{H}, \\ \mp\sum_{n=1}^{\infty} n^{s-\frac{1}{2}} a_n e^{-2\pi i n\tau} & \text{if } \tau \in \mathfrak{H}^-, \end{cases}$$

is holomorphic and 1-periodic. It is not Γ-invariant.

(c) The *period function*

$$\psi(\tau) = f(\tau) - \tau^{-2s} f(-1/\tau),$$

which extends holomorphically to $\mathbb{C}' = \mathbb{C} \smallsetminus (-\infty, 0]$. It satisfies (1.3)

$$\psi(\tau) = \psi(\tau+1) + (\tau+1)^{-2s}\psi\left(\frac{\tau}{\tau+1}\right), \qquad \tau^{-2s}\psi\left(\frac{1}{\tau}\right) = \pm\psi(\tau).$$

The following theorem is not explicitly stated in [3]. Don Zagier pointed it out to me.

Theorem 1. *The periodic function f associated to a Maass cusp form can be extended to $\mathbb{Q}$ such that for all $\xi \in \mathbb{Q}$:*

$$(1.4)\qquad f(\tau) = f(\xi) + o(1) \qquad (\tau \xrightarrow{\text{ga}} \xi).$$

It satisfies

$$f(\xi+1) = f(\xi) \qquad \textit{for all } \xi \in \mathbb{Q},$$
$$\xi^{-2s} f\left(\frac{-1}{\xi}\right) = f(\xi) - \psi(\xi) \qquad \textit{for all } \xi \in \mathbb{Q} \cap (0,\infty).$$

Geodesic approach. Here $\tau \stackrel{\text{ga}}{\to} \xi$ means that τ approaches ξ along a geodesic in $\mathfrak{H}$ or $\mathfrak{H}^-$. So τ tends to ξ along a piece of a vertical line through ξ, or along a euclidean circle with its center on the real axis. The implicit constant in $o(1)$ may depend on the geodesic.

Proof. In §I.4 of [3], the integral representation

$$f(\tau) = \frac{1}{2\pi i} \int_{\operatorname{Re}\rho = C} \tilde{f}_\pm(\rho) e^{\pm i\pi\rho/2} \tau^{-\rho}\, d\rho$$

is given for $\pm\operatorname{Im}\tau > 0$, with $C > 0$. Moving the line of integration to $\operatorname{Re}\rho = -\varepsilon$ for some small $\varepsilon > 0$, we pick up a residue at $\rho = 0$. The residue gives a constant $f(0)$, which can be expressed in terms of the L-function; see (1.16b) in [3]. An estimate of the L-function shows that the integral over $\operatorname{Re}\rho = -\varepsilon$ contributes $o(1)$ as $\tau \stackrel{\text{ga}}{\to} 0$. This gives (1.4) for $\xi = 0$. Repeatedly applying the periodicity and the relation $\psi(\tau) = f(\tau) - \tau^{-2s} f(-1/\tau)$, gives (1.4) at all $\xi \in \mathbb{Q}$. We use that $\tau \mapsto \tau + 1$ and $\tau \mapsto \frac{-1}{\tau}$ map geodesics to geodesics. □

1.1.1 Explicit formula

A direct extension of the reasoning in [3] gives an expression in terms of the twisted L-series

$$L_o(\rho, \xi) = \sum_{n \neq 0} n a_n e^{2\pi i n \xi} |n|^{1-\rho}, \tag{1.5}$$

which has an analytic continuation and a functional equation. For $\xi = \frac{a}{c}$, $a, c \in \mathbb{Z}$, $c > 0$, $(a, c) = 1$, $a\tilde{a} \equiv 1 \bmod c$:

$$f\left(\frac{a}{c}\right) = -\frac{1}{4} \pi^{-s-\frac{1}{2}} \Gamma\left(s + \frac{1}{2}\right) c^{2s} L_0\left(s + \frac{1}{2}, -\frac{\tilde{a}}{c}\right), \tag{1.6}$$

One needs the meromorphic continuation of both L_o and $L_e(\rho, \xi) = \sum_{n \neq 0} a_n e^{2\pi i n \xi} |n|^{-\rho}$. This continuation is obtained from integrals of $u(\xi + iy)$ and $\partial_x u(\xi + iy)$ over $y \in (0, \infty)$. The functional equations relate these L-series for $\xi = \frac{a}{c}$ and $\xi = \frac{\tilde{a}}{c}$, and imply that $L_e\left(\pm(s - \frac{1}{2}), \xi\right) = 0$ for all $\xi \in \mathbb{Q}$.

1.1.2 Automorphic distribution

As we shall discuss in §2.1, the periodic function f represents a hyperfunction on $\mathbb{R}$, which for f coming from a Maass cusp form is actually a distribution. The limit behavior in Theorem 1 indicates that this distribution is rather regular. Schmid and Miller, [7], [6], have studied this distribution as the distribution derivative of a continuous function.

1.1.3 Smooth extension of the period function

The proposition in §III.3, [3] implies that there is $\tilde{\psi} \in C^\infty(\mathbb{R})$ such that $\psi(x) = \tilde{\psi}(x)$ for $x > 0$, and $|x|^{-2s}\tilde{\psi}(-1/x) = \tilde{\psi}(x)$ for $x \in \mathbb{R} \setminus \{0\}$. So $\tilde{\psi}$ is real analytic on $\mathbb{R} \setminus \{0\}$. At $x < 0$, the value of $\tilde{\psi}(x)$ has no direct relation to the values of $\psi(\tau)$ with $\tau \in \mathbb{C}'$ near x.

We have $|\xi|^{-2s} f(-1/\xi) = f(\xi) - \tilde{\psi}(\xi)$ for almost all $\xi \in \mathbb{Q}$.

1.1.4 Principal series

This leads to a reformulation in terms of the smooth vectors in the principal series representation with spectral parameter s.

Let $\mathcal{V}_s^\infty$ be the space of $f \in C^\infty(\mathbb{R})$ for which $x \mapsto |x|^{-2s} f(-1/x)$ extends as an element of $C^\infty(\mathbb{R})$. This space is invariant under the following right action of G:

$$(1.7) \qquad (f|_{2s}g)(x) = |cx+d|^{-2s} f(gx) \qquad \text{for } g = \begin{bmatrix} a & b \\ c & d \end{bmatrix} \in G.$$

The extended period function $\tilde{\psi}$ is an element of $\mathcal{V}_s^\infty$. There is a 1-cocycle $\gamma \mapsto c_\gamma$ on Γ with values in $\mathcal{V}_s^\infty$ such that for all $\gamma \in \Gamma$

$$(1.8) \qquad f(\xi) - (f|_{2s}\gamma)(\xi) = c_\gamma(\xi) \qquad \text{for almost all } \xi \in \mathbb{Q}.$$

So indeed, the function on $\mathbb{Q}$ induced by the periodic function satisfies a modular transformation property, modulo an element of $\mathcal{V}_s^\infty$.

Actually, $\tilde{\psi}$ is real analytic outside 0 and ∞. So the cocycle c takes values in the G-invariant subspace $\mathcal{V}_s^{\omega,\infty}$ of semi-analytic vectors in $\mathcal{V}_s^\infty$: vectors that are real analytic except at finitely many points. A vector f is real analytic at ∞ if $x \mapsto |x|^{-2s} f(-1/x)$ is given by a power series in $\frac{-1}{x}$ that converges on a neighborhood of $\frac{-1}{x} = 0$.

1.1.5 Definition of quantum Maass forms

Let $\mathcal{R}_s$ be the space of (equivalence classes of) functions $\mathbb{Q} \smallsetminus E \to \mathbb{C}$ where $E \subset \mathbb{Q}$ is a finite set. We consider two such functions to be equal in $\mathcal{R}_s$ if their difference is zero outside a finite subset of $\mathbb{Q}$. The formula (1.7) defines a right action of Γ in R_s. By restriction, $\mathcal{V}_s^{\omega,\infty}$ determines a Γ-submodule of $\mathcal{R}_s$. Define the Γ-module Q_s as the quotient:

$$0 \longrightarrow \mathcal{V}_s^{\omega,\infty} \longrightarrow \mathcal{R}_s \longrightarrow Q_s \longrightarrow 0. \tag{1.9}$$

The periodic function f coming from the Maass cusp form u determines an element of $\mathcal{R}_s$. Its class in Q_s is Γ-invariant. The function f is not in $\mathcal{R}_s^\Gamma$. Actually, $\mathcal{R}_s^\Gamma \cong \mathbb{C}$ is not very interesting. Let us define the space of quantum Maass forms with spectral parameter s as

$$\mathsf{qMf} := Q_s^\Gamma / \mathcal{R}_s^\Gamma. \tag{1.10}$$

Theorem 1 shows that there is a map $\mathbf{q}_s : \mathsf{Mf}_s^0 \to \mathsf{qMf}_s$.

1.1.6 Map to the semi-analytic cohomology

To (1.9) is associated a long exact sequence, of which we use the following part:

$$0 \longrightarrow 0 \longrightarrow \mathcal{R}_s^\Gamma \longrightarrow Q_s^\Gamma \longrightarrow H^1(\Gamma, \mathcal{V}_s^\infty). \tag{1.11}$$

(The space $(\mathcal{V}_s^\infty)^\Gamma$ is zero.) This implies that there is an injective map $\mathbf{c}_s : \mathsf{qMf}_s \to H^1(\Gamma, \mathcal{V}_s^{\omega,\infty})$.

In [4] it is shown that there is an injective map $\mathbf{r}_s : \mathsf{Mf}_s^0 \to H^1(\Gamma, \mathcal{V}_s^{\omega,\infty})$. The image is the parabolic cohomology subgroup $H^1_{\text{par}}(\Gamma, \mathcal{V}_s^{\omega,\infty})$. (For the modular group $\Gamma = \mathrm{PSL}_2(\mathbb{Z})$, the parabolic cohomology group $H^1_{\text{par}}(\Gamma, A)$ can be defined as the subgroup of classes in $H^1(\Gamma, A)$ that can be represented by a cocycle c that satisfies $c_T = 0$ for $T = \begin{bmatrix} 1 & 1 \\ 0 & 1 \end{bmatrix} \in \Gamma$.)

It turns out that $\mathbf{r}_s$ is equal to the composition $\mathbf{c}_s \circ \mathbf{q}_s$. So the quantum Maass forms constitute a space through which we can factorize the map from Maass cusp forms to cohomology. The injectivity of $\mathbf{r}_s$ implies that $\mathbf{q}_s$ is injective.

1.2 Eisenstein series, extended quantum Maass forms

The Eisenstein series

$$(1.12)\qquad \begin{aligned} E_s(z) &= \Lambda(2s)y^s + \Lambda(2s-1)y^{1-s} \\ &\quad + 2\sum_{n\neq 0} |n|^{\frac{1}{2}-s}\sigma_{2s-1}(|n|)\sqrt{y}K_{s-1/2}(2\pi|n|y)e^{2\pi inx}\,, \\ \Lambda(u) &= \pi^{-\frac{u}{2}}\Gamma\left(\frac{u}{2}\right)\zeta(u)\,, \qquad \sigma_u(n) = \sum_{d|n} d^u \end{aligned}$$

gives a meromorphic family of elements of Mf_s, complementary to the cusp forms. It is holomorphic in the region $0 < \operatorname{Re} s < 1$, $s \neq \frac{1}{2}$.

A modification of the theory of Lewis and Zagier works for the Eisenstein series; see §IV.1, [3]. The holomorphic periodic function associated to E_s can be chosen as

$$(1.13)\quad f_s(\tau) = \pm\frac{\sqrt{\pi}\Gamma(1-s)\Lambda(2s)}{2\Gamma\left(\frac{1}{2}-s\right)} \pm \pi^s\Gamma(1-s)\sum_{n=1}^{\infty}\sigma_{2s-1}(n)e^{\pm 2\pi in\tau}\,,$$

with the convention $\pm 1 = \operatorname{sign}\operatorname{Im}\tau$. The period function $\psi_s(\tau) = f_s(\tau) - \tau^{-2s}f_s(-1/\tau)$ is holomorphic on $\mathbb{C}'$. However, it has no smooth extension through 0. The function $\tilde{\psi}_s$ defined by $\tilde{\psi}_s(x) = \psi_s(x)$ for $x > 0$, and $\tilde{\psi}_s(x) = |x|^{-2s}\psi_s(-1/x)$ is real analytic on $\mathbb{R}\smallsetminus\{0\}$, but it is not an element of $\mathcal{V}_s^{\omega,\infty}$.

The expression (1.13) is explicit, and implies that at the rational point $\frac{a}{c}$, $a, c \in \mathbb{Z}$, $c > 0$, $(a,c) = 1$, we do not have a limit value, but an expansion:

(1.14)

$$\begin{aligned} f_s(\tau) &= \frac{i}{2}c^{2s-2}\Lambda(2s-1)\frac{1}{\tau-\frac{a}{c}} \\ &\quad \pm\frac{1}{2}\pi^{-1/2}\Gamma(1-s)\Gamma\left(s+\frac{1}{2}\right)c^{-2s}\Lambda(2s)\left(\mp i\left(\tau-\frac{a}{c}\right)\right)^{-2s} \\ &\quad - i\pi^{\frac{1}{2}-s}\Gamma\left(s+\frac{1}{2}\right)\sum_{x=1}^{c-1}\left(\!\left(\frac{dx}{c}\right)\!\right)\zeta(2s,x/c) + o(1) \qquad \tau \xrightarrow{\text{ga}} \frac{a}{c}\,, \end{aligned}$$

where $\pm 1 = \sin\operatorname{Im}\tau$, and where $\zeta(\cdot,\cdot)$ denotes the Hurwitz zeta function. The computations are similar to those for cusp forms. We use the fact that the twisted L-series associated to (1.13) have expressions in terms of the Hurwitz zeta function.

The example of the Eisenstein series shows that there are two difficulties: (1) There are no limit values, but expansions. (2) The cocycle has values in a larger space than $\mathcal{V}_s^{\omega,\infty}$.

If we do not want to end the study of quantum Maass forms here, generalizations are needed.

1.2.1 Type of quantum Maass forms

$\mathcal{V}_s^{\omega,\infty}$ consists of the semi-analytic vectors in $\mathcal{V}^\infty$. It contains the G-invariant subspace $\mathcal{V}_s^\omega$ of vectors that are analytic everywhere. Analytic means analytic on $\mathbb{R}$, and $x \mapsto |x|^{-2s} f(-1/x)$ given by a power series in $-1/x$ converging if $|x|$ is sufficiently large. Another space, containing $\mathcal{V}_s^{\omega,\infty}$, is $\mathcal{V}_s^{\omega,\mathrm{fs}}$ consisting of vectors that are analytic except at finitely many points. The function $\tilde{\psi}_s$ for Eisenstein series is in $\mathcal{V}_s^{\omega,\mathrm{fs}}$, and $\tilde{\psi}$ for cusp forms is in $\mathcal{V}_s^{\omega,\infty} = \mathcal{V}_s^\infty \cap \mathcal{V}_s^{\omega,\mathrm{fs}}$.

Let t be one of ω, (ω,∞), (ω,fs), or let t denote another Γ-invariant space of functions inside the principal series with spectral parameter s. Then we define Q_s^t as the quotient in

$$0 \longrightarrow \mathcal{V}_s^t \longrightarrow \mathcal{R}_s \longrightarrow Q_s^t \longrightarrow 0\,. \tag{1.15}$$

We define the space of quantum Maass forms of type t as $\mathsf{qMf}_s^t = \left(Q_s^t\right)^\Gamma / \mathcal{R}_s^\Gamma$. In §1.1, we have omitted $t = (\omega,\infty)$ from the notations.

The inclusions between space of vectors in the principal series induces the following natural maps

$$\begin{array}{ccccc} & & & \nearrow & \mathsf{qMf}_s^{\omega,\mathrm{fs}} \\ \mathsf{qMf}_s^\omega & \longrightarrow & \mathsf{qMf}_s^{\omega,\infty} & & \\ & & & \searrow & \mathsf{qMf}_s^\infty \end{array}$$

1.2.2 Extended quantum Maass forms

Let $\tilde{\mathcal{R}}_s$ be the space of systems p that associate to almost all $\xi \in \mathbb{Q}$ an expansion $p(\xi,\tau)$ holomorphic in τ near ξ. Two such systems of expansions are equal if $p_1(\xi,\tau) = p_2(\xi,\tau) + o(1)$ $(\tau \xrightarrow{\text{ga}} \xi)$ for almost all ξ. A constant is a holomorphic function, so $\mathcal{R}_s \subset \tilde{\mathcal{R}}_s$. The group Γ acts in $\tilde{\mathcal{R}}_s$ by $(p|_{2s}\gamma)(\xi,\tau) = j_\gamma(\tau)^{-s} p(\gamma\xi, \gamma\tau)$, where

$$(1.16) \qquad j_\gamma(\tau)^{-s} = \left((c\tau+d)^2\right)^{-s} \quad \text{for } \gamma = \begin{bmatrix} a\,b \\ c\,d \end{bmatrix}$$

is a holomorphic extension of $x \mapsto |cx+d|^{-2s}$. For any type t we define $\tilde{Q}^t_s = \tilde{\mathcal{R}}_s/\mathcal{V}^t_s$. This is a huge space. We refrain from giving a name to the elements of $\left(\tilde{Q}^t_s\right)^\Gamma / \tilde{\mathcal{R}}^\Gamma_s$.

To the Eisenstein series E_s, we can associate an element of $\left(\tilde{Q}^{\omega,\text{fs}}_s\right)^\Gamma / \tilde{\mathcal{R}}_s$, represented by the system of expansions in (1.14). It turns out that the middle term in (1.14), with $\pm\,((\mp i(\tau - a/c))^{-2s}$, determines a system of expansions in $\tilde{\mathcal{R}}^\Gamma_s$. So this term may be considered as trivial and should be divided out.

It seems sensible to define $\dot{\mathcal{R}}_s \subset \tilde{\mathcal{R}}_s$ as the systems of expansions of the form

$$(1.17) \qquad p(\xi,\tau) = \frac{d_\xi}{\tau - \xi} + c_\xi + o(1) \quad (\tau \xrightarrow{\text{ga}} \xi)\,.$$

It is a Γ-invariant subspace. We define $\dot{Q}^t_s = \dot{\mathcal{R}}_s/\mathcal{V}^t_s$, and call elements of $\mathsf{eqMf}^t_s = \left(\dot{Q}^t_s\right)^\Gamma / \dot{\mathcal{R}}^\Gamma_s$ *extended quantum Maass forms* of *type t*. If necessary, we shall call elements of qMf^t_s *genuine* quantum Maass forms. By definition, there is an injective map $\mathbf{c}^t_s : \mathsf{eqMf}^t_s \to H^1(\Gamma, \mathcal{V}^t_s)$.

We have seen that there is a map $\mathbf{q}^{\omega,\text{fs}}_s : \mathsf{Mf}_s \to \mathsf{eqMf}^{\omega,\text{fs}}_s$.

2 Quantum Maass forms associated to invariant eigenfunctions

The aim is to attach extended quantum Maass forms to all invariant eigenfunctions in $\mathcal{E}^\Gamma_s$.

2.1 Invariant hyperfunctions

Up till now, we have used the line model of the principal series. When dealing with hyperfunctions it is more comfortable to use the projective model. In that model, $\mathcal{V}_s^\omega$ and $\mathcal{V}_s^\infty$ correspond to analytic, respectively smooth, functions on $\mathbb{P}^1_\mathbb{R}$. The transformation behavior is

(2.1)
$$f(x)|_{2s}^{\mathrm{pr}} g(x) = j_g^{\mathrm{pr}}(x)^{-s} f(gx)\,, \qquad j^{\mathrm{pr}}_{\left[\begin{smallmatrix} a & b \\ c & d \end{smallmatrix}\right]}(x) = \frac{(ax+b)^2 + (cx+d)^2}{1+x^2}\,.$$

If f is in the line model, then $f^{\mathrm{pr}}(x) = \left(x^2+1\right)^s f(x)$ is in the projective model.

The results of §1 can be formulated in the projective model: Principal series functions have to be multiplied by $\left(1+x^2\right)^s$. To an (extended) quantum Maass form represented by p we associate the corresponding object in the projective model, given by $p^{\mathrm{pr}}(\xi,\tau) = \left(1+\tau^2\right)^s p(\xi,\tau)$. When dealing with p^{pr}, there is no problem in also considering $p(\infty,\tau)$. We assume that from now on, this reformulation in the projective model has been done. A drawback of the projective model is that the simple relation $f(\xi+1) = f(\xi)$ in Theorem 1 becomes $\left(\frac{1+\xi^2}{1+(\xi+1)^2}\right)^s f^{\mathrm{pr}}(\xi+1) = f^{\mathrm{pr}}(\xi)$.

Let $\mathcal{E}_s$ be the space of all solutions of $-y^2\left(\partial_x^2+\partial_y^2\right)u = s(1-s)u$. Helgason, [2], has shown that there is an isomorphism of G-spaces $\Pi_s : \mathcal{V}_s^{-\omega} \to \mathcal{E}_s$, the Poisson integral. For further explanations, see [8], [1], or [4]. Under this isomorphism, $(\mathcal{V}_s^{-\omega})^\Gamma$ corresponds to $\mathcal{E}_s^\Gamma$.

Each $u \in \mathcal{E}_s^\Gamma$ has a Fourier expansion

(2.2)
$$u(z) = b_0 y^{1-s} + a_0 y^s + \sum_{n\in\mathbb{Z}\setminus\{0\}} \sqrt{y}\left(a_n K_{s-1/2}(2\pi|n|y) + b_n I_{1/2-s}(2\pi|n|y)\right) e^{2\pi i n x}\,.$$

The convergence of this series implies that $b_n = O\left(e^{-A|n|}\right)$ for each $A > 0$, and $a_n = O\left(e^{\varepsilon|n|}\right)$ for each $\varepsilon > 0$.

The hyperfunction α_u such that $\Pi_s\alpha_u = u$ is represented on $\mathbb{R}$ by the function $f_u^{\mathrm{pr}}(\tau) = \left(1+\tau^2\right)^s f_u(\tau)$, where f_u is given by

(2.3)
$$f_u(\tau) = \pm\frac{\sqrt{\pi}\,\Gamma(1-s)}{2\Gamma\left(\frac12 - s\right)} a_0 \pm \frac{\pi^s}{2}\Gamma(1-s)\sum_{n=1}^\infty n^{s-\frac12} a_{\pm n} e^{\pm 2\pi i n\tau}\,,$$

with $\pm\mathrm{Im}\,\tau > 0$. See Proposition 2.1 in [1].

A representative g of α_u is a holomorphic function on $U \smallsetminus \mathbb{P}^1_{\mathbb{R}}$ for some neighborhood U of $\mathbb{P}^1_{\mathbb{R}}$ in $\mathbb{P}^1_{\mathbb{C}}$. Two representatives differ by a holomorphic function on a full neighborhood of $\mathbb{P}^1_{\mathbb{R}}$. That difference corresponds to an element of $\mathcal{V}_s^{\omega}$.

If $\gamma \in \Gamma$, then $f|_{2s}^{\text{pr}}\gamma$ is an other representative of α_u. So $\gamma \mapsto r_\gamma = g - g|_{2s}^{\text{pr}}\gamma$ is a cocycle with values in $\mathcal{V}_s^{\omega}$. This induces a map $\mathbf{r}_s^{\omega} : \mathcal{E}_s^{\Gamma} \to H^1(\Gamma, \mathcal{V}_s^{\omega})$, which is shown in [4] to be injective.

2.2 Quantum Maass forms of analytic type

Theorem 2. *There is an injective linear map* $\mathbf{q}_s^{\omega} : \mathcal{E}_s^{\Gamma} \to \mathsf{eqMf}_s^{\omega}$ *such that* $\mathbf{c}_s^{\omega} \circ \mathbf{q}_s^{\omega} = \mathbf{r}_s^{\omega}$.

Proof. Let $u \in \mathcal{E}_s^{\Gamma}$, $\alpha \in (\mathcal{V}_s^{-\omega})^{\Gamma}$, $u = \Pi_s\alpha$. A representative g of α determines $p \in \tilde{\mathcal{R}}_s$ by $p(\xi,\tau) = g(\tau) + o(1)$ $(\tau \overset{\text{ga}}{\to} \xi)$. The Γ-invariance of α implies that $p|_{2s}^{\text{pr}}\gamma = p + r_\gamma$, with $r_\gamma \in \mathcal{V}_s^{\omega}$. We have obtained an element of $\left(\tilde{\mathcal{Q}}_s^{\omega}\right)^{\Gamma} / \tilde{\mathcal{R}}_s^{\Gamma}$, which does not depend on the choice of the representative g. We want to subtract an element of $\tilde{\mathcal{R}}_s^{\Gamma}$ from p such that we are left with an element of $\dot{\mathcal{R}}_s$. Then we shall have an element of eqMf_s^{ω} that is mapped under $\mathbf{c}_s^{\omega}$ to $\mathbf{r}_s^{\omega}u$.

To do this, we investigate the behavior of the representative g near ∞. We use §4.3 in [1]. We can decompose α as the sum of three Γ_∞-invariant hyperfunctions, corresponding to various parts of the Fourier expansion (2.2). Γ_∞ denotes the subgroup of Γ generated by $T = \left[\begin{smallmatrix} 1 & 1 \\ 0 & 1 \end{smallmatrix}\right]$.

Lemma 4.2 gives a hyperfunction α^c corresponding to the terms with $K_{s-1/2}$ in (2.2). A representative of α^c is given by an integral involving the terms with $n \neq 0$ in f_u. Its behavior is $c_\infty + o(1)$ as $\tau \overset{\text{ga}}{\to} \infty$.

To $a_0 y^s$ corresponds a hyperfunction α^0 that is near ∞ represented by a multiple of $m_0(\infty,\tau) = \pm\frac{e^{\mp\pi i s}}{2\cos\pi s}\tau^{2s}\left(1+\tau^{-2}\right)^s$, with $\pm\text{Im}\,\tau > 0$. It turns out that this representative is invariant under T. So we can define an element of $\tilde{\mathcal{R}}_s^{\Gamma}$ by defining $m_0(\xi,\gamma\tau) = j_\gamma^{\text{pr}}(\tau)^s m_0(\infty,\tau)$, independent of the choice of $\gamma \in \Gamma$ such that $\xi = \gamma\infty$.

The remaining part α^∞ is a hyperfunction supported in $\{\infty\}$. The term $b_0 y^{1-s}$ corresponds to a hyperfunction that is near ∞ represented by a multiple of $\tau \mapsto \tau$. This has the behavior of the expansion at ∞ of an element of $\dot{\mathcal{R}}_s$.

The terms with $I_{1/2-s}$ correspond to a hyperfunction that is near ∞ represented by

$$(2.4)\quad B(\tau) = -\frac{i\pi^{\frac{1}{2}-s}}{2\Gamma\left(\frac{3}{2}-s\right)} \sum_{n\neq 0} |n|^{\frac{1}{2}-s} b_n \tau \left(1+\tau^{-2}\right)^s {}_1F_1(1;2-2s;2\pi i n\tau)\,.$$

This last contribution is not explicitly given in §4.3 in [1]; so we give more details.

First we have to check that the hyperfunction β induced by B is sent to the $I_{1/2-s}$-part of (2.2). The estimate $b_n = O\left(e^{-A|n|}\right)$ for each $A > 0$ implies the absolute convergence in (2.4). The support of β is contained in $\{\infty\}$. The Poisson map is given by an integral around ∞:

$$\Pi_s\beta(z) = \frac{1}{\pi}\int_{|\tau|=R} B(\tau)\left(\frac{y(1+\tau^2)}{(\tau-z)(\tau-\bar{z})}\right)^{1-s} \frac{d\tau}{1+\tau^2}\,,$$

with $R > 1$, $R > |z|$. After a change of variables:

$$\begin{aligned}\Pi_s\beta(z) &= \frac{-i\pi^{-1/2-s}}{2\Gamma\left(\frac{3}{2}-s\right)} \sum_{n\neq 0} |n|^{s-1/2} b_n y^{1-s} \int_{|\tau|=R} \left(1+\frac{y^2}{\tau^2}\right)^{s-1} \left(1+\frac{x}{\tau}\right)^{1-2s} \\ &\quad \cdot {}_1F_1(1;2-2s;2\pi i n(\tau+x))\,\frac{d\tau}{\tau}\end{aligned}$$

for sufficiently large R. Insert the power series expansion for ${}_1F_1$ and the binomial series for $\left(1+\frac{y^2}{\tau^2}\right)^{s-1}$ and $\left(1+\frac{x}{\tau}\right)^{1-2s}$. Some computations lead to the power series expansion of $I_{1/2-s}$, and give the desired terms in (2.2).

It turns out that $B|_{2s}^{\mathrm{pr}}T(\tau) = B(\tau) + o(1)$ as $\tau \xrightarrow{\mathrm{ga}} \infty$. To check this, we use again the power series expansion of ${}_1F_1$ and the binomial formula for powers of $1+\frac{1}{\tau}$. So the contribution of $B(\tau)$ to $p(\infty,\tau)$ can be transported to all $\xi \in \mathbb{Q}$ to give an element of $\tilde{\mathcal{R}}_s^\Gamma$.

Now we have a system of expansions $p_1 \in \tilde{\mathcal{R}}_s$ given by $p_1(\xi,\tau) = g(\tau) - m(\xi,\tau)$ with $m \in \tilde{\mathcal{R}}_s^\Gamma$, and $p(\infty,\tau) = d_\infty\tau + c_\infty + o(1)$ ($\tau \xrightarrow{\mathrm{ga}} \infty$). Let $\gamma = \left[\begin{smallmatrix} a & b \\ c & d\end{smallmatrix}\right] \in \Gamma$ such that $\xi = \gamma\infty \in \mathbb{Q}$. As $\tau \xrightarrow{\mathrm{ga}} \xi$:

$$\begin{aligned}p_1(\xi,\tau) &= j_\gamma(\gamma^{-1}\tau)^s \left(p_1|_{2s}^{\mathrm{pr}}\right)(\infty,\gamma^{-1}\tau) \\ &= j_\gamma(\gamma^{-1}\tau)^s \left(g|_{2s}^{\mathrm{pr}}\gamma(\gamma^{-1}\tau) - \left(m|_{2s}^{\mathrm{pr}}\gamma\right)(\infty,\gamma^{-1}\tau)\right)\end{aligned}$$

$$\begin{aligned}
&= j_\gamma(\gamma^{-1}\tau)\left(g(\gamma^{-1}\tau) + r_\gamma(\gamma^{-1}\tau) - m(\infty,\gamma^{-1}\tau)\right)\\
&= j_\gamma(\gamma^{-1}\tau)^s\left(r_\gamma(\gamma^{-1}\tau) + p_1(\infty,\gamma^{-1}\tau)\right)\\
&= (\text{Anal}) + (\text{Anal})\left(d_\infty\left(-\frac{d}{c} - \frac{1}{c^2(\tau-\xi)}\right) + c_\infty + o(1)\right),
\end{aligned}$$

where (Anal) means a function that is holomorphic in τ on a neighborhood of $\mathbb{P}^1_\mathbb{R}$. This shows that $p_1 \in \dot{\mathcal{R}}_s$. □

The construction also gives a criterion for obtaining a genuine quantum Maass form:

Proposition 3. *Let $u \in \mathcal{E}_s^\Gamma$. Then $\mathbf{q}_s^\omega u$ is in qMf_s^ω if and only if $b_0 = 0$ in* (2.2).

Parabolic representative. Let us consider the image $\mathbf{q}_s^{\omega,\mathrm{fs}} u$ of $\mathbf{q}_s^\omega u$ under the natural map $\mathsf{eqMf}_s^\omega \to \mathsf{eqMf}_s^{\omega,\mathrm{fs}}$. This image can be represented by $p + \eta$ where η is any element of $\mathcal{V}_s^{\omega,\mathrm{fs}}$.

If the function g represents α_u on $\mathbb{P}^1_\mathbb{R}$, it differs by the representative $f_u^{\mathrm{pr}} : \tau \mapsto \left(1+\tau^2\right)^s f_u(\tau)$ by an element $\eta \in \mathcal{V}_s^{\omega,\mathrm{fs}}$. Indeed, $\eta = g - f_u^{\mathrm{pr}}$ is holomorphic on a neighborhood of $\mathbb{R}$. So $p_1 = p + \eta$ represents $\mathbf{q}_s^{\omega,\mathrm{fs}} u$. It has the special property that $p_1|_{2s}^{\mathrm{pr}} T = p_1$; this we call a *parabolic representative*.

2.3 Quantum Maass forms and analytic cohomology

In [4] it is shown that the injective map $\mathbf{r}_s : \mathcal{E}_s^\Gamma \to H^1(\Gamma, \mathcal{V}_s^\omega)$ is not surjective. So the question arises whether $\mathbf{q}_s^\omega \mathcal{E}_s^\Gamma$ is equal to the whole of eqMf_s^ω. The answer is that eqMf_s^ω is as large as possible:

Theorem 4. *Let $0 < \mathrm{Re}\, s < 1$, $s \neq \frac{1}{2}$. The map $\mathbf{c}_s^\omega : \mathsf{eqMf}_s^\omega \to H^1(\Gamma, \mathcal{V}_s^\omega)$ is bijective.*

Up till now, the assumption $s \neq \frac{1}{2}$ was convenient. Here it is essential.

Proof. Let f be holomorphic on a neighborhood of ∞. For $|\mathrm{Im}\,\tau|$ sufficiently large, we consider the one-sided average

$$f|_{2s}^{\mathrm{pr}}\mathsf{Av}_T^+ = \sum_{n=0}^\infty f|_{2s}^{\mathrm{pr}} T^n : \tau \mapsto \sum_{n=0}^\infty \left(\frac{1+\tau^2}{1+(\tau+n)^2}\right)^s f(\tau+n). \tag{2.5}$$

The sum converges absolutely for $\operatorname{Re} s > \frac{1}{2}$, and defines a holomorphic function that satisfies

$$f|_{2s}^{\mathrm{pr}}\mathsf{Av}_T^+|_{2s}^{\mathrm{pr}}(1-T) \;=\; f|_{2s}^{\mathrm{pr}}(1-T)|_{2s}^{\mathrm{pr}}\mathsf{Av}_T^+ \;=\; f\,. \tag{2.6}$$

If $f(\infty) = 0$, the convergence and relation (2.6) hold for $\operatorname{Re} s > 0$. For $f_0(\tau) = \left(1+\tau^{-2}\right)^s$ we find

$$f_0|_{2s}^{\mathrm{pr}}\mathsf{Av}_T^+(\tau) \;=\; \tau^{2s}\left(1+\tau^{-2}\right)^s\zeta(2s,\tau)\,, \tag{2.7}$$

where $\zeta(\cdot,\cdot)$ is the Hurwitz zeta function. This gives the meromorphic continuation of $f_0|_{2s}^{\mathrm{pr}}\mathsf{Av}_T^+$ to $s \in \mathbb{C}$, with a first order pole at $s = \frac{1}{2}$ as the sole singularity in $0 < \operatorname{Re} s$. Writing a general f as $f = f(\infty)f_0 + f_1$, we have $f_1(\infty) = 0$, and find $f|_{2s}^{\mathrm{pr}}\mathsf{Av}_T^+$ for all $s \neq \frac{1}{2}$, $0 < \operatorname{Re} s < 1$. The relation (2.6) extends as well. The asymptotic behavior of the Hurwitz zeta function implies that there are A_0 and A_1 such that

$$\left(f|_{2s}^{\mathrm{pr}}\mathsf{Av}_T^+\right)(\tau) \;=\; A_1\tau + A_0 + o(1) \qquad (\tau \xrightarrow{\mathrm{ga}} \infty)\,. \tag{2.8}$$

For $\xi = \gamma\infty$, the subgroup $\Gamma_\xi \subset \Gamma$ fixing ξ is generated by $\pi_\xi = \gamma T\gamma^{-1}$. The one-sided average $\mathsf{Av}_{\pi_\xi}^+$ is obtained from Av_T^+ by conjugation.

Let $c \in Z^1\left(\Gamma, \mathcal{V}_s^\infty\right)$. We define $p \in \dot{\mathcal{R}}_s$ by

$$p(\xi,\tau) \;=\; -\left(c_\pi|_{2s}^{\mathrm{pr}}\mathsf{Av}_{\pi_\xi}^+\right)(\tau) + o(1) \qquad (\tau \xrightarrow{\mathrm{ga}} \xi)\,. \tag{2.9}$$

For $\delta \in \Gamma$, we find

$$\begin{aligned}
(p|_{2s}^{\mathrm{pr}}\delta)(\xi,\tau) &= j_\delta(\tau)^{-s}p(\delta\xi,\delta\tau)\\
&= -j_\delta(\tau)^{-s}c_{\delta\pi_\xi\delta^{-1}}|_{2s}^{\mathrm{pr}}\mathsf{Av}_{\delta\pi_\xi\delta^{-1}}^+(\delta\tau) + o(1)\\
&= -c_{\delta\pi_\xi\delta^{-1}}|_{2s}^{\mathrm{pr}}\delta|\mathsf{Av}_{\pi_\xi}^+|_{2s}^{\mathrm{pr}}\delta^{-1}|_{2s}^{\mathrm{pr}}\delta(\tau) + o(1)\\
&= -c_\delta|_{2s}^{\mathrm{pr}}(\pi_\xi - 1)|_{2s}^{\mathrm{pr}}\mathsf{Av}_{\pi_\xi}^+(\tau) - c_{\pi_\xi}|_{2s}^{\mathrm{pr}}\mathsf{Av}_{\pi_\xi}^+(\tau) + o(1)\\
&= c_\delta(\tau) + p(\xi,\tau)\,.
\end{aligned}$$

So p represents $[p] \in \mathsf{eqMf}_s^\omega$ and $\mathbf{c}_s^\omega[p] = [c]$. □

References

[1] R.W.Bruggeman: *Automorphic forms, hyperfunction cohomology, and period functions;* J. reine angew. Math. **492** (1997) 1–39

[2] S.Helgason: *Topics in harmonic analysis on homogeneous spaces;* Progr. in Math. **13**, Birkhäuser, 1981

[3] J.Lewis, D.Zagier: *Period functions for Maass wave forms. I;* Ann. of Math. **153** (2001) 191–258

[4] R.Bruggeman, J.Lewis, D.Zagier: *Period functions for Maass wave forms. II;* in preparation

[5] H.Maass: *Modular functions of one complex variable;* Notes by S.Lal, Tata Institue of Fundamental research; reprint Springer-Verlag, 1983

[6] S.D.Miller, W.Schmid: *The highly oscillatory behavior of automorphic distributions for* SL(2)*;* Letter in Math. Phys. **69** (2004) 265–286

[7] W.Schmid: *Automorphic distributions for* $\mathrm{SL}(2,\mathbb{R})$*;* p. 345–387 in Conférence Moshé Flato 1999, Quantization and Symmetries, vol. 1, Math. Phys. Studies **21** (2000), Kluwer Acad. Publ.

[8] H.Slichtkrull: *Hyperfunctions and harmonic analysis on symmetric spaces;* Birkhäuser, 1984

[9] D.Zagier: *Vassiliev invariants and a strange identity related to the eta function;* Topology **40** (2001) 945–960

Roelof Bruggeman
Mathematisch Instituut,
Universiteit Utrecht, Postbus 80010,
NL-3508 TA Utrecht, Nederland
Email: bruggeman@math.uu.nl

$\mathcal{L}$-Invariant of p-Adic L-Functions

H. HIDA[1]

Contents

Introduction[2]

Let $\overline{\mathbb{Q}} \subset \mathbb{C}$ be the field of all algebraic numbers. We fix a prime $p > 2$ and a p-adic absolute value $|\cdot|_p$ on $\overline{\mathbb{Q}}$. Then $\mathbb{C}_p$ is the completion of $\overline{\mathbb{Q}}$ under $|\cdot|_p$. We write $W = \left\{x \in K \middle| |x|_p < 1\right\}$ for the p-adic integer ring of

[1]The author is partially supported by the NSF grant: DMS 0244401 and DMS 0456252.

[2]This introduction is the note of the lecture delivered at the conference on L-functions at Kyushu University on February in 2006, and the following four sections are the notes of the lectures at Harvard University, Spring 2006, in their eigenvariety semester organized by Mazur and Taylor. We would like to thank the organizers of the conference at Kyushu University and the semester at Harvard University.

sufficiently large extension $K/\mathbb{Q}_p$ inside $\mathbb{C}_p$. We write $\overline{\mathbb{Q}}_p$ for the field of all numbers in $\mathbb{C}_p$ algebraic over $\mathbb{Q}_p$. Start with a strictly compatible system $\{\rho_{\mathfrak{l}}\}$ of semi-simple Galois representations $\rho_{\mathfrak{l}} : \mathrm{Gal}(\overline{\mathbb{Q}}/\mathbb{Q}) \to GL_d(T_{\mathfrak{l}})$ for primes $\mathfrak{l}$ of the coefficient field $T \subset \overline{\mathbb{Q}}$. We suppose that ρ is associated to a pure (absolute Hodge) motive in the sense of Deligne (see [D]). We assume that ρ does not contain the trivial representation as a subquotient. We write S for the finite set of ramification of ρ and $\rho_{\mathfrak{l}}$ is unramified outside $S \cup \{\infty, \ell\}$, where ℓ is the residual characteristic of $\mathfrak{l}$. We write $\mathfrak{p} = \left\{\xi \in O_T \middle| |\xi|_p < 1\right\}$ and often write $W := O_{T,\mathfrak{p}}$, where O_T is the integer ring of T. Often we just write ρ for $\rho_{\mathfrak{p}}$ which acts on $V = T_{\mathfrak{p}}^d$.

For simplicity, we assume that $p \notin S$. Assume that $\mathfrak{q} \nmid \ell$, and let $E_\ell(X) = \det(1 - \rho_{\mathfrak{q}}(Frob_\ell)|_{V_{I_\ell}} X) \in T[X]$. We always assume that $\rho_{\mathfrak{p}}$ is ordinary in the following sense: ρ restricted to $\mathrm{Gal}(\overline{\mathbb{Q}}_p/\mathbb{Q}_p)$ is upper triangular with diagonal characters $\mathcal{N}^{a_j}$ on the inertia I_p for the p-adic cyclotomic character $\mathcal{N}$ ordered from top to bottom as $a_1 \geq a_2 \geq \cdots \geq 0 \geq \cdots \geq a_d$. Thus

$$\rho|_{I_p} = \begin{pmatrix} \mathcal{N}^{a_1} & * & \cdots & * \\ 0 & \mathcal{N}^{a_2} & \cdots & * \\ \vdots & \vdots & \ddots & \vdots \\ 0 & 0 & \cdots & \mathcal{N}^{a_d} \end{pmatrix}.$$

In other words, we have a decreasing filtration $\mathcal{F}^{i+1}\rho \subset \mathcal{F}^i\rho$ stable under $\mathrm{Gal}(\overline{\mathbb{Q}}_p/\mathbb{Q}_p)$ such that $gr^i(\rho)(-i) := (\mathcal{F}^i\rho/\mathcal{F}^{i+1}\rho)(-i)$, the Tate twists, is unramified. Define

$$H_p(X) = \prod_i \det(1 - Frob_p|_{gr^i(\rho)(-i)} p^i X) = \prod_{j=1}^d (1 - \alpha_j X).$$

Then it is believed to be $E_p(X) = H_p(X)$ if $p \notin S$ and $E_p(X)|H_p(X)$ otherwise. In any case, $\mathrm{ord}_p(\alpha_j) \in \mathbb{Z}$. Let us define

$$\beta_j = \begin{cases} \alpha_j & \text{if } \mathrm{ord}_p(\alpha_j) \geq 1, \\ p\alpha_j^{-1} & \text{if } \mathrm{ord}_p(\alpha_j) \leq 0 \end{cases}$$

and put $e = |\{j|\beta_j = p\}|$.

$$\mathcal{E}(\rho) = \prod_{j=1}^d (1 - \beta_j p^{-1}) \text{ and } \mathcal{E}^+(\rho) = \prod_{j=1, \beta_j \neq p}^d (1 - \beta_j p^{-1}).$$

Then the complex L-function is defined by $L(s,\rho) = \prod_\ell E_\ell(\ell^{-s})^{-1}$. We assume that the value at 1 is critical for $L(s,\rho)$ (in the motivic sense of Deligne in [D]). We suppose to have an algebraicity result (conjectured by Deligne) that for a well defined period $c^+(\rho)(1)) \in \mathbb{C}^\times$ such that $\frac{L(s,\rho\otimes\varepsilon)}{c^+(\rho(1))} \in \overline{\mathbb{Q}}$ for all finite order characters $\varepsilon : \mathbb{Z}_p^\times \to \mu_{p^\infty}(\overline{\mathbb{Q}})$. Then we should have

Conjecture 0.1. *Suppose that $s = 1$ is critical for ρ. Then there exist a power series $\Phi^{an}(X) \in W[[X]]$ and a p-adic L-function $L_p^{an}(s,\rho) = \Phi_\rho^{an}(\gamma^{1-s}-1)$ interpolating $L(1,\rho\otimes\varepsilon)$ for p-power order character ε such that $\Phi_\rho^{an}(\varepsilon(\gamma)-1) \sim \mathcal{E}(\rho\otimes\varepsilon)\frac{L(1,\rho\otimes\varepsilon)}{c^+(\rho(1))}$ with the modifying p-factor $\mathcal{E}(\rho)$ as above (putting $\mathcal{E}(\rho\otimes\varepsilon) = 1$ if $\varepsilon \neq 1$). The L-function $L_p^{an}(s,\rho)$ has zero of order $e + \mathrm{ord}_{s=1} L(s,\rho)$ for a nonzero constant $\mathcal{L}^{an}(\rho) \in \mathbb{C}_p^\times$ (called the analytic $\mathcal{L}$-invariant), we have*

$$\lim_{s\to 1}\frac{L_p^{an}(s,\rho)}{(s-1)^e} = \mathcal{L}^{an}(\rho)\mathcal{E}^+(\rho)\frac{L(1,\rho)}{c^+(\rho(1))},$$

where "$\lim_{s\to1}$" is the p-adic limit, $c^+(\rho(1))$ is the transcendental factor of the critical complex L-value $L(1,\rho)$, and $\mathcal{E}^+(\rho)$ is the product of nonvanishing modifying p-factors.

When $e > 0$, we call that $L_p^{an}(s,\rho)$ has an *exceptional zero* at $s = 1$. Here is an example. Start with a Dirichlet character $\chi : (\mathbb{Z}/N\mathbb{Z}) \to \overline{\mathbb{Q}}^\times$ with $\chi(-1) = -1$. Then $c(\rho^+(1)) = (2\pi i)$. If we suppose $\chi = \left(\frac{-D}{\cdot}\right)$ for a square free positive integer D, the modifying Euler factor vanishes at $s = 1$ if the Legendre symbol $\left(\frac{-D}{p}\right) = 1 \Leftrightarrow (p) = \mathfrak{p}\overline{\mathfrak{p}}$ in $O_{\mathbb{Q}[\sqrt{-D}]}$ with $\mathfrak{p} = \left\{x \in O_{\mathbb{Q}[\sqrt{-D}]} \middle| |x|_p < 1\right\}$. By a work of Kubota–Leopoldt and Iwasawa, we have a p-adic analytic L-function $L_p^{an}(s,\chi) = \Phi^{an}(\gamma^{1-s}-1)$ for a power series $\Phi^{an}(X) \in \Lambda = W[[X]]$ and $\gamma = 1 + p$ such that for $\mathcal{E}(\chi\mathcal{N}^m) = (1-\chi(p)p^{m-1})$

$$L_p^{an}(m,\chi) = \Phi^{an}(\gamma^{1-m}-1) = \mathcal{E}(\chi\mathcal{N}^m)L(1-m,\chi) \sim \mathcal{E}(\chi\mathcal{N}^m)\frac{L(m,\chi)}{(2\pi i)^m}$$

for all positive integer m as long as $|n^m - n|_p < 1$ for all n prime to p. If we have an exceptional zero at 1, it appears that we lose the exact connection of the p-adic L-value and the corresponding complex L-value. However, the conjecture says we can recover the complex L-value via an appropriate derivative of the p-adic L-function as long as we can compute $\mathcal{L}^{an}(\rho)$. We may regard χ as a Galois character $\mathrm{Gal}(\mathbb{Q}[\mu_N]/\mathbb{Q}) = (\mathbb{Z}/N\mathbb{Z})^\times \xrightarrow{\chi} \{\pm 1\}$, and we remark that $\chi(Frob_p) = 1$ to have the exceptional zero. For our later use, for the class number h of $\mathbb{Q}[\sqrt{-D}]$, we write the generator of $\overline{\mathfrak{p}}^h$ as ϖ; so, $\overline{\mathfrak{p}}^h = (\varpi)$ for $\varpi \in \mathbb{Q}[\sqrt{-D}]$.

Though we formulated the conjecture for $p \notin S$, if $\rho_\mathfrak{p}$ is ordinary semi-stable, we have the same phenomena and can formulate a conjecture similarly. Here is such an example. Start with an elliptic curve $E_{/\mathbb{Q}}$, which yields a compatible system $\rho_E := \{T_\ell E\}$ given by the ℓ-adic Tate module $T_\ell E$. Suppose that E has split multiplicative reduction at p. In this case, $H_p(X) = (1-X)(1-pX)$ and $E_p(X) = (1-X)$, $\mathcal{E}(\rho_E) = 0$ and $\mathcal{E}^+(\rho) = 1$. Then by the solution of the Shimura-Taniyama conjecture by Wiles et al, this L-function has p-adic analogue constructed by Mazur such that we have $\Phi_E^{an}(X) \in \Lambda$ with $\Phi_E^{an}(\varepsilon(\gamma)-1) = \mathcal{E}(\rho_E \otimes \varepsilon)\frac{G(\varepsilon^{-1})L(1,E,\varepsilon)}{\Omega_E}$ for all p-power order character $\varepsilon : \mathbb{Z}_p^\times \to W^\times$; in other words, $L_p^{an}(s,E) = \Phi_E^{an}(\gamma^{1-s}-1)$. Here Ω_E is the period of the Néron differential of E. Thus if $Frob_p$ has eigenvalue 1 on $T_\ell E$, the exceptional zero appears at $s = 1$ as in the case of Dirichlet character. The $Frob_p$ has eigenvalue 1 if and only if E has multiplicative reduction mod p.

The problem of $\mathcal{L}$-invariant is to compute explicitly the $\mathcal{L}$-invariant $\mathcal{L}^{an}(\rho)$. The $\mathcal{L}$-invariant in the cases where $\rho = \chi = \left(\frac{-D}{\cdot}\right)$ as above and $\rho = \rho_E$ for E with split multiplicative reduction is computed in the 1970s to 90s, and the results are

Theorem 0.2. *Let the notation and the assumption be as above.*

1. $\mathcal{L}^{an}(\chi) = \frac{\log_p(q)}{\mathrm{ord}_p(q)} = -\frac{\log_p(q)}{h}$ *for* $q \in \mathbb{C}_p$ *given by* $q = \varpi/\overline{\varpi}$ $(\overline{\mathfrak{p}}^h = (\varpi))$ *and the class number* h *of* $\mathbb{Q}[\sqrt{-D}]$*;*
2. *For* E *split multiplicative at* p*, writing* $E(\mathbb{C}_p) = \mathbb{C}_p^\times/q^{\mathbb{Z}}$ *for the Tate period* $q \in \mathbb{Q}_p^\times$*, we have* $\mathcal{L}^{an}(\rho_E) = \frac{\log_p(q)}{\mathrm{ord}_p(q)}$.

Here $\log_p$ *is the Iwasawa logarithm and* $|x|_p = p^{-\mathrm{ord}_p(x)}$.

The first assertion is due to Gross–Koblitz [GsK] and Ferrero–Greenberg [FG], and the second was conjectured by Mazur–Tate–Teitelbaum [MTT] and later proven by Greenberg–Stevens [3] and by some others further later.

In the first formula of the theorem, the sign "–" of $-\frac{\log_p(q)}{h}$ is correct as explained in [G] (10) (and if we evaluate the value at $s = 0$, the minus sign should be removed as was done in [FG].

For an ordinary p-adic Galois representation $\rho : \mathrm{Gal}(\overline{\mathbb{Q}}/\mathbb{Q}) \to GL_d(W)$, there is a systematic way to create many Galois representations whose eigenvalues of $Frob_p$ contain 1. Indeed, let ρ acts on the $d \times d$ matrices $M_d(W)$ by conjugation. Since

$$M_d(W) = Ad(W) \oplus \{\text{scalar matrices}\}$$

for the trace 0 space $Ad(W)$ which is stable under the conjugation. Then the action of $Ad(\rho)(Frob_p)$ on $Ad(W)$ has eigenvalue 1 with multiplicity $\geq d-1$. However it is easy to check that the system $Ad(\rho)$ is not critical if $d > 2$. Thus we assume that $d = 2$. Now require that $\rho_F : \mathrm{Gal}(\overline{\mathbb{Q}}/F) \to GL_2(W)$ be a Galois representation associated to a p-ordinary Hilbert Hecke eigenform (belonging to a discrete series at ∞) over a totally real field F. We make $Ad(\rho_F)$ and consider the induced representation $\mathrm{Ind}_F^{\mathbb{Q}} Ad(\rho_F)$ whose eigenvalues of $Frob_p$ has 1 with multiplicity e for the number e of prime factors of p in F. The system $\mathrm{Ind}_F^{\mathbb{Q}} Ad(\rho_F)$ is critical at $s = 1$.

Arithmetic $\mathcal{L}$-invariant. Returning to a general ordinary representation $\rho = \rho_{\mathfrak{p}}$, we describe an arithmetic way of constructing p-adic L-function due to Iwasawa and others. We define Galois cohomologically the Selmer group

$$\mathrm{Sel}(\rho) \subset H^1(\mathrm{Gal}(\overline{\mathbb{Q}}/\mathbb{Q}_\infty), \rho \otimes \mathbb{Q}_p/\mathbb{Z}_p)$$

for the $\mathbb{Z}_p$-extension $\mathbb{Q}_\infty/\mathbb{Q}$ inside $\mathbb{Q}(\mu_{p^\infty})$ by the subgroup of cohomology classes unramified outside p whose image vanishes in $H^1(I_p, (\rho/\mathcal{F}^+\rho) \otimes_{\mathbb{Z}_p} \mathbb{Q}_p/\mathbb{Z}_p)$. Here $\mathcal{F}^+\rho$ is the middle filtration $\mathcal{F}^1\rho$ and I_p is each inertia group at p. The Galois group $\Gamma = \mathrm{Gal}(\mathbb{Q}_\infty/\mathbb{Q})$ acts on $H^1(\mathrm{Gal}(\overline{\mathbb{Q}}/\mathbb{Q}_\infty), \rho \otimes \mathbb{Q}_p/\mathbb{Z}_p)$ and hence on $\mathrm{Sel}(\rho)$, making it as a discrete module over the group algebra $W[[\Gamma]] = \varprojlim_n W[\Gamma/\Gamma^{p^n}]$. Identifying Γ with $1 + p\mathbb{Z}_p$ by the cyclotomic character, we may regard $\gamma \in \Gamma$. Then $W[[\Gamma]] \cong \Lambda$ by $\gamma \mapsto 1 + X$. By the classification theory of compact Λ-modules of finite type, the Pontryagin dual $\mathrm{Sel}^*(\rho)$ has a Λ-linear map into $\prod_{f \in \Omega} \Lambda/f\Lambda$ with finite kernel and cokernel for a finite set $\Omega \subset \Lambda$. The power series $\Phi_\rho = \prod_{f \in \Omega} f(X)$

is uniquely determined up to unit multiple. We then define $L_p(s,\rho) = \Phi_\rho(\gamma^{1-s}-1)$. Greenberg gave a recipe of defining $\mathcal{L}(\rho)$ for this $L_p(s,\rho)$ and verified in 1994 the conjecture for this $L_p(s,\rho)$ except for the nonvanishing of $\mathcal{L}(\rho)$ (under some restrictive conditions). For the adjoint square $Ad(\rho_F)$ for ρ_F associated to a Hilbert modular form, the conjecture (except for the nonvanishing of $\mathcal{L}(\rho)$) was again proven in my paper [H00] in the Israel journal (in 2000) under the condition that $\overline{\rho}_F = (\rho_F \mod \mathfrak{m}_W)$ is absolutely irreducible over $\mathrm{Gal}(\overline{\mathbb{Q}}/F[\mu_p])$ and the p-distinguishedness condition for $\overline{\rho}_F|_{\mathrm{Gal}(\overline{F}_\mathfrak{p}/F_\mathfrak{p})}$ for all $\mathfrak{p}|p$ (which we recall later). If there exists an analytic p-adic L-function $L_p^{an}(s,\rho) = \Phi_\rho^{an}(\gamma^{1-s}-1)$ interpolating complex L-values, the main conjecture of Iwasawa's theory confirms $\Phi_\rho = \Phi_\rho^{an}$ up to unit multiple.

Suppose now that ρ_F is associated to a Hilbert modular Hecke eigenform of weight $k \geq 2$ over a totally real field F. Following Greenberg's recipe, we try to compute $\mathcal{L}(Ad(\rho_F)) = \mathcal{L}(\mathrm{Ind}_F^{\mathbb{Q}} Ad(\rho_F))$. By ordinarity, we have $\rho_F|_{\mathrm{Gal}(\overline{\mathbb{Q}}_p/F_\mathfrak{p})} \cong \begin{pmatrix} \beta_\mathfrak{p} & * \\ 0 & \alpha_\mathfrak{p} \end{pmatrix}$ with two distinct diagonal characters $\alpha_\mathfrak{p}$ and $\beta_\mathfrak{p}$ factoring through $I_\mathfrak{p} \to \mathrm{Gal}(F_\mathfrak{p}[\mu_{p^\infty}]/F_\mathfrak{p})$ for the inertia group $I_\mathfrak{p}$ for all $\mathfrak{p}|p$. We consider the universal nearly ordinary deformation $\rho : \mathrm{Gal}(\overline{\mathbb{Q}}/\mathbb{Q}) \to GL_2(R)$ over K with the pro-Artinian local universal K-algebra R. This means that for any Artinian local K-algebra A with maximal ideal $\mathfrak{m}_A$ and any Galois representation $\rho_A : \mathrm{Gal}(\overline{\mathbb{Q}}/F) \to GL_2(A)$ such that

1. unramified outside ramified primes for ρ_F;

2. $\rho_A|_{\mathrm{Gal}(\overline{\mathbb{Q}}_p/F_\mathfrak{p})} \cong \begin{pmatrix} * & * \\ 0 & \alpha_{A,\mathfrak{p}} \end{pmatrix}$ with $\alpha_{A,\mathfrak{p}} \equiv \alpha_\mathfrak{p} \mod \mathfrak{m}_A$ such that the diagonal characters factor through $I_\mathfrak{p} \to \mathrm{Gal}(F_\mathfrak{p}[\mu_{p^\infty}]/F_\mathfrak{p})$ for all $\mathfrak{p}|p$;

3. $\det(\rho_A) = \det \rho_F$;

4. $\rho_A \equiv \rho_F \mod \mathfrak{m}_A$,

there exists a unique K-algebra homomorphism $\varphi : R \to A$ such that $\varphi \circ \rho \cong \rho_A$. Write $\Gamma_\mathfrak{p} \cong \mathbb{Z}_p$ for the p-profinite part of $\mathrm{Gal}(F_\mathfrak{p}[\mu_{p^\infty}]/F_\mathfrak{p})$. Choose a generator $\gamma_\mathfrak{p}$ of $\Gamma_\mathfrak{p}$ and identify $W[[\Gamma_\mathfrak{p}]]$ with $W[[X_\mathfrak{p}]]$ by $\gamma_\mathfrak{p} \leftrightarrow 1+X_\mathfrak{p}$. Since $\rho|_{\mathrm{Gal}(\overline{\mathbb{Q}}_p/F_\mathfrak{p})} \cong \begin{pmatrix} * & * \\ 0 & \delta_\mathfrak{p} \end{pmatrix}$, $\delta_\mathfrak{p}\alpha_\mathfrak{p}^{-1} : \mathrm{Gal}(F_\mathfrak{p}[\mu_{p^\infty}]/F_\mathfrak{p}) \to R$ factors through $\Gamma_\mathfrak{p}$ and induces an algebra structure on R over $W[[X_\mathfrak{p}]]$. Thus R is an algebra over

$K[[X_\mathfrak{p}]]_{\mathfrak{p}|p}$. If we write $\varphi_\rho : R \to K$ for the morphism with $\varphi_\rho \circ \boldsymbol{\rho} \cong \rho_F$, by our construction, $\mathrm{Ker}(\varphi_\rho) \supset (X_\mathfrak{p})_{\mathfrak{p}|p} = (X)$. We state a conjecture:

Conjecture 0.3. *We have* $R \cong K[[X_\mathfrak{p}]]_{\mathfrak{p}|p}$.

By the results of Wiles, Taylor-Wiles, Fujiwara, Kisin and Skinner-Wiles, the conjecture holds at least if either $F = \mathbb{Q}$ (see [Ki], [Ki1] and [Ki2]) or the residual representation $\overline{\rho}_F = (\rho \mod \mathfrak{m}_W)$ is absolutely irreducible and $\overline{\rho}_F|_{\mathrm{Gal}(\overline{F}_\mathfrak{p}/F_\mathfrak{p})})^{ss} \cong \overline{\alpha}_\mathfrak{p} \oplus \overline{\beta}_\mathfrak{p}$ with $\overline{\alpha}_\mathfrak{p} \neq \overline{\beta}_\mathfrak{p}$ (see [Fu] and [Fu1]). Here is a theorem:

Theorem 0.4. *Assume* Conjecture 0.3. *Then for the local Artin symbol* $[p, F_\mathfrak{p}] = Frob_\mathfrak{p}$, *we have*

$$\mathcal{L}(\mathrm{Ind}_F^{\mathbb{Q}} Ad(\rho_F)) = \det\left(\frac{\partial \delta_\mathfrak{p}([p, F_\mathfrak{p}])}{\partial X_{\mathfrak{p}'}}\right)_{\mathfrak{p},\mathfrak{p}'} \Big|_{X=0} \prod_\mathfrak{p} \log_p(\gamma_\mathfrak{p}) \alpha_\mathfrak{p}([p, F_\mathfrak{p}])^{-1},$$

where $\gamma_\mathfrak{p}$ *is the generator of* $\Gamma_\mathfrak{p}$ *by which we identify the group algebra* $W[[\Gamma_\mathfrak{p}]]$ *with* $W[[X_\mathfrak{p}]]$.

This result is proved as [HMI] Theorem 3.73 under some redundant hypothesis in order to make the book [HMI] self-contained. In this note, we will sketch the proof in the general case assuming that p totally splits in $F/\mathbb{Q}$ and that ρ_F is unramified outside p and ∞. The proof in the nonsplit case is more complicated, and we will give a full proof in [H06] along with conjectures predicting $\mathcal{L}$-invariant of symmetric powers of ρ_F.

Here are some examples showing usefulness of this theorem: Take a totally imaginary quadratic extension M/F in which all prime factors $\mathfrak{p}|p$ in F splits as $\mathfrak{P}\overline{\mathfrak{P}}$. Take a set $\Sigma = \{\mathfrak{P}|p\}$ so that $\Sigma \sqcup \overline{\Sigma}$ is the set of all prime factors of p in M. Write h for the class number of M and choose $\varpi(\mathfrak{P}) \in M$ so that $\mathfrak{P}^h = (\varpi(\mathfrak{P}))$ for $\mathfrak{P} \in \overline{\Sigma}$. For any Galois character $\psi : \mathrm{Gal}(\overline{\mathbb{Q}}/M) \to W^\times$ of M with $\psi(\sigma) \neq \psi^c(\sigma)$ for $\psi^c(\sigma) = \psi(c\sigma c^{-1})$ and a complex conjugation $c \in \mathrm{Gal}(\overline{\mathbb{Q}}/F)$, we have $Ad(\mathrm{Ind}_M^F \psi) = \chi \oplus \mathrm{Ind}_M^F \psi^{1-c}$ for $\chi = \left(\frac{M/F}{\cdot}\right)$, and we can easily show $\mathcal{L}(\chi) = \mathcal{L}(Ad(\mathrm{Ind}_M^F \psi))$. The arithmetic p-adic L-function $L_p(s, \chi)$ for $\chi = \left(\frac{M/F}{\cdot}\right)$ constructed à la Iwasawa has an exceptional zero of order $\geq e$ for $e = |\Sigma|$. Since we can compute explicitely the universal deformation $\boldsymbol{\rho}$ of $\rho = \mathrm{Ind}_M^F \psi$, we get from the theorem

Corollary 0.5. *We have*

$$\mathcal{L}(\mathrm{Ind}_F^{\mathbb{Q}}\chi) = (-1)^e \frac{\det\left(\log_p(N_{\mathfrak{P}'}(\varpi(\mathfrak{P})^{(1-c)}))\right)_{\mathfrak{P},\mathfrak{P}'\in\overline{\Sigma}}}{\prod_{\mathfrak{P}\in\overline{\Sigma}}(h/e(\mathfrak{P}))},$$

where $N_{\mathfrak{P}}$ *is the local norm* $N_{M_{\mathfrak{P}}/\mathbb{Q}_p}$ *and* $e(\mathfrak{P})$ *is the ramification index of* $\mathfrak{P}/p$.

A proof of this is given in [HMI] Corollary 5.39, though the sign $(-1)^e$ is erroneously omitted there. We will revisit briefly the proof of this corollary in Lecture 3 and correct the sign error.

If $E_{/F}$ is an elliptic curve with split multiplicative reduction at all $\mathfrak{p}|p$, we write $E_{/F_{\mathfrak{p}}}(\overline{F}_{\mathfrak{p}}) \cong \overline{F}_{\mathfrak{p}}^{\times}/q_{\mathfrak{p}}^{\mathbb{Z}}$ for the Tate period $q_{\mathfrak{p}} \in F_{\mathfrak{p}}^{\times}$. Then we have directly from the theorem the following

Corollary 0.6. *Assume that* $E_{/F}$ *is associated to a Hilbert modular form on* $GL(2)_{/F}$. *Then the* $\mathcal{L}$*-invariant* $\mathcal{L}(\mathrm{Ind}_F^{\mathbb{Q}} Ad(\rho_E))$ *is given by* $\prod_{\mathfrak{p}} \frac{\log_p(N_{\mathfrak{p}}(q_{\mathfrak{p}}))}{\mathrm{ord}_p(N_{\mathfrak{p}}(q_{\mathfrak{p}}))}$, *where* $N_{\mathfrak{p}}$ *is the local norm* $N_{F_{\mathfrak{p}}/\mathbb{Q}_p}$.

The above two corollaries are obtained by explicitly computing the universal representation $\boldsymbol{\rho}$. The case where $F = \mathbb{Q}$ is treated in [H04]. We will generalize this corollary to Theorem 2.1 dealing with elliptic curves having multiplicative reduction at some prime factors of p and ordinary good reduction at the other prime factors of p and give a proof of Theorem 2.1 limiting ourselves to E having good reduction everywhere outside p and to $F/\mathbb{Q}$ in which p totally splits. The general case will be treated in [H06].

1 Lecture 1: Galois deformation and $\mathcal{L}$-invariant

In the following three lectures delivered at the eigenvariety semester at Harvard University in Spring 2006, we give a sketch of the proof of the results mentioned in the introduction, assuming as a simplifying assumption that p completely splits in $F/\mathbb{Q}$ (the general case is treated in [HMI] Chapters 3 and 5 and also in [H06]).

Let $p > 2$ be a prime, and fix a totally real finite extension $F/\mathbb{Q}$. As we wrote, for **simplicity**, hereafter, we assume always that p splits completely in $F/\mathbb{Q}$. We start with a Galois representation $\rho_F : \mathrm{Gal}(\overline{\mathbb{Q}}/F) \to GL_2(W)$

associated to a discrete series ($\Leftrightarrow k \geq 2$) Hilbert modular form f (over F) with coefficients in a finite extension $W/\mathbb{Z}_p$ (a DVR). We assume the ordinarity of ρ_F:

$$\rho_F|_{D_\mathfrak{p}} \cong \begin{pmatrix} \epsilon_\mathfrak{p} & * \\ 0 & \alpha_\mathfrak{p} \end{pmatrix} \text{ with } \epsilon_\mathfrak{p} \neq \alpha_\mathfrak{p},\ \epsilon_\mathfrak{p}|_{I_\mathfrak{p}} = \mathcal{N}^{k-1} \text{ and } \alpha_\mathfrak{p}(I_\mathfrak{p}) = 1$$

on the decomposition group and the inertial group $I_\mathfrak{p} \subset D_\mathfrak{p} \subset \mathrm{Gal}(\overline{\mathbb{Q}}/F)$ for all prime factor $\mathfrak{p}$ of p in F. Here $\mathcal{N}(\sigma) \in \mathbb{Z}_p^\times$ is the p-adic cyclotomic character with $\exp(\frac{2\pi i}{p^n})^\sigma = \exp(\frac{\mathcal{N}(\sigma)2\pi i}{p^n})$ for all $n > 0$ and $k > 1$ is an integer. Again for **simplicity**, we assume that ρ is unramified outside p and ∞. Thus for any prime $\mathfrak{l} \nmid p$, writing $f|T(\mathfrak{l}) = a_\mathfrak{l} f$, we have $\mathrm{Tr}(\rho(Frob_\mathfrak{l})) = a_\mathfrak{l} \in W$. Let K be the quotient field of W (so, $K/\mathbb{Q}_p$ is a finite extension).

We consider the **universal** nearly ordinary deformation $\boldsymbol{\rho} : \mathrm{Gal}(\overline{\mathbb{Q}}/F) \to GL_2(R)$ over K with the pro-Artinian local universal K-algebra R. This means that for any Artinian local K-algebra A with maximal ideal $\mathfrak{m}_A$ and any Galois representation $\rho_A : \mathrm{Gal}(\overline{\mathbb{Q}}/F) \to GL_2(A)$ such that

(K1) unramified outside p;

(K2) $\rho_A|_{\mathrm{Gal}(\overline{\mathbb{Q}}_p/F_\mathfrak{p})} \cong \left(\begin{smallmatrix} * & * \\ 0 & \alpha_{A,\mathfrak{p}} \end{smallmatrix}\right)$ with $\alpha_{A,\mathfrak{p}} \equiv \alpha_\mathfrak{p} \mod \mathfrak{m}_A$;

(K3) $\det(\rho_A) = \det \rho_F$;

(K4) $\rho_A \equiv \rho_F \mod \mathfrak{m}_A$,

there exists a unique K-algebra homomorphism $\varphi : R \to A$ such that $\varphi \circ \boldsymbol{\rho} \cong \rho_A$. Note that $\mathcal{N} : \mathrm{Gal}(F_\mathfrak{p}[\mu_{p^\infty}]/F_\mathfrak{p}) \cong \mathbb{Z}_p^\times$ (by splitting of p in $F/\mathbb{Q}$). Let $\Gamma_\mathfrak{p} = 1 + p\mathbb{Z}_p \subset \mathrm{Gal}(F_\mathfrak{p}[\mu_{p^\infty}]/F_\mathfrak{p})$. Choose a generator $\gamma_\mathfrak{p}$ of $\Gamma_\mathfrak{p}$ and identify $W[[\Gamma_\mathfrak{p}]]$ with $W[[X_\mathfrak{p}]]$ by $\gamma_\mathfrak{p} \leftrightarrow 1 + X_\mathfrak{p}$. Since $\boldsymbol{\rho}|_{\mathrm{Gal}(\overline{\mathbb{Q}}_p/F_\mathfrak{p})} \cong \left(\begin{smallmatrix} * & * \\ 0 & \delta_\mathfrak{p} \end{smallmatrix}\right)$, $\delta_\mathfrak{p}\alpha_\mathfrak{p}^{-1} : \Gamma_\mathfrak{p} \to R$ induces an algebra structure on R over $W[[X_\mathfrak{p}]]$. Thus R is an algebra over $K[[X_\mathfrak{p}]]_{\mathfrak{p}|p}$. If we write $\varphi : R \to K$ for the morphism with $\varphi \circ \boldsymbol{\rho} \cong \rho_F$, by our construction, $\mathrm{Ker}(\varphi) \supseteq (X_\mathfrak{p})_{\mathfrak{p}|p}$.

Here is the theorem we have seen in the first lecture:

Theorem 1.1. *Suppose $R \cong K[[X_\mathfrak{p}]]_{\mathfrak{p}|p}$. Then, if $\varphi \circ \boldsymbol{\rho} \cong \rho_F$, for the local Artin symbol $[p, F_\mathfrak{p}] = Frob_\mathfrak{p}$, we have*

$$\mathcal{L}(Ad(\rho_F)) = \det\left(\frac{\partial \delta_\mathfrak{p}([p, F_\mathfrak{p}])}{\partial X_{\mathfrak{p}'}}\right)_{\mathfrak{p},\mathfrak{p}'}\Big|_{X=0} \prod_\mathfrak{p} \log_p(\gamma_\mathfrak{p})\alpha_\mathfrak{p}([p, F_\mathfrak{p}])^{-1}.$$

Greenberg proposed a conjectural formula of the $\mathcal{L}$-invariant for a general p-adic p-ordinary Galois representation V with an exceptional zero. When $V = Ad(\rho_F)$, his definition goes as follows. Under some hypothesis, he found a unique subspace $\mathbb{H} \subset H^1(F, Ad(\rho_F))$ of dimension $e = |\{\mathfrak{p}|p\}|$ represented by cocycles $c : \mathrm{Gal}(\overline{\mathbb{Q}}/F) \to Ad(\rho_F)$ such that

1. c is unramified outside p;

2. c restricted to $D_\mathfrak{p}$ is upper triangular after conjugation for all $\mathfrak{p}|p$.

By the condition (2), $c|_{I_\mathfrak{p}}$ modulo upper nilpotent matrices factors through the cyclotomic Galois group $\mathrm{Gal}(\mathbb{Q}_p[\mu_{p^\infty}]/\mathbb{Q}_p)$ because $F_\mathfrak{p} = \mathbb{Q}_p$, and hence $c|_{D_\mathfrak{p}}$ modulo upper nilpotent matrices becomes unramified everywhere over the cyclotomic $\mathbb{Z}_p$-extension F_∞/F. In other words, the cohomology class $[c]$ is in $\mathrm{Sel}_{F_\infty}(Ad(\rho_F))$ but not in $\mathrm{Sel}_F(Ad(\rho_F))$.

Take a basis $\{c_\mathfrak{p}\}_{\mathfrak{p}|p}$ of $\mathbb{H}$ over K. Write

$$c_\mathfrak{p}(\sigma) \sim \begin{pmatrix} -a_\mathfrak{p}(\sigma) & * \\ 0 & a_\mathfrak{p}(\sigma) \end{pmatrix} \quad \text{for } \sigma \in D_{\mathfrak{p}'} \text{ with any } \mathfrak{p}'|p.$$

Then $a_\mathfrak{p} : D_{\mathfrak{p}'} \to K$ is a homomorphism. His $\mathcal{L}$-invariant is defined by

$$\begin{aligned}&\mathcal{L}(Ad(\rho_F))\\ &= \det\left((a_\mathfrak{p}([p, F_{\mathfrak{p}'}])_{\mathfrak{p},\mathfrak{p}'|p} \left(\log_p(\gamma_{\mathfrak{p}'})^{-1} a_\mathfrak{p}([\gamma_{\mathfrak{p}'}, F_{\mathfrak{p}'}]))_{\mathfrak{p},\mathfrak{p}'|p} \right)^{-1} \right).\end{aligned}$$

The above value is independent of the choice of the basis $\{c_\mathfrak{p}\}_\mathfrak{p}$. When $F = \mathbb{Q}$, by a result of Kisin [Ki] 9.10, [Ki1] and [Ki2] 3.4 (generalizing those of Wiles [W] and Taylor-Wiles [TW]), we always have $R \cong K[[X_p]]$. In general, assuming the following two conditions:

(ai) $\overline{\rho} = (\rho_F \mod \mathfrak{m}_W)$ is absolutely irreducible over $\mathrm{Gal}(\overline{\mathbb{Q}}/F[\mu_p])$;

(ds) $\overline{\rho}^{ss}$ has a non-scalar value over $\mathrm{Gal}(\overline{F}_\mathfrak{p}/F_\mathfrak{p})$ for all prime factors $\mathfrak{p}|p$,

by using a result of Fujiwara (see [Fu] and [Fu1]), we can prove $R \cong K[[X_\mathfrak{p}]]_{\mathfrak{p}|p}$. The following conjecture for the arithmetic L-function is a theorem under the condition (ai) and semi-stability of ρ_F over O except for the nonvanishing $\mathcal{L}(Ad(\rho_F)) \neq 0$ (see [HMI] Theorem 5.27 combined with (5.2.6) there):

Conjecture 1.2 (Greenberg). *Suppose* (ds) *and that* $\overline{\rho}$ *is absolutely irreducible. For* $L_p^{arith}(s, Ad(\rho_F)) = \Phi^{arith}(\gamma^{1-s} - 1)$, *then* $L_p^{arith}(s, Ad(\rho_F))$ *has zero of order equal to* $d = [F : \mathbb{Q}]$ *and for the constant* $\mathcal{L}(Ad(\rho_F)) \in K^\times$ *specified by the determinant as in the theorem, we have*

$$\lim_{s\to 1} \frac{L_p^{arith}(s, Ad(\rho_F))}{(s-1)^d} = \mathcal{L}(Ad(\rho_F)) \left| |\mathrm{Sel}_{\mathbb{Q}}(\mathrm{Ind}_F^{\mathbb{Q}} Ad(\rho_F))| \right|_p^{-1/[K:\mathbb{Q}_p]}$$

up to units.

The factor $\mathcal{E}^+(Ad(\rho))$ does not show up in the above formula, because if we write $S_F(Ad(\rho_F))$ for the Bloch-Kato Selmer group $H^1_f(F, Ad(\rho) \otimes \mathbb{Q}_p/\mathbb{Z}_p)$, we have, as explained in [MFG] page 284,

$$\begin{aligned} &\left| |\mathrm{Sel}_{\mathbb{Q}}(\mathrm{Ind}_F^{\mathbb{Q}} Ad(\rho_F))| \right|_p^{-1/[K:\mathbb{Q}_p]} \\ &\qquad = \mathcal{E}^+(Ad(\rho_F)) \left| |S_F(Ad(\rho_F))| \right|_p^{-1/[K:\mathbb{Q}_p]} \quad \text{up to units,} \end{aligned}$$

and the value $\left| |S_F(Ad(\rho_F))| \right|_p^{-1/[K:\mathbb{Q}_p]}$ is directly related the primitive complex L-value $L(1, Ad(\rho_F))$ up to a period. In the following section, we describe the Selmer group and how to define $\mathbb{H}$.

1.1 Selmer Groups

We recall Greenberg's definition of Selmer groups. Write $F^{(p)}/F$ for the maximal extension unramified outside p and ∞. Put $\mathfrak{G} = \mathrm{Gal}(F^{(p)}/F)$ and $\mathfrak{G}_M = \mathrm{Gal}(F^{(p)}/M)$. Let $V = Ad(\rho_F)$ with a continuous action of $\mathfrak{G}$. We fix a W-lattice T in V stable under $\mathfrak{G}$.

Write $D = D_\mathfrak{p} \subset \mathfrak{G}$ for the decomposition group of each prime factor $\mathfrak{p}|p$. Choosing a basis of ρ_F so that $\rho_F|_D$ is upper triangular. We have a 3-step filtration:

(ord) $$V \supset \mathcal{F}_\mathfrak{p}^- V \supset \mathcal{F}_\mathfrak{p}^+ V \supset \{0\},$$

where taking a basis so that $\rho_F|_D$ is upper triangular, $\mathcal{F}_\mathfrak{p}^- V$ is made up of upper triangular matrices, and $\mathcal{F}_\mathfrak{p}^+ V$ is made up of upper nilpotent matrices, and on $\mathcal{F}_\mathfrak{p}^- V/\mathcal{F}_\mathfrak{p}^+ V$, D acts trivially (getting eigenvalue 1 for $Frob_\mathfrak{p}$). Since V is self-dual, its dual $V^*(1) = \mathrm{Hom}_K(V, K) \otimes \mathcal{N}$ again satisfies (ord).

Let M/F be a subfield of $F^{(p)}$, and put $\mathfrak{G}_M = \mathrm{Gal}(F^{(p)}/M)$. We write $\mathfrak{p}$ for a prime of M over p and $\mathfrak{q}$ for general primes of M. We put

$$L_\mathfrak{p}(V) = \mathrm{Ker}(\mathrm{Res} : H^1(M_\mathfrak{p}, V) \to H^1(I_\mathfrak{p}, \frac{V}{\mathcal{F}_\mathfrak{p}^+(V)})).$$

Then for a $\mathfrak{G}_M$-stable W-lattice T of V, we define for the image $L_\mathfrak{p}(V/T)$ of $L_\mathfrak{p}(V)$ in $H^1(M_\mathfrak{p}, V/T)$

$$\text{(1.1)} \quad \mathrm{Sel}_M(A) = \mathrm{Ker}(H^1(\mathfrak{G}_M, A) \to \prod_\mathfrak{p}(\frac{H^1(M_\mathfrak{p}, A)}{L_\mathfrak{p}(A)})) \text{ for } A = V, V/T.$$

The classical Selmer group of V is given by $\mathrm{Sel}_M(V/T)$, equipped with discrete topology. Write F_∞ for the cyclotomic $\mathbb{Z}_p$–extension of F. We define "–" Selmer group $\mathrm{Sel}_M^-(V/T)$ replacing $L_\mathfrak{p}(A)$ by

$$L_\mathfrak{p}^-(V) = \mathrm{Ker}(\mathrm{Res} : H^1(M_\mathfrak{p}, V) \to H^1(I_\mathfrak{p}, \frac{V}{\mathcal{F}_\mathfrak{p}^-(V)})).$$

Lemma 1.3. *Suppose $R \cong K[[X_\mathfrak{p}]]_{\mathfrak{p}|p}$. Then we have* $\mathrm{Sel}_F^-(V) \cong \mathrm{Hom}_K(\mathfrak{m}_R/\mathfrak{m}_R^2, K)$ *and* $\mathrm{Sel}_F(V) = 0$.

Proof. Consider the space $Der_K(R, K)$ of continuous K-derivations. Let $K[\varepsilon] = K[t]/(t^2)$ for the dual number $\varepsilon = (t \mod t^2)$. Then writing K-algebra homomorphism $\phi : R \to K[\varepsilon]$ as $\phi(r) = \phi_0(r) + \phi_1(r)\varepsilon$ and sending ϕ to $\phi_1 \in Der_K(R, K)$, we have $\mathrm{Hom}_{K\text{-alg}}(R, K[\varepsilon]) \cong Der_K(R, K) = \mathrm{Hom}_K(\mathfrak{m}_R/\mathfrak{m}_R^2, K)$. By the universality of (R, ρ), we have

$$\mathrm{Hom}_{K\text{-alg}}(R, K[\varepsilon]) \cong \frac{\{\rho : \mathfrak{G} \to GL_2(K[\varepsilon]) | \rho \text{ satisfies the condtions (K1–4)}\}}{\cong}.$$

Pick ρ as above. Write $\rho(\sigma) = \rho_0(\sigma) + \rho_1(\sigma)\varepsilon$. Then $c_\rho = \rho_1\rho_F^{-1}$ can be easily checked to be a 1-cocycle having values in $M_2(K) \supset V$. Since $\det(\rho) = \det(\rho_F) \Rightarrow \mathrm{Tr}(c_\rho) = 0$, c_ρ has values in V. By the reducibility condition (K2), $[c_\rho] \in \mathrm{Sel}_F^-(V)$. We see easily that $\rho \cong \rho' \Leftrightarrow [c_\rho] = [c_{\rho'}]$. We can reverse the above argument starting a cocycle c giving an element of $\mathrm{Sel}_F^-(V)$ to construct a deformation ρ_c with values in $K[\varepsilon]$. Thus we have

$$\frac{\{\rho : \mathrm{Gal}(\overline{\mathbb{Q}}/F) \to GL_2(K[\varepsilon]) | \rho \text{ satisfies the condtions (K1–4)}\}}{\cong} \cong \mathrm{Sel}_F^-(V).$$

Since the algebra structure of R over $W[[X_{\mathfrak{p}}]]_{\mathfrak{p}|p}$ is given by $\delta_{\mathfrak{p}}\alpha_{\mathfrak{p}}^{-1}$, the K-derivation $\delta : R \to K$ corresponding to a $K[\varepsilon]$-deformation ρ is a $W[[X_{\mathfrak{p}}]]$-derivation if and only if $\rho_1|_{\mathrm{Gal}(\overline{F}_{\mathfrak{p}}/F_{\mathfrak{p}})} \sim \left(\begin{smallmatrix} * & * \\ 0 & 0 \end{smallmatrix}\right)$, which is equivalent to $[c_\rho] \in \mathrm{Sel}_F(V)$, because we already knew that $\mathrm{Tr}(c_\rho) = 0$. Thus we have $\mathrm{Sel}_F(V) \cong Der_{W[[X_{\mathfrak{p}}]]}(R, K) = 0$. □

We also have

Lemma 1.4.

$$\text{(V)} \qquad \mathrm{Sel}_F(V) = 0 \Rightarrow H^1(\mathfrak{G}, V) \cong \prod_{\mathfrak{p}} \frac{H^1(F_{\mathfrak{p}}, V)}{L_{\mathfrak{p}}(V)}.$$

Indeed, by the Poitou-Tate exact sequence, the following sequence is exact:

$$\mathrm{Sel}_F(V) \to H^1(\mathfrak{G}_M, V) \to \prod_{\mathfrak{p}} \frac{H^1(F_{\mathfrak{p}}, V)}{L_{\mathfrak{p}}(V)} \to \mathrm{Sel}_F(V^*(1))^*.$$

An old theorem of Greenberg gives $\dim \mathrm{Sel}_F(V) = \dim \mathrm{Sel}_F(V^*(1))^*$ (see [G] Proposition 2); so, we have the assertion (V). □

1.2 Greenberg's $\mathcal{L}$-invariant

Here is Greenberg's definition of $\mathcal{L}(V)$: The long exact sequence of $\mathcal{F}_{\mathfrak{p}}^- V/\mathcal{F}_{\mathfrak{p}}^+ V \hookrightarrow V/\mathcal{F}_{\mathfrak{p}}^+ V \twoheadrightarrow V/\mathcal{F}_{\mathfrak{p}}^- V$ gives a homomorphism, noting $F_{\mathfrak{p}} = \mathbb{Q}_p$,

$$H^1(F_{\mathfrak{p}}, \mathcal{F}_{\mathfrak{p}}^- V/\mathcal{F}_{\mathfrak{p}}^+ V) = \mathrm{Hom}(\mathfrak{G}_{\mathbb{Q}_p}^{ab}, \mathcal{F}_{\mathfrak{p}}^- V/\mathcal{F}_{\mathfrak{p}}^+ V) \xrightarrow{\iota_{\mathfrak{p}}} H^1(F_{\mathfrak{p}}, V)/L_{\mathfrak{p}}(V).$$

Note that

$$\mathrm{Hom}(G_{\mathbb{Q}_p}^{ab}, \mathcal{F}_{\mathfrak{p}}^- V/\mathcal{F}_{\mathfrak{p}}^+ V) \cong (\mathcal{F}_{\mathfrak{p}}^- V/\mathcal{F}_{\mathfrak{p}}^+ V)^2 \cong K^2$$

canonically by $\phi \mapsto (\frac{\phi([\gamma, F_{\mathfrak{p}}])}{\log_p(\gamma)}, \phi([p, F_{\mathfrak{p}}]))$. Here $[x, F_{\mathfrak{p}}] = [x, \mathbb{Q}_p]$ is the local Artin symbol (suitably normalized). Since

$$L_p(\mathcal{F}_{\mathfrak{p}}^- V/\mathcal{F}_{\mathfrak{p}}^+ V) = \mathrm{Ker}(H^1(F_{\mathfrak{p}}, \mathcal{F}_{\mathfrak{p}}^- V/\mathcal{F}_{\mathfrak{p}}^+ V) \xrightarrow{\mathrm{Res}} H^1(I_{\mathfrak{p}}, \mathcal{F}_{\mathfrak{p}}^- V/\mathcal{F}_{\mathfrak{p}}^+ V)),$$

the image of $\iota_\mathfrak{p}$ is isomorphic to $\mathcal{F}_\mathfrak{p}^- V/\mathcal{F}_\mathfrak{p}^+ V \cong K$. By (V), we have a unique subspace $\mathbb{H}$ of $H^1(\mathfrak{G}, V)$ projecting down onto

$$\prod_\mathfrak{p} \mathrm{Im}(\iota_\mathfrak{p}) \hookrightarrow \prod_\mathfrak{p} \frac{H^1(F_\mathfrak{p}, V)}{L_\mathfrak{p}(V)}.$$

Then by the restriction, $\mathbb{H}$ gives rise to a subspace L of

$$\prod_\mathfrak{p} \mathrm{Hom}(G_{F_\mathfrak{p}}^{ab}, \mathcal{F}_\mathfrak{p}^- V/\mathcal{F}_\mathfrak{p}^+ V) \cong \prod_\mathfrak{p} (\mathcal{F}_\mathfrak{p}^- V/\mathcal{F}_\mathfrak{p}^+ V)^2$$

isomorphic to $\prod_\mathfrak{p}(\mathcal{F}_\mathfrak{p}^- V/\mathcal{F}_\mathfrak{p}^+ V)$. If a cocycle c representing an element in $\mathbb{H}$ is unramified, it gives rise to an element in $\mathrm{Sel}_F(V)$. By the vanishing of $\mathrm{Sel}_F(V)$ (Lemma 1.3), this implies $c = 0$; so, the projection of L to the first factor $\prod_\mathfrak{p} \frac{\mathcal{F}_\mathfrak{p}^- V}{\mathcal{F}_\mathfrak{p}^+ V}$ (via $\phi \mapsto (\phi([\gamma, F_\mathfrak{p}])/\log_p(\gamma))_\mathfrak{p}$) is surjective. Thus this subspace L is a graph of a K–linear map $\mathcal{L} : \prod_\mathfrak{p} \mathcal{F}_\mathfrak{p}^- V/\mathcal{F}_\mathfrak{p}^+ V \to \prod_\mathfrak{p} \mathcal{F}_\mathfrak{p}^- V/\mathcal{F}_\mathfrak{p}^+ V$. We then define $\mathcal{L}(V) := \det(\mathcal{L}) \in K$.

Let $\boldsymbol{\rho} : \mathfrak{G}_F \to GL_2(R)$ be the universal nearly ordinary deformation with $\boldsymbol{\rho}\big|_D = \begin{pmatrix} * & * \\ 0 & \boldsymbol{\delta} \end{pmatrix}$. Then $c_\mathfrak{p} = \frac{\partial \boldsymbol{\rho}}{\partial X_\mathfrak{p}}|_{X=0}\rho_F^{-1}$ is a 1-cocycle (by the argument proving Lemma 1.3) giving rise to a class of $\mathbb{H}$. By Lemma 1.3, $\mathbb{H} = \mathrm{Sel}_F^-(V)$, and $\{c_\mathfrak{p}\}_\mathfrak{p}$ gives a basis of $\mathbb{H}$ over K. We have $\boldsymbol{\delta}([u, F_\mathfrak{p}]) = (1 + X_\mathfrak{p})^{\log_p(u)/\log_p(\gamma)}$ for $u \in O_\mathfrak{p}^\times = \mathbb{Z}_p^\times$. Writing

$$c_\mathfrak{p}(\sigma) = \begin{pmatrix} -a_\mathfrak{p}(\sigma) & * \\ 0 & a_\mathfrak{p}(\sigma) \end{pmatrix} \rho_F(\sigma)^{-1},$$

we have $a_\mathfrak{p} = \boldsymbol{\delta}^{-1}\frac{d\boldsymbol{\delta}}{dX_\mathfrak{p}}|_{X=0}$, and from this we get the desired formula of $\mathcal{L}(Ad(\rho_F))$. This finishes the proof of Theorem 0.4. □

If one restricts $c \in \mathbb{H}$ to $\mathfrak{G}_\infty = \mathrm{Gal}(F^{(p)}/F_\infty)$, its ramification is exhausted by $\Gamma = \mathrm{Gal}(F_\infty/F)$ (because $F_\mathfrak{p} = \mathbb{Q}_p$) giving rise to a class $[c] \in \mathrm{Sel}_{F_\infty}(V)$. The kernel of the restriction map: $H^1(\mathfrak{G}, V) \to H^1(\mathfrak{G}_\infty, V)$ is given by $H^1(\Gamma, H^0(\mathfrak{G}_\infty, V)) = 0$ because $H^0(\mathfrak{G}_\infty, V) = 0$. Thus the image of $\mathbb{H}$ in $\mathrm{Sel}_{F_\infty}(V/T)$ gives rise to the order d exceptional zero of $L^{arith}(s, Ad(\rho_F))$ at $s = 1$. We have proved

Proposition 1.5. *For the number of prime factors $e = [F : \mathbb{Q}]$ of p in F, we have*

$$\mathrm{ord}_{s=1} L_p^{arith}(s, Ad(\rho_F)) \geq e.$$

2 Lecture 2: Elliptic curves with multiplicative reduction

Let p be an odd prime. Order the prime factors of p in F as $\mathfrak{p}_1, \dots, \mathfrak{p}_e$. In this lecture, we describe the computation of the $\mathcal{L}$-invariant of $Ad(T_pE)$ for a modular elliptic curve $E_{/F}$ with split multiplicative reduction at $\mathfrak{p}_j|p > 2$ for $j = 1, 2, \dots, k$ and ordinary good reduction at $\mathfrak{p}_j|p$ for $j > k$.

Theorem 2.1. *Assume that $R \cong \mathbb{Q}_p[[X_\mathfrak{p}]]_{\mathfrak{p}|p}$. Suppose that the Hilbert-modular elliptic curve E has split multiplicative reduction at $\mathfrak{p}_j$ for $j = 1, 2, \dots, k$ $(k \le e)$ with Tate period q_j at $\mathfrak{p}_j$ for $j \le k$ and has ordinary good reduction at $\mathfrak{p}_i$ with $i > k$. Then for the local Artin symbol $[p, F_\mathfrak{p}] = Frob_\mathfrak{p}$ and the norm $Q_j = N_{F_{\mathfrak{p}_j}/\mathbb{Q}_p}(q_j)$, we have for $\rho_E = T_pE$*

$$\mathcal{L}(\mathrm{Ind}_F^{\mathbb{Q}} Ad(\rho_E)) \\ = \left(\prod_{j=1}^{k} \frac{\log_p(Q_j)}{\mathrm{ord}_p(Q_j)}\right) \cdot \det\left(\frac{\partial \delta_\mathfrak{p}([p, F_i])}{\partial X_j}\right)_{i>k, j>k} \Big|_{X=0} \prod_{i>k} \frac{\log_p(\gamma_{\mathfrak{p}_i})}{\alpha_\mathfrak{p}([p, F_i])},$$

where, with the identification of the group algebra $W[[\Gamma_\mathfrak{p}]]$ and $W[[X_\mathfrak{p}]]$, $\gamma_\mathfrak{p}$ is the generator of the p-profinite part $\Gamma_\mathfrak{p}$ of $\mathcal{N}(\mathrm{Gal}(F_\mathfrak{p}[\mu_{p^\infty}]/F_\mathfrak{p}))$.

In the proof, for simplicity, as before, we assume that p is completely split in $F/\mathbb{Q}$. Also, again for simplicity, in the following proof, we assume E has good reduction outside p and $k = 1$. We put $\Gamma_F = \prod_\mathfrak{p} \Gamma_\mathfrak{p}$.

2.1 Hecke algebras for quaternion algebras

We make some preparation for the proof, gathering known facts. We assume that $F \neq \mathbb{Q}$ (otherwise the theorem is known by Greenberg-Stevens). For simplicity, p splits completely in $F/\mathbb{Q}$. Take first a quaternion algebra $B_{0/F}$ central over F unramified everywhere such that $B_0 \otimes_\mathbb{Q} \mathbb{R} \cong M_2(\mathbb{R})^r \times \mathbb{H}^{d-r}$ with $0 \le r \le 1$ (so $r \equiv d \mod 2$). Then we consider the automorphic variety (either a Shimura curve $(r = 1)$ or a 0-dimensional point set $(r = 0)$) given by

$$X_{11}(p^n) = B_0^\times \backslash B_{0,\mathbb{A}}^\times / S_{11}(p^n) Z_\mathbb{A} C_\infty,$$

where $Z_{\mathbb{A}} \cong F_{\mathbb{A}}^{\times}$ is the center of $B_{\mathbb{A}}^{\times}$, C_{∞} is a maximal compact subgroup of the identity component of $B_{0,\infty}^{\times}$ and identifying $B_{0,\mathfrak{l}}^{(\infty)} = B_0 \otimes_{\mathbb{Q}} F_{\mathfrak{l}}$ with $M_2(F_{\mathfrak{l}})$ for all primes $\mathfrak{l}$,

$$S_{11}(p^n) = \left\{ \left(\begin{smallmatrix} a & b \\ c & d \end{smallmatrix}\right) \in GL_2(\widehat{O}) \middle| \left(\begin{smallmatrix} a & b \\ c & d \end{smallmatrix}\right) \equiv \left(\begin{smallmatrix} 1 & * \\ 0 & 1 \end{smallmatrix}\right) \mod p^n \right\}$$

for $\widehat{O} = \prod_{\mathfrak{l}} O_{\mathfrak{l}}$. Consider $M_n \cong H_r(X_{11}(p^n), \mathbb{Z}_p)$ which is the Pontryagin dual of $H^r(X_{11}(p^n), \mathbb{Q}_p/\mathbb{Z}_p)$ which is a finite rank free $\mathbb{Z}_p$-module with Hecke operator action of $T(\mathfrak{n})$ for all prime ideals outside p and $U(p_{\mathfrak{p}}^n) = U(p_{\mathfrak{p}})^n$ and the diamond operator action $\langle z \rangle$ coming from $\left(\begin{smallmatrix} z & 0 \\ 0 & 1 \end{smallmatrix}\right)$ for $z \in O_p$. Let $e = \lim_{n \to \infty} U(p)^{n!}$ as an operator acting on M_n ($U(p) = \prod_{\mathfrak{p}} U(p_{\mathfrak{p}})$). Let M_n^{ord} be the direct summand eM_n. We have natural trace map $M_m \twoheadrightarrow M_n$ for $m > n$ compatible with all Hecke operators and all diamond operators. By the diamond operator action, $M_{\infty}^{ord} = \varprojlim_n M_n^{ord}$ naturally become a $W[[\Gamma_F]]$-module. Here is an old theorem of mine:

Theorem 2.2. *The $W[[\Gamma_F]]$-module M_{∞}^{ord} is free of finite rank over $W[[\Gamma_F]]$.*

Let $\mathbf{h}$ be the $W[[\Gamma_F]]$-algebra generated over $W[[\Gamma_F]]$ by $T(\mathfrak{n})$ for all $\mathfrak{n}$ prime to p and all $U(\mathfrak{p})$. Then we have

Corollary 2.3. *The $W[[\Gamma_F]]$-module $\mathbf{h}$ is torsion free of finite type with $\mathbf{h}_F/(X_{\mathfrak{p}})_{\mathfrak{p}|p}\mathbf{h}_F$ pseudo isomorphic to the Hecke algebra associated with $H_r(X_{11}(p), W)$.*

Actually if $p \geq 5$, $\mathbf{h}$ is known to be free over $W[[\Gamma_F]]$ and the pseudo isomorphism as above is actually an isomorphism.

Let $\mathbb{T}$ be the local ring of the universal nearly ordinary Hecke algebra $\mathbf{h}$ acting nontrivially on the Hecke eigenform associated to E. Let $P \in \mathrm{Spf}(\mathbb{T})(\mathbb{Q}_p)$ corresponding to ρ_E, that is, $\rho_{\mathbb{T}} \mod P \sim \rho_E$. Let $\widehat{\mathbb{T}}_P = \varprojlim_n \mathbb{T}_P/P^n\mathbb{T}_P$ for the localization $\mathbb{T}_P$. Since $\rho_E = T_pE \otimes \mathbb{Q}_p$ is absolutely irreducible, by the technique of pseudo representation, we can construct the modular deformation $\rho_{\mathbb{T}} : \mathfrak{G} \to GL_2(\widehat{\mathbb{T}}_P)$ which satisfies (K1–4); in particular, $\det \rho_{\mathbb{T}} = \mathcal{N}$, because the central character is trivial. Since E is modular over F, we have the surjective $\mathbb{Q}_p$-algebra homomorphism $R \to \widehat{\mathbb{T}}_P$ for the localization-completion $\widehat{\mathbb{T}}_P$. Since $\widehat{\mathbb{T}}_P$ is integral and of dimension d, we have

Corollary 2.4. *If $R \cong K[[X_{\mathfrak{p}}]]_{\mathfrak{p}|p}$, then $R \cong \widehat{\mathbb{T}}_P$.*

Under absolute irreducibility of $\overline{\rho}_F$ over $\mathrm{Gal}(\overline{F}/F[\mu_p])$ with non-scalar semi-simplification of $\overline{\rho}_F|_{\mathrm{Gal}(\overline{F}_{\mathfrak{p}}/F_{\mathfrak{p}})}$ for all $\mathfrak{p}|p$, the isomorphism $R \cong K[[X_{\mathfrak{p}}]]_{\mathfrak{p}|p}$ will be proven by showing $R \cong \widehat{\mathbb{T}}_P$ first (see Appendix).

Take a quaternion algebra $B_{1/F}$ such that $B_1 \otimes_{\mathbb{Q}} \mathbb{R} \cong M_2(\mathbb{R})^q \times \mathbb{H}^{d-q}$ with $q \leq 1$ and B is ramified only at $\mathfrak{p}_1$ (among finite places). Then at $\mathfrak{p}_1$, we have a unique maximal order R_1 in $B_{\mathfrak{p}_1}$. Then we define $U_{11}(p^n)$ to be the product of $S_{11}(p^n)^{(\mathfrak{p}_1)}$ and $R_1^\times$ and define

$$Y_{11}(p^n) = B_1^\times \backslash B_{1,\mathbb{A}}^\times / U_{11}(p^n) Z_{\mathbb{A}} C_\infty.$$

Then we define $e_1 = \lim_{n\to\infty} U(p^{(\mathfrak{p}_1)})^{n!}$ acting on $N_n = H_q(Y_{11}(p^n), \mathbb{Z}_p)$, the dual of the cohomology group $H^q(Y_{11}(p^n), \mathbb{Q}_p/\mathbb{Z}_p)$. Let $\Gamma_1 = \prod_{\mathfrak{p}\neq\mathfrak{p}_1} \Gamma_{\mathfrak{p}}$. We go through all the above process and define $\mathbf{h}_1 \subset \mathrm{End}_{W[[\Gamma_1]]}(\varprojlim_n e_1 N_n)$. Since ρ_E (or corresponding automorphic representation π_E) is Steinberg at $\mathfrak{p}_1$, by the Jacquet-Langlands correspondence, we have a Hecke eigenvector f_1 in $H^q(Y_{11}(p), \mathbb{Z}_p)$ giving rise to E. Then we define $\mathbb{T}_1$ to be the local ring of $\mathbf{h}_1$ acting nontrivially on f_1. Let $P_1 \in \mathrm{Spf}(\mathbb{T}_1)(W)$ be the point associated to ρ_E. We then have a deformation $\rho_{\mathbb{T}_1} : \mathfrak{G} \to GL_2(\widehat{\mathbb{T}}_{1,P})$ of ρ_E. Since the central character is trivial, we have $\det \rho_{\mathbb{T}_1} = \mathcal{N}$.

Theorem 2.5. *We have*

1. $\mathbf{h}_1$ *is torsion-free of finite rank over* $W[[\Gamma_1]]$, *and*
$$\widehat{\mathbb{T}}_{1,P_1} \cong K[[X_{\mathfrak{p}_2}, \ldots, X_{\mathfrak{p}_d}]];$$

2. $\rho_{\mathbb{T}_1}$ *restricted to* $\mathrm{Gal}(\overline{F}_{\mathfrak{p}_1}/F_{\mathfrak{p}_1})$ *is isomorphic to* $\begin{pmatrix} \varepsilon\mathcal{N} & * \\ 0 & \varepsilon \end{pmatrix}$, *where* $\varepsilon = \pm 1$ *is the eigenvalue of* $Frob_{\mathfrak{p}_1}$ *on the étale quotient of* T_pE;

3. *There is a surjective algebra homomorphism* $\mathbb{T}/X_{\mathfrak{p}_1}\mathbb{T} \twoheadrightarrow \mathbb{T}_1$ *inducing an isomorphism* $\widehat{\mathbb{T}}_P/X_{\mathfrak{p}_1}\widehat{\mathbb{T}}_P \cong \widehat{\mathbb{T}}_{1,P_1}$;

4. *There is a surjective algebra homomorphism* $\mathbb{T}/(U(\mathfrak{p}_1) - \varepsilon)\mathbb{T} \twoheadrightarrow \mathbb{T}_1$ *sending* $T(\mathfrak{n})$ *to* $T(\mathfrak{n})$, *where* $U(\mathfrak{p}_1) = U(p_{\mathfrak{p}_1})$.

Here is a sketch of proof. The first assertion follows from construction; in other words, it can be proven by the same way as the proof of Corollary 2.3. By the Jacquet-Langlands correspondence, $\mathbb{T}$ covers $\mathbb{T}_1$. Any automorphic representation π corresponding to a point of $\mathrm{Spf}(\mathbb{T}_1)(\overline{\mathbb{Q}}_p)$ is Steinberg at $\mathfrak{p}_1$ because B_1 ramifies at $\mathfrak{p}_1$. Since points corresponding classical automorphic representation is Zariski dense in $\mathrm{Spf}(\mathbb{T}_1)$, the Galois representation has to have the form as in (2). Thus the eigenvalue of $U(\mathfrak{p}_1)$ of π is ± 1 and the corresponding Galois representation has the form as in (2). The assertion (1) implies (3). By (2), $U(\mathfrak{p}_1)$ is either ± 1. Since $U(\mathfrak{p}_1)$ is a formal function on the connected $\mathrm{Spf}(\mathbb{T}_1)$, $U(\mathfrak{p}_1) = \varepsilon$ is a constant, which implies (4). □

2.2 Proof of Theorem 2.1

Write for simplicity, $X_j := X_{\mathfrak{p}_j}$, $F_j = F_{\mathfrak{p}_j}$ and $p_j = p_{\mathfrak{p}_j}$. By (3) and (4) of Theorem 2.5, $U(\mathfrak{p}_1) \equiv \varepsilon \mod X_1$ is a constant independent of $X_j := X_{\mathfrak{p}_j}$ for all $j \geq 2$. Thus $\frac{\partial U(\mathfrak{p}_1)}{\partial X_j}|_{X_1=0} = 0$ for all $j \geq 2$. Thus

$$\det\left(\frac{\partial U(\mathfrak{p}_i)}{\partial X_j}\right)\Big|_{X=0} = \frac{\partial U(\mathfrak{p}_1)}{\partial X_1}\Big|_{X_1=0} \times \det\left(\frac{\partial U(\mathfrak{p}_i)}{\partial X_j}\right)_{i\geq 2, j\geq 2}\Big|_{X=0}.$$

Since $\delta_{\mathfrak{p}_i}([p, F_i]) = U(\mathfrak{p}_i)$, we get from the formula we stated in the first lecture:

$$\begin{aligned}&\mathcal{L}(\mathrm{Ind}_F^{\mathbb{Q}} Ad(\rho_E))\\ &\quad= \det\left(\frac{\partial \delta_{\mathfrak{p}}([p, F_{\mathfrak{p}}])}{\partial X_{\mathfrak{p}'}}\right)_{\mathfrak{p},\mathfrak{p}'}\Big|_{X=0} \prod_{\mathfrak{p}} \log_p(\gamma_{\mathfrak{p}})\alpha_{\mathfrak{p}}([p, F_{\mathfrak{p}}])^{-1},\end{aligned}$$

the following new formula:

$$\begin{aligned}(2.1)\quad \mathcal{L}(\mathrm{Ind}_F^{\mathbb{Q}} Ad(\rho_E)) &= \frac{\partial \delta_{\mathfrak{p}}([p, F_1])}{\partial X_1}\Big|_{X_1=0} \log_p(\gamma_{\mathfrak{p}_1})\alpha_{\mathfrak{p}_1}([p, F_1])^{-1}\\ &\times \det\left(\frac{\partial \delta_{\mathfrak{p}}([p, F_i])}{\partial X_j}\right)_{i\geq 2, j\geq 2}\Big|_{X=0} \prod_{j\geq 2} \log_p(\gamma_{\mathfrak{p}_j})\alpha_{\mathfrak{p}}([p, F_j])^{-1}.\end{aligned}$$

Thus the result follows from the following lemma of Greenberg-Stevens:

Lemma 2.6. *Let us write $\gamma = \gamma_{\mathfrak{p}_1}$. We have, for $\delta_1 = \delta_{\mathfrak{p}_1}$,*

$$\frac{\partial \delta_1([p, F_1])}{\partial X_1}\Big|_{X_1=0} \log_p(\gamma)\alpha_{\mathfrak{p}_1}([p, F_1])^{-1} = \frac{\log_p(q_1)}{\mathrm{ord}_p(q_1)}$$

Proof. Since $\alpha_{\mathfrak{p}_1}([p, F_1]) = 1$ (split multiplicative reduction), we can forget about this factor. Since the linear operator $\mathcal{L} : \prod_{\mathfrak{p}} \mathcal{F}_{\mathfrak{p}}^- V/\mathcal{F}_{\mathfrak{p}}^+ V \to \prod_{\mathfrak{p}} \mathcal{F}_{\mathfrak{p}}^- V/\mathcal{F}_{\mathfrak{p}}^+ V$ induces $\mathcal{L}_1 : \mathcal{F}_{\mathfrak{p}_1}^- V/\mathcal{F}_{\mathfrak{p}_1}^+ V \to \mathcal{F}_{\mathfrak{p}_1}^- V/\mathcal{F}_{\mathfrak{p}_1}^+ V$ by our diagonalization of its matrix. This $\mathcal{L}_1$ comes from the subspace

$$L_1 \subset \mathrm{Hom}(D_1^{ab}, \mathcal{F}_{\mathfrak{p}_1}^- V/\mathcal{F}_{\mathfrak{p}_1}^+ V) \cong \mathrm{Hom}(D_1^{ab}, \mathbb{Q}_p)$$

for $D_1 = \mathrm{Gal}(\overline{\mathbb{Q}}_p/F_1)$ has a generator $\phi_0 = \delta_1^{-1} \frac{\partial \delta_1}{\partial X_1}\Big|_{X_1=0} : D_1^{ab} \to \mathbb{Q}_p$. Thus by definition

$$\frac{\partial \delta_1([p, F_1])}{\partial X_1}\Big|_{X_1=0} \log_p(\gamma) = \log_p(\gamma) \frac{\phi_0([p, F_1])}{\phi_0([\gamma, F_1])}.$$

Let $\rho_E = T_p E \otimes_{\mathbb{Z}_p} \mathbb{Q}_p$, $\widetilde{\rho}_E = (\rho \mod (X_1^2, X_2, \ldots, X_d))$, and write $\widetilde{\mathbb{Q}_p} = \mathbb{Q}_p[X_1]/(X_1^2)$. The character $(\delta_1 \mod X_1^2)$ is an infinitesimal deformation of the trivial character fitting into the following commutative diagram of D_1-modules:

$$\begin{array}{ccccc} \widetilde{\mathbb{Q}_p}(\epsilon_1) & \xrightarrow{\hookrightarrow} & \widetilde{\rho}_E & \xrightarrow{\twoheadrightarrow} & \widetilde{\mathbb{Q}_p}(\delta_1) \\ \downarrow & & \downarrow & & \downarrow \\ \mathbb{Q}_p(1) & \longrightarrow & \rho_E & \longrightarrow & \mathbb{Q}_p. \end{array}$$

Twist this diagram by $\epsilon_1^{-1}\mathcal{N} = \delta_1$, getting a new diagram

$$\begin{array}{ccccc} \widetilde{\mathbb{Q}_p}(1) & \xrightarrow{\hookrightarrow} & \widetilde{\rho}_E & \xrightarrow{\twoheadrightarrow} & \widetilde{\mathbb{Q}_p}(\delta_1^2) \\ \downarrow & & \downarrow & & \downarrow \\ \mathbb{Q}_p(1) & \longrightarrow & \rho_E & \longrightarrow & \mathbb{Q}_p. \end{array}$$

Once this type of diagram is obtained (with leftmost column given by $\widetilde{\mathbb{Q}_p}(1) \twoheadrightarrow \mathbb{Q}_p(1)$), by a general result of Greenberg-Stevens in such a situation (see [GS1] (2.3.4)), we get

$$\frac{\partial \delta_1^2}{\partial X_1}([q_1, \mathbb{Q}_p])\Big|_{X_1=0} = 0 \Rightarrow \frac{\partial \delta_1}{\partial X_1}([q_1, \mathbb{Q}_p])\Big|_{X_1=0} = 0.$$

Write $q_1 = p^a u$ for $a = \mathrm{ord}_p(q_1)$ and $u \in \mathbb{Z}_p^\times$. Then $\log_p(u) = \log_p(q_1)$. We have

$$\begin{aligned} \delta_1([q_1, \mathbb{Q}_p]) &= \delta_1([p, \mathbb{Q}_p])^a \delta_1([u, \mathbb{Q}_p]) \\ &= \delta_1([p, \mathbb{Q}_p])^a (1 + X_1)^{-\log_p(u)/\log_p(\gamma)} \end{aligned}$$

(because $\mathcal{N}([u,\mathbb{Q}_p]) = u^{-1}$). Differentiating this identity with respect to X_1, we get from $\delta_1([u,\mathbb{Q}_p])|_{X_1=0} = \delta_1([p,\mathbb{Q}_p])|_{X_1=0} = 1$

$$a\frac{\partial\delta_1([p_1,\mathbb{Q}_p])}{\partial X_1}\Big|_{X_1=0} - \frac{\log_p(q_1)}{\log_p(\gamma)} = a\frac{\partial\delta_1([p_1,\mathbb{Q}_p])}{\partial X_1}\Big|_{X_1=0} - \frac{\log_p(u)}{\log_p(\gamma)} = 0.$$

From this, we conclude

$$\log_p(\gamma)\frac{\partial\delta_1([p_1,\mathbb{Q}_p])}{\partial X_1}\Big|_{X_1=0} = \frac{\log_p(q_1)}{\mathrm{ord}_p(q_1)}.$$

□

The fixed field of the kernel of ϕ_0 is a $\mathbb{Z}_p$-extension $M_\infty/\mathbb{Q}_p$ ($F_1 = \mathbb{Q}_p$). Since $L_1 \ni \phi \mapsto \frac{\phi([\gamma,\mathbb{Q}_p])}{\log_p \gamma} \in \mathbb{Q}_p$ is surjective, M_∞ ramifies fully. Then by local class field theory, $\bigcap_{n=1}^{\infty} N_{M_n/\mathbb{Q}_p}(M_n^\times)$ has a rank 1 torsion-free part, which contains $q_0 = p^b v$ with $a \neq 0$ and $v \in \mathbb{Z}_p^\times$. The quantity $\frac{\log_p(q_0)}{\mathrm{ord}_p(q_0)} \in \mathbb{Q}_p$ is determined uniquely independent of the choice of q_0, and we now prove

Proposition 2.7.

$$\log_p(\gamma)\frac{\partial\delta_1([p_1,\mathbb{Q}_p])}{\partial X_1}\Big|_{X_1=0} = \frac{\log_p(q_0)}{\mathrm{ord}_p(q_0)}.$$

Proof. Let $\phi_0 = \delta_1^{-1}\frac{\partial\delta_1}{\partial X_1} : D_1^{ab} \to \mathbb{Q}_p$. Let $\mathbf{M}_\infty/\mathbb{Q}_p$ be the composite of all $\mathbb{Z}_p$-extensions of $\mathbb{Q}_p$; so, by local class field theory, $\mathrm{Gal}(\mathbf{M}_\infty/\mathbb{Q}_p) \cong \mathbb{Z}_p^2$. Then we have $[q_0,\mathbb{Q}_p] \in \mathrm{Gal}(\mathbf{M}_\infty/M_\infty)$ again by local class field theory, and by definition, $\phi_0([q_0,\mathbb{Q}_p]) = 0$. Since $[q_0,\mathbb{Q}_p] = [v,\mathbb{Q}_p][p,\mathbb{Q}_p]^b$ ($b = \mathrm{ord}_p(q_0)$) we have $0 = \phi_0([q_0,\mathbb{Q}_p]) = \phi_0([v,\mathbb{Q}_p]) + b\phi_0([p,\mathbb{Q}_p])$. Writing $M_\infty^{ur}/\mathbb{Q}_p$ for the unique unramified $\mathbb{Z}_p$-extension and $M_\infty^+/\mathbb{Q}_p$ for the cyclotomic $\mathbb{Z}_p$-extension, the restriction of ϕ_0 to $\Gamma^+ := \mathrm{Gal}(M_\infty^+/\mathbb{Q}_p)$ is a constant multiple of $\log_p \circ \mathcal{N}_p$ for the cyclotomic character $\mathcal{N}_p$; i.e., $\phi_0|_{\Gamma^+} = x(\log_p \circ \mathcal{N}_p)$ for $x \in \mathbb{Q}_p^\times$. Since $\log_p(\mathcal{N}_p([v,\mathbb{Q}_p])) = \log_p(v^{-1}) = -\log_p(q_0)$, we have $x\log_p(v^{-1}) + b\phi_0([p,\mathbb{Q}_p]) = 0$. Thus $\mathcal{L}(Ad(T_pE)) = \phi_0([p,\mathbb{Q}_p])/x = \frac{\log_p(q_0)}{\mathrm{ord}_p(q_0)}$. □

3 Lecture 3: $\mathcal{L}$-invariants of CM fields

Let p be an odd prime. Let M/F be a totally imaginary quadratic extension of the base totally real field F. We study the adjoint square Selmer group when the Galois representation is an induction of a Galois character of $\mathfrak{G}_M := \mathrm{Gal}(M^{(p)}/M)$. Put $\mathfrak{G}_F := \mathrm{Gal}(M^{(p)}/F)$. For simplicity, we assume that $p > 2$ totally splits in $M/\mathbb{Q}$. We relate the Selmer group with a more classical Iwasawa module of a quadratic extension of F, and from the torsion property of the Selmer group already proven, we deduce some (new) torsion property of such classical Iwasawa modules.

3.1 Ordinary CM fields and their Iwasawa modules

Let O_M be the integer ring of M. We consider $Z = \varprojlim_n Cl_M(p^n)$ for the ray class group $Cl_M(p^n)$ of M modulo p^n. Let Δ be the maximal torsion subgroup of Z, and put $\Gamma_M = Z/\Delta$, which has a natural action of $\mathrm{Gal}(M/F)$. We split $Z = \Delta \times \Gamma_M$. We define $\Gamma^+ = H^0(\mathrm{Gal}(M/F), \Gamma_M)$ and $\Gamma^- = \Gamma_M/\Gamma^+$. Since $p > 2$, the action of $\mathrm{Gal}(M/F)$ splits the extension $\Gamma^+ \hookrightarrow \Gamma_M \twoheadrightarrow \Gamma^-$, and we have a canonical decomposition $\Gamma_M = \Gamma^+ \times \Gamma^-$. Write $\pi^- : Z \to \Gamma^-$, $\pi^+ : \Gamma_M \to \Gamma^+$ and $\pi_\Delta : Z \to \Delta$ for the three projections. Take a character $\varphi : \Delta \to \overline{\mathbb{Q}}^\times$, and regard it as a character of Z through the projection: $Z \twoheadrightarrow \Delta$.

Let M_∞ be the composite of all $\mathbb{Z}_p$-extensions of M. Then by class field theory, M_∞ is the subfield of the ray class filed of M modulo p^∞ fixed by Δ. Let $\mathbb{Q}_\infty/\mathbb{Q}$ be the cyclotomic $\mathbb{Z}_p$-extension. Let M_∞^{cyc} be the composite $M\mathbb{Q}_\infty/M$. Define M_∞^- (resp. M_∞^+) for the fixed subfield of Γ^- (resp. Γ^+). Since M_∞^{cyc} is abelian over F, we have $M_\infty^{cyc} \subset M_\infty^+$ and a projection $\pi_{cyc} : \Gamma^+ \to \mathrm{Gal}(M_\infty^{cyc}/M) \subset 1 + p\mathbb{Z}_p$. The Leopoldt conjecture for F asserts that π_{cyc} is an isomorphism; in other words, $M_\infty^+ = M_\infty^{cyc}$. The extension M_∞^-/M is called the anticyclotomic tower over M. Thus if the Leopoldt conjecture holds for F, M_∞ is the composite of the cyclotomic $\mathbb{Z}_p$-extension M_∞^{cyc} and the anticyclotomic $\mathbb{Z}_p^{[F:\mathbb{Q}]}$–extension M_∞^-.

To introduce Iwasawa modules for the multiple $\mathbb{Z}_p$-extensions $M_\infty^?/M$, we fix a CM type Σ, which is a set of embeddings of M into $\overline{\mathbb{Q}}$ such that $I_M = \Sigma \sqcup \Sigma c$ for the generator c of $\mathrm{Gal}(M/F)$. Over $\mathbb{C}$, an abelian variety with complex multiplication by M has $\mathbb{C}$-points isomorphic to $\mathbb{C}^\Sigma/\Sigma(\mathfrak{a})$ for a lattice $\mathfrak{a}$ in M (see [ACM] 5.2), where $\Sigma(\mathfrak{a}) = \{(\sigma(a))_{\sigma\in\Sigma} \in \mathbb{C}^{\Sigma} | a \in \mathfrak{a}\}$. By composing i_p, we write Σ_p for the set of p-adic places induced by $i_p \circ \sigma$ for $\sigma \in \Sigma$. We assume

(spt) $$\Sigma_p \cap \Sigma_p c = \emptyset.$$

This is to guarantee the abelian variety of CM type Σ to have ordinary good reduction modulo p (whose Galois representation is hence ordinary at all $\mathfrak{p}|p$).

Writing $M(p^\infty)$ for the ray class field over M modulo p^∞, we identify Z with $\mathrm{Gal}(M(p^\infty)/M)$ via the Artin reciprocity law. Fix a character φ of Δ. We then define M_Δ by the fixed field of Γ in $M(p^\infty)$; so, $\mathrm{Gal}(M_\Delta/M) = \Delta$.

Since φ is a character of Δ, φ factors through $\mathrm{Gal}(M^?_\infty M_\Delta/M)$ for ? indicating one of $+, -, cyc$ or "nothing". When nothing is attached, it refers to the object for the full multiple $\mathbb{Z}_p$-extension M_∞. Let $L^?_\infty/M^?_\infty M_\Delta$ be the maximal p-abelian extension unramified outside Σ_p. Each $\gamma \in \mathrm{Gal}(L_\infty/M)$ acts on the normal subgroup $X^? = \mathrm{Gal}(L^?_\infty/M^?_\infty M_\Delta)$ continuously by conjugation, and by the commutativity of $X^?$, this action factors through $\mathrm{Gal}(M_\Delta M^?_\infty/M)$. Then we look into the compact p-profinite $\Gamma^?$–module: $X^?[\varphi] = X^? \otimes_{\mathbb{Z}_p[\Delta],\varphi} W$, where $\Gamma^? = \mathrm{Gal}(M^?_\infty/M)$. We study when $X^?[\varphi]$ is a torsion Iwasawa module over $\Lambda^? = W[[\Gamma^?]]$. The module $X^?[\varphi]$ is generally expected to be torsion of finite type over $\Lambda^?$ for the naturally defined multiple $\mathbb{Z}_p$-extensions $M^?_\infty$.

The torsion property of $X^{cyc}[\varphi]$ over Λ^{cyc} is classically known (e.g., [HT2] Theorem 1.2.2). This implies

Theorem 3.1. *The modules $X[\varphi], X^+[\varphi]$ and $X^{cyc}[\varphi]$ are torsion modules over the corresponding Iwasawa algebra Λ, Λ^+ and Λ^{cyc}, respectively.*

We refer this result to [HT2] Theorem 1.2.2 (which was originally due to R. Greenberg). We study the anticyclotomic Iwasawa module $X^-[\varphi]$ over Λ^- from our new view point of Galois deformation theory. As is well known, $X^-[\varphi]$ is a Λ^--module of finite type, and under mild assumptions (including anticyclotomy of φ), we will prove the torsion property of $X^-[\varphi]$ in Theorem 3.3.

The Σ-Leopoldt conjecture for abelian extensions of M is almost equivalent to the torsion property of $X^-[\varphi]$ over Λ^- for all possible φ (see [HT2] Theorem 1.2.2). Here, for an abelian extension L/M with integer ring O_L, the Σ-Leopoldt conjecture asserts the closure $\overline{O_L^\times}$ of $O_L^\times$ in $L_\Sigma = \prod_{\mathfrak{p}\in\Sigma_p} L_{\mathfrak{p}}$ satisfies

$$\dim_{\mathbb{Q}}(O_L^\times \otimes_{\mathbb{Z}} \mathbb{Q}) = \dim_{\mathbb{Q}_p}(\overline{O_L^\times} \otimes_{\mathbb{Z}_p} \mathbb{Q}_p).$$

If $X^-[\varphi]$ is a torsion Λ^--module, we can think of the characteristic element $\mathcal{F}^-(\varphi) \in \Lambda^-$ of the module $X^-[\varphi]$. The anticyclotomic main conjecture (cf. [HT] Conjecture 2.2) predicts the identity (up to units) of $\mathcal{F}^-(\varphi)$ and the projection of (the φ-branch of) the Katz p-adic L-function (constructed in [K] and [HT1]) under π^-.

3.2 Anticyclotomic Iwasawa modules

A character ψ of Δ is called *anticyclotomic* if $\psi(c\sigma c^{-1}) = \psi^{-1}(\sigma)$ for a complex conjugation $c \in \mathrm{Gal}(\overline{\mathbb{Q}}/F)$. Fix an algebraic closure $\overline{F}$ of F. Regarding φ as a Galois character, we define $\varphi^-(\sigma) = \varphi(c\sigma c^{-1}\sigma^{-1})$ for $\sigma \in \mathrm{Gal}(\overline{F}/M)$. Then $\psi := \varphi^-$ is anticyclotomic.

We define a Galois character $\widetilde{\varphi} : \mathfrak{G}_F \to W[[\Gamma^-]]$ by setting $\widetilde{\varphi}(\sigma) = \varphi(\sigma)(\sigma|_{M_\infty^-})^{1/2}$, where $(\sigma|_{M_\infty^-})^{1/2}$ is the unique square root of $(\sigma|_{M_\infty^-})$ in Γ^- and $(\sigma|_{M_\infty^-}) \in \Gamma^-$ is regarded as a group element in $\Gamma^- \subset W[[\Gamma^-]]$. Note that $\widetilde{\varphi}^-(\sigma) = \widetilde{\varphi}(c\sigma c^{-1}\sigma^{-1}) = \psi(\sigma)\sigma|_{M_\infty^-}$. Then we consider $\mathrm{Ind}_M^F(\widetilde{\varphi}) : \mathrm{Gal}(\overline{F}/F) \to GL_2(W[[\Gamma^-]])$. We write $\alpha_{M/F}$ for the quadratic character of $\mathrm{Gal}(\overline{F}/F)$ identifying $\mathrm{Gal}(M/F)$ with $\{\pm 1\}$.

Lemma 3.2. *We have*

1. $\det(\mathrm{Ind}_M^F \chi) = \alpha_{M/F}\chi|_{F_{\mathbb{A}}^\times}$ *and* $\mathrm{Tr}(\mathrm{Ind}_M^F \chi(Frob_{\mathfrak{l}})) = \sum_{\mathfrak{b}\subset O_M, N_{M/F}(\mathfrak{b})=\mathfrak{l}} \chi(\mathfrak{b})$ *for a prime* $\mathfrak{l}$ *of* F *unramified for* $\mathrm{Ind}_M^F \chi$, *identifying a character* χ *of* $\mathrm{Gal}(\overline{F}/M)$ *with a character of* $M_{\mathbb{A}^{(\infty)}}^\times/M^\times$ *by the Artin symbol,*

2. $Ad(\mathrm{Ind}_M^F(\widetilde{\varphi})) \cong \alpha_{M/F} \oplus \mathrm{Ind}_M^F(\widetilde{\varphi}^-)$ *as* $\mathfrak{G}_F$*-modules.*

Since $\mathrm{Ind}_M^F(\widetilde{\varphi})|_{\mathfrak{G}_M} = \widetilde{\varphi} \oplus \widetilde{\varphi}_c$ with $\widetilde{\varphi}_c(\sigma) = \widetilde{\varphi}(c\sigma c^{-1})$, we then define $\mathcal{F}_{\mathfrak{p}}^+ \mathrm{Ind}_M^F \widetilde{\varphi} = \widetilde{\varphi}$ for $\mathfrak{p} \in \Sigma_p$. In Lecture 1, we have already defined $\mathcal{F}_{\mathfrak{p}}^\pm Ad(\mathrm{Ind}_M^F \widetilde{\varphi})$ and the Selmer group $\mathrm{Sel}_F(Ad(\mathrm{Ind}_M^F \widetilde{\varphi}) \otimes_{\mathbb{Z}_p} (W[[\Gamma^-]])^*)$. Since the image of $\mathcal{F}_{\mathfrak{p}}^+(Ad(\mathrm{Ind}_M^F \widetilde{\varphi}))$ in $\alpha_{M/F}$ is trivial in the above decomposition in Lemma 3.2 and the image of $\mathcal{F}_{\mathfrak{p}}^+(Ad(\mathrm{Ind}_M^F \widetilde{\varphi}))$ is given by

$\mathcal{F}_{\mathfrak{p}}^{+}(\mathrm{Ind}_{M}^{F}(\widetilde{\varphi}^{-}))$, we get (cf. [HMI] Exercise 1.12 and Corollary 3.81)

$$\begin{aligned}&\mathrm{Sel}_F(Ad(\mathrm{Ind}_M^F(\widetilde{\varphi})\otimes_{W[[\Gamma^-]]}(W[[\Gamma^-]])^*)\\&=\mathrm{Sel}_F\left(\alpha_{M/F}\otimes_{\mathbb{Z}_p}(W[[\Gamma^-]])^*\right)\oplus\mathrm{Sel}_F(\mathrm{Ind}_M^F(\widetilde{\varphi}^-)\otimes_{W[[\Gamma^-]]}(W[[\Gamma^-]])^*)\\&=\mathrm{Hom}(Cl_M^-\otimes_{\mathbb{Z}}W[[\Gamma^-]],\mathbb{Q}_p/\mathbb{Z}_p)\oplus\mathrm{Sel}_M((\widetilde{\varphi}^-)\otimes_{W[[\Gamma^-]]}W[[\Gamma^-]]^*),\end{aligned}$$

where Cl_M^- is the quotient of CL_M by the image of Cl_F (the order of Cl_M^- is equal to the order of the $\alpha_{M/F}$-eigenspace of Cl_M up to a power of 2). By the definition of the Selmer group, we note that

$$\mathrm{Sel}_M(\widetilde{\varphi}^-\otimes_{W[[\Gamma^-]]}(W[[\Gamma^-]])^*)\cong\mathrm{Hom}(X^-[\varphi^-],\mathbb{Q}_p/\mathbb{Z}_p),\tag{3.1}$$

which shows

Theorem 3.3. *Let the notation be as above. Then we have*

$$\mathrm{Sel}_F^*(Ad(\mathrm{Ind}_M^F(\widetilde{\varphi})))\cong\left(Cl_M^-\otimes_{\mathbb{Z}}W[[\Gamma^-]]\right)\oplus X^-[\varphi^-]$$

as $W[[\Gamma^-]]$-modules. Moreover $X^-[\varphi^-]$ is a torsion $W[[\Gamma^-]]$-module without exceptional zero if $\psi:=\varphi^-$ satisfies the following conditions:

(at1) *The character ψ has order prime to p.*

(at2) *The local character $\psi_{\mathfrak{P}}$ is non-trivial for all $\mathfrak{P}\in\Sigma_p$.*

(at3) *The restriction ψ^* of ψ to $\mathrm{Gal}(\overline{F}/M^*)$ for the composite M^* of M and the unique quadratic extension inside $F[\mu_p]$ is non-trivial.*

The first assertion follows from the argument given as above. The torsion property follows from the theorem of Taylor–Wiles and Fujiwara and the propositions in the following appendix. In [HMI] Theorem 5.33, it is checked under assumptions milder than (at1–3) imply the assumption of the theorem of Taylor–Wiles and Fujiwara.

3.3 The $\mathcal{L}$-invariant of CM fields

Consider the universal $(\mathcal{R}_F,\varrho)$ deforming $\rho_F=\mathrm{Ind}_M^F\varphi \mod \mathfrak{m}_W$ among W-deformations ρ_A into $GL_2(A)$ for proartinian W-algebras A with residue field $W/\mathfrak{m}_W$ satisfying the following conditions

(W1) unramified outside p;

(W2) $\rho_A|_{\mathrm{Gal}(\overline{\mathbb{Q}}_p/F_\mathfrak{p})} \cong \left(\begin{smallmatrix} * & * \\ 0 & \alpha_{A,\mathfrak{p}} \end{smallmatrix}\right)$ with $\alpha_{A,\mathfrak{p}} \equiv \alpha_\mathfrak{p} \mod \mathfrak{m}_A$ and $\alpha_{A,\mathfrak{p}}\alpha_\mathfrak{p}^{-1}|_{I_\mathfrak{p}}$ factoring through $\mathrm{Gal}(F_\mathfrak{p}[\mu_{p^\infty}]/F_\mathfrak{p})$;

(W3) $\det(\rho_A) = \det\rho$;

(W4) $\rho_A \equiv \overline{\rho} \mod \mathfrak{m}_A$,

We know $\mathcal{R}_F \cong \mathbb{T}$ by Fujiwara (see Appendix). Since $\dim\mathrm{Spf}(W[[\Gamma^-]]) = \dim\mathrm{Spf}(\mathbb{T})$, $\mathrm{Spf}(W[[\Gamma^-]])$ gives an irreducible component of $\mathrm{Spf}(\mathcal{R}_F)$. Write $\mathbb{I} = W[[\Gamma^-]]$ simply. Let $\pi_\mathbb{I} : \mathbb{T} = \mathcal{R}_F \twoheadrightarrow \mathbb{I}$ be the projection (which factors through π^{cyc}). We would like to compute the $\mathcal{L}$-invariant of the component $\mathbb{I}$. Thus we need to compute $\mathbf{a}(p_\mathfrak{p}) = \pi_\mathbb{I}(U(p_\mathfrak{p}))$. The following fact follows from the fact $\mathrm{Ind}_M^F \phi|_{\mathfrak{G}_M} = \phi \oplus \phi^c$.

Lemma 3.4. *Let the notation be as above, we have* $\mathbf{a}(p_\mathfrak{p}) = \widetilde{\varphi}([p_\mathfrak{P}, M_\mathfrak{P}])$ *for the prime factor* $\mathfrak{P} \in \Sigma_p^c$ *of* $\mathfrak{p}$.

Define the character $\kappa : \mathrm{Gal}(\overline{F}/M) \to (\Lambda^-)^\times$ by $\sigma \mapsto (\sigma|_{M_\infty^-})^{1/2}$. Then $\widetilde{\varphi} = \lambda\kappa$, and we write $\kappa_\mathbb{I} = \pi_\mathbb{I} \circ \kappa : \mathrm{Gal}(\overline{F}/M) \to \mathbb{I}^\times$. Then, $\kappa_\mathbb{I}$ restricted to the inertia group $I_\mathfrak{P}$ at $\mathfrak{P}$ factors through the projection: $I_\mathfrak{P} \to \mathrm{Gal}(\mathbb{Q}_p[\mu_{p^\infty}]/\mathbb{Q}_p) \cong \mathbb{Z}_p^\times$. Since the $W[[\Gamma_F]]$-algebra structure of $\mathbb{I}$ is induced by the nearly ordinary character of $\mathrm{Ind}_M^F \kappa_\mathbb{I}$ (restricted to the inertia group $I_\mathfrak{P}$), for $u_\mathfrak{P} \in O_{M,\mathfrak{P}}^\times$ ($\mathfrak{P} \in \Sigma_p^c$), we have

$$\kappa_\mathbb{I}([u_\mathfrak{P}, M_\mathfrak{P}]) = (1 + X_\mathfrak{p})^{\log_p(N_\mathfrak{p}(u_\mathfrak{P}))/\log_p(\gamma_\mathfrak{p})}, \tag{3.2}$$

where $\mathfrak{p} = \mathfrak{P} \cap O$, $N_\mathfrak{p} : M_\mathfrak{P} = F_\mathfrak{p} \to \mathbb{Q}_p$ is the norm map and $\gamma_\mathfrak{p}$ is the generator of $\Gamma_\mathfrak{p} := (1 + p\mathbb{Z}_p) \cap N_\mathfrak{p}(O_\mathfrak{p}^\times)$. Choose an element $\varpi(\mathfrak{P}) \in M$ so that $\mathfrak{P}^h = (\varpi(\mathfrak{P}))$ for each $\mathfrak{P} \in \Sigma_p^c$, where $h = |Cl_M|$ (the class number of M). Then $p_\mathfrak{P}^h = u_\mathfrak{P}\varpi(\mathfrak{P})_\mathfrak{P}^{e(\mathfrak{p})}$ with $u_\mathfrak{P} \in O_{M,\mathfrak{P}}^\times$ for the absolute ramification index $e(\mathfrak{p})$ of $\mathfrak{p}$ (which is the absolute ramification index of $\mathfrak{P}$ also). Regarding $\kappa_\mathbb{I}$ as a character of $M_{\mathbb{A}^{(\infty)}}^\times/M^\times$ by class field theory, we have $\kappa_\mathbb{I}(\varpi(\mathfrak{P})) = 1 = \kappa_\mathbb{I}(\varpi(\mathfrak{P})_\mathfrak{l})$ with the $\mathfrak{l}$-component $\varpi(\mathfrak{P})_\mathfrak{l} \in M_\mathfrak{l}^\times$ for any prime $\mathfrak{l}$ outside p, because $\kappa_\mathbb{I}(O_\mathfrak{l}^\times) = 1$ and $\varpi(\mathfrak{P}) \in M^\times$. Then we have

$$\kappa_\mathbb{I}(p_\mathfrak{P}^h) = \kappa_\mathbb{I}(p_\mathfrak{P}^h\varpi(\mathfrak{P})^{-e(\mathfrak{p})}) = \kappa_\mathbb{I}(u_\mathfrak{P}) \prod_{\mathfrak{P}'|p, \mathfrak{P}' \neq \mathfrak{P}} \kappa_\mathbb{I}(\varpi(\mathfrak{P})_{\mathfrak{P}'}^{-e(\mathfrak{p})}),$$

where $\varpi(\mathfrak{P})_{\mathfrak{P}'}$ is the $\mathfrak{P}'$-component of $\varpi(\mathfrak{P}) \in M^\times \subset M_{\mathbb{A}}^\times$. By (3.2), we get

$$\kappa_{\mathbb{I}}(p_{\mathfrak{P}}^h) = (1+X_{\mathfrak{p}})^{\frac{\log_p(N_{\mathfrak{p}}(\varpi(\mathfrak{P})^{e(\mathfrak{p})c}u_{\mathfrak{p}}))}{\log_p(\gamma_{\mathfrak{p}})}} \prod_{\mathfrak{P}' \in \Sigma_p^c - \{\mathfrak{P}\}} (1+X_{\mathfrak{p}'})^{\frac{e(\mathfrak{p})\log_p(N_{\mathfrak{p}'}(\varpi(\mathfrak{P})_{\mathfrak{P}'}^{c-1}))}{\log_p(\gamma_{\mathfrak{p}'})}},$$

where $\mathfrak{p}' = \mathfrak{P}' \cap O$. Here $\log_p$ is the Iwasawa p-adic logarithm defined over $\overline{\mathbb{Q}}_p^\times$ characterized by $\log_p(p) = 0$. In particular, we have

$$\log_p(N_{\mathfrak{p}}(u_{\mathfrak{P}})) = \log(N_{\mathfrak{p}}(p_{\mathfrak{P}}^h \varpi(\mathfrak{P})_{\mathfrak{P}}^{-e(\mathfrak{p})})) = -e(\mathfrak{p})\log_p(N_{\mathfrak{p}}(\varpi(\mathfrak{P})_{\mathfrak{P}})).$$

Thus we have

Lemma 3.5. *Let the notation be as above. Then we have, for primes $\mathfrak{P}' \in \Sigma_p$ and $\mathfrak{p}' = \mathfrak{P}' \cap O$,*

$$\frac{\partial \kappa(p_{\mathfrak{P}})}{\partial X_{\mathfrak{p}'}} = \frac{e(\mathfrak{p})\log_p(N_{\mathfrak{p}'}(\varpi(\mathfrak{P})_{\mathfrak{P}'}^{(c-1)}))}{h\log_p(\gamma_{\mathfrak{p}'})}\kappa(p_{\mathfrak{p}})(1+X_{\mathfrak{p}'})^{-1}.$$

We have $\mathbf{a}(p_{\mathfrak{p}}) = c_{\mathfrak{p}}\kappa(p_{\mathfrak{p}})$ for a nonzero constant $c_{\mathfrak{p}} \in W^\times$, because the nearly ordinary character of $\mathrm{Ind}_M^F \widetilde{\varphi}$ is κ times a character of $D_{\mathfrak{P}}$ with values in $W^\times$. We do not need to pay much attention to the constant $c_{\mathfrak{p}}$, because the formula of the $\mathcal{L}$-invariant only involve

$$\Big(\prod_{\mathfrak{p}|p} \mathbf{a}(p_{\mathfrak{p}})^{-1}\delta_{\mathfrak{p}}([\gamma_{\mathfrak{p}}, F_{\mathfrak{p}}])\Big)\det\left(\left(\frac{\partial \mathbf{a}(p_{\mathfrak{p}})}{\partial X_{\mathfrak{p}'}}\right)_{\mathfrak{p},\mathfrak{p}'}\right)$$

in which the constant $c_{\mathfrak{p}}$ cancels out. Specializing the above formula to the locally cyclotomic point P, we get

Theorem 3.6. *Let the notation and the assumption be as above and as in* Theorem 3.3, *including* (at1–4). *Then we have, for any specialization $\widetilde{\varphi}_P$ of $\widetilde{\varphi}$ modulo a locally cyclotomic point $P \in \mathrm{Spf}(\mathbb{I})(W)$,*

$$\mathcal{L}(Ad(\mathrm{Ind}_M^F \widetilde{\varphi}_P)) = (-1)^e \det\left(\left(\log_p(N_{\mathfrak{p}'}(\varpi(\mathfrak{P})_{\mathfrak{P}'}^{(1-c)}))\right)_{\mathfrak{P},\mathfrak{P}' \in \Sigma_p^c}\right)\prod_{\mathfrak{p}|p}\frac{e(\mathfrak{p})}{h},$$

where $\mathfrak{p} = O \cap \mathfrak{P}$ and $\mathfrak{p}' = O \cap \mathfrak{P}'$.

Note here $\mathrm{ord}_p(\varpi(\mathfrak{P})^{(1-c)}_{\mathfrak{P}'}) = -h/e(\mathfrak{p})$, taking the valuation ord_p associated to $\overline{\mathfrak{P}} \in \Sigma$. By Lemma 3.2 (2) and Theorem 3.3, we see

$$\mathcal{L}(Ad(\mathrm{Ind}_M^F \widetilde{\varphi}_P)) = \mathcal{L}(\alpha_{M/F}),$$

and this is the reason for the independence of $\mathcal{L}(Ad(\mathrm{Ind}_M^F \widetilde{\varphi}_P))$ on the choice of the locally cyclotomic points P. If $F = \mathbb{Q}$, we have $\varpi(\mathfrak{P})\varpi(\mathfrak{P})^c = p^h$ and hence $\log_p(\varpi(\mathfrak{P})) = -\log_p(\varpi(\mathfrak{P})^c)$. Consequently, $\log_p(\varpi(\mathfrak{P})^{1-c}) = 2\log_p(\varpi(\mathfrak{P}))$, and therefore the above formula coincides with the classical analytic $\mathcal{L}$-invariant formula for $\alpha_{M/F}$ of Ferrero–Greenberg.

For a given ordinary CM type (M, Σ_p), we can choose ψ satisfying the assumptions of Theorems 3.3 and 3.6. Then through the above process, we can compute $\mathcal{L}(\alpha_{M/F})$ as follows:

Corollary 3.7. *Suppose that M/F is an ordinary CM-quadratic extension of M satisfying* (spt). *Choose a p-ordinary CM-type Σ of M. Then the $\mathcal{L}$-invariant $\mathcal{L}(\alpha_{M/F})$ of Greenberg for the quadratic Galois character $\alpha_{M/F} = \left(\frac{M/F}{\cdot}\right)$ is given by*

$$(-1)^e \det\left(\left(\log_p(N_{\mathfrak{p}'}(\varpi(\mathfrak{P})^{(1-c)}_{\mathfrak{P}'}))\right)_{\mathfrak{P},\mathfrak{P}'\in\Sigma_p^c}\right)\prod_{\mathfrak{p}|p}\frac{e(\mathfrak{p})}{h},$$

where h is the class number of M, $\mathfrak{p}' = \mathfrak{P}' \cap O$ and $\varpi(\mathfrak{P})$ is a generator of $\mathfrak{P} \in \Sigma_p^c$. If the prime p does not split in $F/\mathbb{Q}$, the $\mathcal{L}$-invariant of $\alpha_{M/F}$ does not vanish.

A regulator similar to the above determinant was introduced long ago in [FeG] (3.8) in the context of (classical) cyclotomic Iwasawa's theory.

Suppose that p does not split in $F/\mathbb{Q}$. Then for $\varpi = \varpi(\mathfrak{P})$, $N_{\mathfrak{P}}(\varpi) = \prod_{\sigma\in\Sigma}\varpi^\sigma = \varpi^\Sigma$. Then we have $|\varpi^\Sigma|_p = 1$ and $|\varpi^{c\Sigma}|_p < 1$ and hence $N_{\mathfrak{P}}(\varpi^{1-c})$ cannot be of the form ζp^α for $\alpha \in \mathbb{Q}$ and a root of unity ζ. Thus $\log_p(N_{\mathfrak{P}}(\varpi^{1-c})) \neq 0$ as claimed in the corollary. Of course, by Baker's argument (exploited by Brumer in the p-adic case), if $M/\mathbb{Q}$ is abelian, we can also confirm the nonvanishing of the determinant in the corollary.

4 Appendix: Differential and adjoint square Selmer group

Recall the **universal** nearly ordinary deformation $\boldsymbol{\rho} : \mathrm{Gal}(\overline{\mathbb{Q}}/F) \to GL_2(R)$ over K with the pro-Artinian local universal K-algebra R. This means that for any Artinian local K-algebra A with maximal ideal $\mathfrak{m}_A$ and any Galois representation $\rho_A : \mathrm{Gal}(\overline{\mathbb{Q}}/F) \to GL_2(A)$ such that

(K1) unramified outside p;

(K2) $\rho_A|_{\mathrm{Gal}(\overline{\mathbb{Q}}_p/F_\mathfrak{p})} \cong \left(\begin{smallmatrix} * & * \\ 0 & \alpha_{A,\mathfrak{p}} \end{smallmatrix}\right)$ with $\alpha_{A,\mathfrak{p}} \equiv \alpha_\mathfrak{p} \mod \mathfrak{m}_A$;

(K3) $\det(\rho_A) = \det \rho$;

(K4) $\rho_A \equiv \rho \mod \mathfrak{m}_A$,

there exists a unique K-algebra homomorphism $\varphi : R \to A$ such that $\varphi \circ \boldsymbol{\rho} \cong \rho_A$. We write $\Phi_K(A)$ the collection of the isomorphism classes of the deformations ρ_A.

Let $\overline{\rho} = (\rho \mod \mathfrak{m}_W)$, and consider a similar deformation changing base ring from K to W. Then we have a universal couple $(\mathcal{R}, \varrho)$ as long as (ai_F) $\overline{\rho}$ is absolutely irreducible and (ds) $\overline{\rho}^{ss}$ is not scalar-values over $D_\mathfrak{p}$ for all $\mathfrak{p}|p$ (these assumptions we always assume). This means that for any pro-Artinian local W-algebra A with $A/\mathfrak{m}_A = W/\mathfrak{m}_W = \mathbb{F}$ for the maximal ideal $\mathfrak{m}_A$ and any Galois representation $\rho_A : \mathrm{Gal}(\overline{\mathbb{Q}}/F) \to GL_2(A)$ such that

(W1) unramified outside p;

(W2) $\rho_A|_{\mathrm{Gal}(\overline{\mathbb{Q}}_p/F_\mathfrak{p})} \cong \left(\begin{smallmatrix} * & * \\ 0 & \alpha_{A,\mathfrak{p}} \end{smallmatrix}\right)$ with $\alpha_{A,\mathfrak{p}} \equiv \alpha_\mathfrak{p} \mod \mathfrak{m}_A$;

(W3) $\det(\rho_A) = \det \rho$;

(W4) $\rho_A \equiv \overline{\rho} \mod \mathfrak{m}_A$,

there exists a unique W-algebra homomorphism $\varphi : \mathcal{R} \to A$ such that $\varphi \circ \varrho \cong \rho_A$. We write $\Phi(A)$ the collection of the isomorphism classes of this finer deformations ρ_A. Thus $\Phi(A) \cong \mathrm{Hom}_{W\text{-}alg}(R, A)$.

Let $\widetilde{\rho} \in \Phi(A)$ acting on $\widetilde{L}$. Define

$$\widetilde{T} = \left\{\phi \in \mathrm{End}_A(\widetilde{L}) \middle| \mathrm{Tr}(\phi) = 0\right\}. \tag{4.1}$$

We let $\sigma \in \mathfrak{G}_F$ act on $v \in \widetilde{T}$ by conjugation $v \mapsto \widetilde{\rho}(\sigma)v\widetilde{\rho}(\sigma)^{-1}$. As before, $\widetilde{T}$ has the following three step filtration stable under $D_\mathfrak{p}$ for each prime ideal $\mathfrak{p}|p$ of F:

$$\widetilde{T} \supset \mathcal{F}_\mathfrak{p}^-\widetilde{T} \supset \mathcal{F}_\mathfrak{p}^+\widetilde{T} \supset \{0\}. \tag{4.2}$$

Let $\mathbb{Z}_p^* = \mathbb{Q}_p/\mathbb{Z}_p = \mathrm{Hom}(\mathbb{Z}_p.\mathbb{Q}_p/\mathbb{Z}_p)$ and $A^* = \mathrm{Hom}(A, \mathbb{Q}_p/\mathbb{Z}_p)$. We thus have $\mathrm{Sel}_F(V/T) = \mathrm{Sel}_F(T \otimes \mathbb{Z}_p^*)$ for $V/T := T \otimes_{\mathbb{Z}_p} \mathbb{Q}_p/\mathbb{Z}_p$, and we also have

$$\begin{aligned}\mathrm{Sel}_F(Ad(\widetilde{\rho})) &= \mathrm{Sel}_F(\widetilde{T} \otimes_A A^*)\\ &= \mathrm{Ker}(H^1(\mathfrak{G}, \widetilde{T} \otimes_A A^*) \to \prod_{\mathfrak{p}|p} H^1(I_\mathfrak{p}, \frac{\widetilde{T} \otimes_A A^*}{\mathcal{F}_\mathfrak{p}^+\widetilde{T} \otimes_A A^*}))\end{aligned}$$

for $\widetilde{\rho} \in \Phi(A)$ and the inertia subgroup $I_\mathfrak{p} \subset \mathfrak{G}$. Note that $D_\mathfrak{p}$ acts trivially on $\mathcal{F}_\mathfrak{p}^- V/\mathcal{F}_\mathfrak{p}^+ V$. We often indicate this fact by writing $\mathcal{F}_\mathfrak{p}^- V/\mathcal{F}_\mathfrak{p}^+ V \cong K$ as $D_\mathfrak{p}$-modules.

Proposition 4.1 (B. Mazur). *Suppose Φ has a universal couple $(\mathcal{R}_F, \varrho_F)$. Then the Pontryagin dual $\mathrm{Sel}_F^*(V/T)$ is canonically isomorphic to the module of 1-differentials $\Omega_{\mathcal{R}_F/W[[\Gamma_F]]} \otimes_{\mathcal{R}_F,\varphi} W$, where $\varphi : \mathcal{R}_F \to W$ is the W-algebra homomorphism such that $\rho \cong \varphi \circ \varrho_F$. More generally, for any $\widetilde{T} \in \Phi(A)$, we have*

$$\mathrm{Sel}_F^*(\widetilde{T} \otimes_A A^*) = \mathrm{Hom}(\mathrm{Sel}_F(\widetilde{T} \otimes_A A^*), \mathbb{Z}_p^*) \cong \Omega_{\mathcal{R}_F/W[[\Gamma_F]]} \otimes_{\mathcal{R}_F,\phi} A,$$

where $\phi : \mathcal{R}_F \to A$ is the W-algebra homomorphism such that $\widetilde{\rho} \cong \phi \circ \varrho_F$.

This proposition is from [MFG] Theorem 5.14. Here we assume that *Kähler 1-differentials* are continuous with respect to the profinite topology.

Here is a sketch of a **proof** due to Mazur: Write simply $(\mathcal{R}, \varrho)$ for $(\mathcal{R}_F, \varrho_F)$. Let $\Phi = \Phi^{n.ord,\nu}$; so, $\Phi(A) \cong \mathrm{Hom}_{W\text{-alg}}(\mathcal{R}, A)$. For simplicity, we assume that X be a profinite $\mathcal{R}$-module, (in general, we take an inductive limit of such modules). Then $\mathcal{R}[X]$ is an object in CL_W. We consider the W-algebra homomorphism $\xi : \mathcal{R} \to \mathcal{R}[X]$ with $\xi \mod X = \mathrm{id}$. Then we can write $\xi(r) = r \oplus d_\xi(r)$ with $d_\xi(r) \in X$. By the above definition of the product, we get $d_\xi(rr') = rd_\xi(r') + r'd_\xi(r)$ and $d_\xi(W) = 0$. Thus d_ξ is a W-derivation, i.e., $d_\xi \in Der_W(\mathcal{R}, X)$. For any derivation $d : \mathcal{R} \to X$ over

W, $r \mapsto r \oplus d(r)$ is obviously a W-algebra homomorphism, and we get

$$\begin{aligned}(4.3)\quad &\left\{\widetilde{\rho} \in \Phi(\mathcal{R}[X])\middle|\widetilde{\rho} \mod X = \varrho\right\} / \approx_X \\ &\cong \left\{\widetilde{\rho} \in \Phi(\mathcal{R}[X])\middle|\widetilde{\rho} \mod X \approx \varrho\right\} / \approx \\ &\cong \left\{\xi \in \mathrm{Hom}_{W\text{-alg}}(\mathcal{R}, \mathcal{R}[X])\middle|\xi \mod X = \mathrm{id}\right\} \\ &\cong Der_W(\mathcal{R}, X) \cong \mathrm{Hom}_{\mathcal{R}}(\Omega_{\mathcal{R}/W}, X),\end{aligned}$$

where "$\approx_X$" is conjugation under $(1 \oplus M_n(X)) \cap GL_2(\mathcal{R}[X])$, and "$\approx$" is conjugation by elements in $GL_2(\mathcal{R}[X])$.

Let $\widetilde{\rho}$ be the deformation in the left-hand side of (4.3). Then we may write $\widetilde{\rho}(\sigma) = \varrho(\sigma) \oplus u'(\sigma)$ (here $u'(\sigma)$ is a "derivative" of $\widetilde{\rho}(\sigma)$). We see

$$\begin{aligned}\varrho(\sigma\tau) \oplus u'(\sigma\tau) &= (\varrho(\sigma) \oplus u'(\sigma))(\varrho(\tau) \oplus u'(\tau)) \\ &= \varrho(\sigma\tau) \oplus (\varrho(\sigma)u'(\tau) + u'(\sigma)\varrho(\tau)).\end{aligned}$$

Define $u(\sigma) = u'(\sigma)\varrho(\sigma)^{-1}$, which is a cocycle with values on $M_2(X)$ by the above formula. Since $\det\widetilde{\rho} = \det\rho = \det\varrho$, $x(\sigma) = \widetilde{\rho}(\sigma)\varrho(\sigma)^{-1}$ has values in $SL_2(\mathcal{R}[X])$, u has values in $Ad(X) = L(Ad(\varrho)) \otimes_{\mathcal{R}} X$. Hence $u : \mathfrak{G}_F^S \to Ad(X)$ is a 1-cocycle. It is a straightforward computation to see the injectivity of the map:

$$\left\{\widetilde{\rho} \in \Phi(\mathcal{R}[X])\middle|\widetilde{\rho} \mod X \approx \varrho\right\} / \approx_X \hookrightarrow H^1(\mathfrak{G}_F^S, Ad(X))$$

given by $\widetilde{\rho} \mapsto [u]$. We put $\mathcal{F}_{\mathfrak{p}}^{\pm}(Ad(X)) = \mathcal{F}_{\mathfrak{p}}^{\pm}L(Ad(\varrho)) \otimes_{\mathcal{R}} X$. Since $\varrho|_{I_{\mathfrak{p}}}$ is upper-triangular (up to conjugation), we have $u|_{I_{\mathfrak{p}}}$ has values in $\mathcal{F}_{\mathfrak{p}}^{-}Ad(X)$.

If further we insist on $d_\xi(W[[\Gamma_F]]) = 0$, since $W[[\Gamma_F]]$-algebra structure is given by $\delta_{\mathfrak{p}}\alpha_{\mathfrak{p}}^{-1}$ which is the character of lower right corner of ϱ (restricted to $I_{\mathfrak{p}}$) this means the corresponding cocycle $u|_{I_{\mathfrak{p}}}$ has values in $\{\left(\begin{smallmatrix} * & * \\ 0 & 0 \end{smallmatrix}\right)\}$. Since $\mathrm{Tr}(u) = 0$, we conclude $u|_{I_{\mathfrak{p}}} \in \mathcal{F}_{\mathfrak{p}}^{+}Ad(X)$. If $\widetilde{L}$ is finite, we may tale $X = \widetilde{L}$, and this gives the desired isomorphism, because $\widetilde{T} \otimes_A A^* = Ad(\widetilde{L})$. If $\widetilde{L}$ is not finite, then $\widetilde{L} \otimes_A A^*$ can be written as a union of $Ad(X)$ for finite X, and by taking the inductive limit, we get the assertion. □

If we replace $\mathcal{F}^{+}\widetilde{?}$ in the definition of the Selmer group by $\mathcal{F}^{-}\widetilde{?}$, we get the "minus" Selmer group $\mathrm{Sel}_F^-(?)$, and by the same argument

$$\mathrm{Sel}_F^-(\widetilde{T} \otimes_A A^*)^* \cong \Omega_{\mathcal{R}_F/W} \otimes_{\mathcal{R}} A.$$

We can apply the above argument to (R, ρ). If $\rho \in \Phi(W)$, we have a unique $P \in \mathrm{Spf}(\mathcal{R}_F)(W)$ such that $\varrho \mod P = \rho$. Then R is canonically isomorphic to the P-adic completion-localization $\widehat{\mathcal{R}}_P$ of $\mathcal{R}$ at P and $\rho : \mathfrak{G}_F \xrightarrow{\varrho} GL_2(\mathcal{R}) \to GL_2(\widehat{\mathcal{R}}_P) = GL_2(R)$. Thus we get

Corollary 4.2. *Assume $R \cong K[[X_\mathfrak{p}]]$. Then we have $\Omega_{R/K} \otimes_{R,\varphi_\rho} K \cong \mathrm{Sel}^-_F(V)$ which is isomorphic to $\bigoplus_{\mathfrak{p}|p} KdX_\mathfrak{p}$.*

Under the conjecture, the Selmer group $\mathrm{Sel}^-_F(V)$ is exactly $\mathbb{H} \subset H^1(\mathfrak{G}, V)$ discussed in the second lecture, and the restriction map takes $\mathbb{H} = \mathrm{Sel}^-_F(V)$ into $\mathrm{Sel}_{F_\infty}(V)$ as we have seen. Recall Greenberg's formula for a base $a_\mathfrak{p}$ of $\mathbb{H}$:

$$\mathcal{L}(Ad(\rho)) = \det\left(a_\mathfrak{p}\left([p, F_{\mathfrak{p}'}]\right)_{\mathfrak{p},\mathfrak{p}'|p}\left(\log_p(\gamma_\mathfrak{p})^{-1} a_\mathfrak{p}([\gamma_{\mathfrak{p}'}, F_{\mathfrak{p}'}])_{\mathfrak{p},\mathfrak{p}'|p}\right)^{-1}\right).$$

Then by the above corollary, putting $c_\mathfrak{p} = \frac{\partial \rho}{\partial X_\mathfrak{p}} \rho^{-1}\Big|_{X=0}$, $\{c_\mathfrak{p}\}_{\mathfrak{p}|p}$ is a basis of $\mathbb{H}$. Then writing $c_\mathfrak{p} \sim \begin{pmatrix} -a_\mathfrak{p} & * \\ 0 & a_\mathfrak{p} \end{pmatrix}$, we can compute the above formula. Note that $a_\mathfrak{p} = \delta_\mathfrak{p}^{-1} \frac{\partial \rho}{\partial \delta_\mathfrak{p}}\Big|_{X=0}$, and $\delta_\mathfrak{p}([\gamma_\mathfrak{p}, F_\mathfrak{p}]) = (1 + X_\mathfrak{p})$ and $\delta_\mathfrak{p}([p_\mathfrak{p}, F_\mathfrak{p}]) = U(p_\mathfrak{p})$. From this we get the formula we stated in the first lecture.

We add the following condition to the deformations $\widetilde{L}$ satisfying (W1–4) to make the universal ring small enough to prove $\mathrm{Sel}_F(V/T)$ is finite (and $\mathrm{Sel}_F(V) = 0$). Let Σ_p be the set of all prime factors of p in O. Fix a pair of integers $(\kappa_{1,\mathfrak{p}}, \kappa_{2,\mathfrak{p}})$ for each $\mathfrak{p} \in \Sigma_p$, and write κ for the tuple $(\kappa_{1,\mathfrak{p}}, \kappa_{2,\mathfrak{p}})_\mathfrak{p}$. We assume that $[\kappa] = \kappa_{1,\mathfrak{p}} + \kappa_{2,\mathfrak{p}}$ is independent of $\mathfrak{p} \in \Sigma_p$. As an extra condition, we now consider

(W5) On $\widetilde{T}/\mathcal{F}^+_\mathfrak{p}\widetilde{T}$, $\mathrm{Gal}(F^{ur}_\mathfrak{p}[\mu_{p^\infty}]/F^{ur}_\mathfrak{p})$ acts by the character $\mathcal{N}^{\kappa_{1,\mathfrak{p}}}$ for all $\mathfrak{p}|p$, and $\det(\widetilde{T}) = \mathcal{N}^{[\kappa]}$ on an open subgroup of $I_\mathfrak{p}$.

We write $\Phi_\kappa(A)$ for the set of isomorphism classes of deformations $\widetilde{\rho} : \mathfrak{G}_F \to GL_2(A)$ of $\overline{\rho}$ satisfying (W1–5). Under (ai_F) or (ds), we have the universal couple $(R_{\kappa,F}, \varrho_{\kappa,F})$ among the deformations satisfying (W1–5). We call $c \in \mathfrak{G}_F$ a complex conjugation, if c is in the conjugacy class of a complex conjugation in $\mathrm{Gal}(\overline{\mathbb{Q}}/\mathbb{Q})$.

Conjecture 4.3. *Suppose* (ds) *and* (ai_F) *for $\overline{\rho}$ and that F is totally real. If $\det(\rho)(c) = -1$ for any complex conjugation c, the universal ring $R_{\kappa,F}$ is free of finite rank over W, and $R_{\kappa,F}$ is a reduced local complete intersection if $\kappa_{2,\mathfrak{p}} - \kappa_{1,\mathfrak{p}} \geq 1$ for all $\mathfrak{p} \in \Sigma_p$.*

Here a reduced algebra A free of finite rank over $W[[x_1,\dots,x_t]]$ is a *local complete intersection* over $R = W[[x_1,\dots,x_t]]$ if A is isomorphic to $R[[T_1,\dots,T_r]]/(f_1(T),\dots,f_r(T))$ for r power series $f_i(T)$, where r is the number of variables in $R[[T_1,\dots,T_r]]$. Though the assertion of $R_{\kappa,F}$ being a local complete intersection is technical, as we will see later, this claim is a key to relating the size of the Selmer group with the corresponding L-value. In the classical setting of Galois representations associated to elliptic modular forms of weight k (in $S_k(\Gamma_1(N))$), we have $\kappa = (0, k-1)$. Thus the condition $\kappa_{2,\mathfrak{p}} - \kappa_{1,\mathfrak{p}} \geq 1$ is equivalent to requiring $k \geq 2$.

Theorem 4.4 (Wiles, Taylor, Fujiwara). *Suppose that the initial representation ρ is associated to a Hilbert modular form of p-power level (in this case, we call ρ modular). If* (ai$_M$) *holds for $M = F[\mu_p]$,* Conjecture 4.3 *holds.*

See Fujiwara's papers [Fu] and [Fu1]. A more general version of this theorem is proven as Theorem 3.67 and Corollary 3.42 in [HMI].

Proposition 4.5. *Assume* Conjecture 4.3. *Then*

1. $\mathrm{Sel}_F(Ad(\widetilde{\rho}) \otimes_W W^*)$ *is finite for any $\widetilde{\rho} \in \Phi_k(W)$ and $R \cong K[[X_\mathfrak{p}]]_{\mathfrak{p}\in\Sigma_p}$ if $\kappa_{2,\mathfrak{p}} - \kappa_{1,\mathfrak{p}} \geq 1$ for all $\mathfrak{p} \in \Sigma_p$,*
2. *$\mathcal{R}_F$ is a reduced local complete intersection free of finite rank over $W[[\Gamma_F]]$,*
3. $\mathrm{Sel}^*_F(Ad(\varrho_F) \otimes_{\mathcal{R}_F} \mathcal{R}^*_F)$ *is a torsion $\mathcal{R}_F$-module,*
4. *For an irreducible component* $\mathrm{Spf}(\mathbb{I})$ *of* $\mathrm{Spf}(\mathcal{R}_F)$*, write $\rho_\mathbb{I} = \pi \circ \varrho_F$ for the projection $\pi : \mathcal{R}_F \twoheadrightarrow \mathbb{I}$. Then* $\mathrm{Sel}^*_F(Ad(\rho_\mathbb{I}) \otimes_\mathbb{I} \mathbb{I}^*)$ *is a torsion $\mathbb{I}$-module.*

Proof. If R is reduced and free of finite rank over W, $\Omega_{R/W}$ is a finite module. Thus the first assertion follows. Note that $P_\kappa = \mathrm{Ker}(\kappa : W[[\Gamma_F]] \to W)$ is generated by $((1 + x_\mathfrak{p}) - \mathcal{N}(\gamma_\mathfrak{p})^{\kappa_{1,\mathfrak{p}}})$ for $\mathfrak{p} \in S$. Thus $\cap_\kappa P_\kappa = \{0\}$. Since $\mathcal{R}_F/P_\kappa\mathcal{R}_F \cong R_{\kappa,F}$ which is free of finite rank s over W, by Nakayama's lemma, $\mathcal{R}_F$ is generated by s elements $r_1,\dots,r_s$ over $W[[\Gamma_F]]$ which give a basis of $R_{\kappa,F}$ over W. Thus we have a surjective $W[[\Gamma_F]]$-linear map $\iota : W[[\Gamma_F]]^s \to \mathcal{R}_F$ sending $(a_1,\dots,a_s)$ to $\sum_j a_j r_j$. Taking another κ', we find that $\mathcal{R}_F/P_{\kappa'}\mathcal{R}_F \cong R_{\kappa',F}$ which is free over W; so, it has to be free of rank s over W. Thus $\mathrm{Ker}(\iota) \subset P_{\kappa'}^s$ for all κ'; so, ι has to be an isomorphism. This shows the freeness in the second assertion.

Let C be the set of all $\kappa = (\kappa_{\mathfrak{p}})_{\mathfrak{p}}$ such that $\kappa_{2,\mathfrak{p}} - \kappa_{1,\mathfrak{p}} \geq 1$ for all $\mathfrak{p}$. Then we still have $\bigcap_{\kappa \in C} P_\kappa = \{0\}$. Thus the natural W-algebra homomorphism $\mathcal{R}_F \to \prod_{\kappa \in C} R_{\kappa,F}$ is an injection. The right-hand side is reduced (i.e., no nilpotent radical), and $\mathcal{R}_F$ is reduced.

We write

$$R_\kappa = \mathcal{R}/P_\kappa = \frac{W[[T_1, \ldots, T_r]]}{(\overline{f}_1(T), \ldots, \overline{f}_r(T))}.$$

Write $\overline{t}_j \in \mathfrak{m}_{R_\kappa}$ for the image of T_j in R_κ. Take a lift t_j in $\mathfrak{m}_{\mathcal{R}}$ of $\overline{t}_j$ so that $\overline{t}_j = (t_j \mod P_\kappa \mathcal{R})$. Define $\varphi : W[[\Gamma_F]][[T_1, \ldots, T_r]] \twoheadrightarrow \mathcal{R}_F$ by $\varphi(f(T_1, \ldots, T_r)) = f(t_1, \ldots, t_r)$. Since $\mathcal{R}_F$ is $W[[\Gamma_F]]$-free, $\mathrm{Ker}(\varphi) \otimes_{W[[\Gamma_F]],\kappa} W = (\overline{f}_1, \ldots, \overline{f}_r)$; so, taking a lift $f_j \in \mathrm{Ker}(\varphi)$ of $\overline{f}_j$, we find $\mathrm{Ker}(\varphi) = (f_1, \ldots, f_r)$ by Nakayama's lemma, and hence $\mathcal{R}_F$ is a local complete intersection over $W[[\Gamma_F]]$.

Since $\mathcal{R}_F$ is reduced and finite over $W[[\Gamma_F]]$, $\Omega_{\mathcal{R}_F/W[[\Gamma_F]]}$ is a torsion $\mathcal{R}_F$-module. From this, the last two assertions follow. As $R_{\kappa,F} = \mathcal{R}_F/P_\kappa \mathcal{R}_F$ is reduced, $\mathrm{Spf}(\mathcal{R}_F)$ is étale over $\mathrm{Spf}(W[[\Gamma_F]])$ around $\rho = P$; so, $R = \widehat{\mathcal{R}}_P \cong K[[X_{\mathfrak{p}}]]$. This finishes the proof. □

Since $\mathcal{R}_F$ is reduced and free of finite rank over $W[[\Gamma_F]]$, its total quotient ring Q is a product of fields of finite dimension over the field $\mathcal{K}$ of fractions of $W[[\Gamma_F]]$. For simplicity, we assume that $\mathbb{I} = W[[\Gamma_F]]$. In particular, writing $\mathbb{K}$ for the field of fractions of $\mathbb{I}$, we have $Q = \mathbb{K} \oplus X$ for a complementary ring direct summand X. Let $\mathbb{I}'$ be the projection of $\mathcal{R}_F$ to X. Then $\mathrm{Spf}(\mathcal{R}_F) = \mathrm{Spf}(\mathbb{I}) \cup \mathrm{Spf}(\mathbb{I}')$ (and $\mathrm{Spf}(\mathbb{I}')$ is the union of irreducible components other than $\mathrm{Spf}(\mathbb{I})$). We take the intersection $\mathrm{Spf}(C_0) = \mathrm{Spf}(\mathbb{I}) \cap \mathrm{Spf}(\mathbb{I}')$; so, $C_0 = \mathbb{I} \otimes_{\mathcal{R}_F} \mathbb{I}'$, which is a torsion $\mathbb{I}$-module called the *congruence module* of $\mathbb{I}$ (or of $\mathrm{Spf}(\mathbb{I})$). It is easy to see that $\mathbb{I} \otimes_{\mathcal{R}_F} \mathbb{I}' \cong \mathbb{I}/((\mathbb{K} \oplus 0) \cap \mathcal{R}_F)$ (cf. [H88] 6.3). By the above expression, the $W[[\Gamma_F]]$-freeness tells us that $\mathrm{char}_{\mathbb{I}}(C_0) = (\mathbb{K} \oplus 0) \cap \mathcal{R}_F$ is an intersection of a power of prime divisors (cf. [BCM] 7.4.2). Since $\mathbb{I} = W[[\Gamma_F]]$ is regular, and hence $\mathrm{char}(C_0)$ is a principal ideal generated by $h \in \mathbb{I}$. For this conclusion, we do not need the isomorphism $\mathbb{I} \cong W[[\Gamma_F]] = W[[x_{\mathfrak{p}}]]_{\mathfrak{p}}$ but a milder condition that $\mathbb{I}$ is a Gorenstein ring over $W[[\Gamma_F]]$ is enough (that is, $\mathrm{Hom}_{W[[\Gamma_F]]}(\mathbb{I}, W[[\Gamma_F]]) \cong \mathbb{I}$ as $\mathbb{I}$-modules; see [H88] Theorem 6.8). Note that a local complete intersection over $W[[\Gamma_F]]$ is a Gorenstein ring (e.g., [CRT] Theorem 21.3). Now by a theorem of Tate (e.g., [MFG] 5.3.4),

$$\mathrm{char}(\Omega_{\mathcal{R}_F/W[[\Gamma_F]]} \otimes_{\mathcal{R}_F} \mathbb{I}) = \mathrm{char}(C_0) = (h).$$

We have for any prime ideal $P \in \mathrm{Spf}(\mathbb{I})$ with $\iota : \mathbb{I}/P \cong W$, writing $\rho_P = \iota \circ \rho_{\mathbb{I}} : \mathfrak{G}_F \to GL_2(W)$

$$\begin{aligned}\mathrm{Sel}^*_F(Ad(\rho_P) \otimes_W W^*) &\cong \Omega_{\mathcal{R}_F/W[[\Gamma_F]]} \otimes_{\mathcal{R}_F,P} W \\ &\cong \mathrm{Sel}^*_F(Ad(\rho_{\mathbb{I}}) \otimes_{\mathbb{I}} \mathbb{I}^*) \otimes_{\mathbb{I}} \mathbb{I}/P.\end{aligned}$$

This shows that if $\mathrm{char}(\mathrm{Sel}^*_F(Ad(\rho_{\mathbb{I}}) \otimes_{\mathbb{I}} \mathbb{I}^*)) = (h)$ for $h \in \mathbb{I}$, we have

$$\mathrm{char}(\mathrm{Sel}^*_F(Ad(\rho_P) \otimes_W W^*)) = (h(P)),$$

where $h(P) = (h \mod P) \in W$. Thus we get

Corollary 4.6. *We have* $|\mathrm{Sel}^*_F(Ad(\rho_P) \otimes_W W^*)| = |h(P)|_p^{-[K:\mathbb{Q}_p]}$ *for all* $P \in \mathrm{Spf}(\mathbb{I})(W)$.

In this corollary, we do not preclude the case where $\mathrm{Sel}^*_F(Ad(\rho_P) \otimes_W W^*)$ is infinite. In such an extreme case, simply $h(P) = 0$ and, hence, $|h(P)|_p^{-[K:\mathbb{Q}_p]} = \infty$.

References

Books

[ACM] G. Shimura, *Abelian Varieties with Complex Multiplication and Modular Functions*, Princeton University Press, Princeton, NJ, 1998.

[BCM] N. Bourbaki, *Algèbre Commutative*, Hermann, Paris, 1961–83.

[CRT] H. Matsumura, *Commutative Ring Theory*, Cambridge studies in advanced mathematics **8**, Cambridge Univ. Press, 1986.

[HMI] H. Hida, *Hilbert Modular Forms and Iwasawa Theory*, Oxford University Press, 2006.

[MFG] H. Hida, *Modular Forms and Galois Cohomology*, Cambridge Studies in Advanced Mathematics **69**, 2000, Cambridge University Press.

Articles

[D] P. Deligne, Valeurs des fonctions L et périodes d'intégrales, Proc. Symp. Pure Math. **33.2** (1979), 313–346.

[FeG] L. Federer and B. H. Gross, Regulators and Iwasawa Modules, Inventiones Math. **62** (1981), 443–457.

[FG] B. Ferrero and R. Greenberg, On the behavior of p-adic L-functions at $s = 0$, Inventiones Math. **50** (1978), 91–102.

[Fu] K. Fujiwara, Deformation rings and Hecke algebras in totally real case, preprint, 1999 (arXiv.math.NT/0602606).

[Fu1] K. Fujiwara, Galois deformations and arithmetic geometry of Shimura varieties, ICM proceedings, 2006.

[G] R. Greenberg, Trivial zeros of p–adic L–functions, Contemporary Math. **165** (1994), 149–174.

[GS] R. Greenberg and G. Stevens, p-adic L–functions and p–adic periods of modular forms, Inventiones Math. **111** (1993), 407–447.

[GS1] R. Greenberg and G. Stevens, On the conjecture of Mazur, Tate, and Teitelbaum, Contemporary Math. **165** (1994), 183–211.

[GsK] B. H. Gross and N. Koblitz, Gauss sums and the p-adic Γ-function, Ann. of Math. **109** (1979), 569–581.

[H88] H. Hida, Modules of congruence of Hecke algebras and L-functions associated with cusp forms, Amer. J. Math. **110** (1988) 323–382.

[H00] H. Hida, Adjoint Selmer groups as Iwasawa modules, Israel Journal of Math. **120** (2000), 361–427 (a preprint version downloadable at `www.math.ucla.edu/ hida`).

[H04] H. Hida, Greenberg's $\mathcal{L}$-invariants of adjoint square Galois representations, IMRN. **59** (2004), 3177–3189 (a preprint version downloadable at `www.math.ucla.edu/ hida`).

[H06] H. Hida, $\mathcal{L}$-invariants of Tate curves, preprint, 2006, (downloadable at `www.math.ucla.edu/ hida`).

[HT] H. Hida and J. Tilouine, Katz p-adic L-functions, congruence modules and deformation of Galois representations, LMS Lecture notes **153** (1991), 271–293.

[HT1] H. Hida and J. Tilouine, Anticyclotomic Katz p–adic L–functions and congruence modules, Ann. Sci. Ec. Norm. Sup. 4th series **26** (1993), 189–259.

[HT2] H. Hida and J. Tilouine, On the anticyclotomic main conjecture for CM fields, Inventiones Math. **117** (1994), 89–147.

[K] N. M. Katz, p-adic L-functions for CM fields, Inventiones Math. **49** (1978), 199–297.

[Ki] M. Kisin, Overconvergent modular forms and the Fontaine-mazur conjecture, Inventiones Math. **153** (2003), 373–454.

[Ki1] M. Kisin, Geometric deformations of modular Galois representations, Inventiones Math. **157** (2004), 275–328.

[Ki2] M. Kisin, Moduli of finite flat group schemes, and modularity, preprint, 2005.

[MTT] B. Mazur, J. Tate and J. Teitelbaum, On p-adic analogues of the conjectures of Birch and Swinnerton-Dyer, Inventiones Math. **84** (1986), 1–48.

[TW] R. Taylor and A. Wiles, Ring theoretic properties of certain Hecke algebras, Ann. of Math. **141** (1995), 553–572.

[W] A. Wiles, Modular elliptic curves and Fermat's last theorem, Ann. of Math. **141** (1995), 443–551.

Haruzo Hida
Department of Mathematics,
UCLA,
Los Angeles, CA 90095-1555,
U.S.A.
Email: hida@math.ucla.edu

Siegel Modular Forms of Weight Three and Conjectural Correspondence of Shimura Type and Langlands Type

T. IBUKIYAMA

In this short article, we give conjectures on explicit dimensions of Siegel modular forms of degree two of weight 3 in several cases. We also explain much deeper conjectures, a conjecture on correspondence between Siegel modular forms of half-integral weight and integral weight of degree two, and a conjecture on correspondence between the split group of type C_2 and its compact twist, from which the conjectures on dimensions come. Also some relations of these conjectures to super-singular abelian surfaces will be explain.

The dimension of the space of Siegel modular forms can be calculated, in principle, either by Riemann-Roch theorem or by Selberg trace formula, as far as the weight is big enough. Of course in actual calculation, there are difficult technical details, and there is no explicit dimension formula for general degree except for conjectural one [12], [13], but at least there is a general way to calculate it, and it is often well handled when the degree is two. On the other hand, if the weight is very small, e.g. if weight (= the representation at the real place) is not integrable, then there is no known general way to calculate it. For example, for Siegel modular forms in the case of degree two, i.e. for automorphic forms of $Sp(2,\mathbb{R})$ (split rank 2), the weight 1, 2, or 3 is a difficult weight of this type. On the other hand, these are interesting weights for geometry. The weight three case corresponds to the canonimal divisor, so this is very interesting case. The weight two case has some conjectural connection with a degree two version of Shimura-Taniyama conjecture as conjectured by Hiroyuki Yoshida. As for weight one, it rarely exists (cf. [15]). This article concerns about weight 3.

1 Definition of Siegel modular forms

1.1

We briefly review the definitions and notation. Let H_n be the Siegel upper half space of degree n and $Sp(n,\mathbb{R})$ the symplectic group of rank n.

$$H_2 = \{Z = {}^tZ \in M_2(\mathbb{C}); Im(Z) > 0\}$$
$$Sp(2,\mathbb{R}) = \{g = \begin{pmatrix} A & B \\ C & D \end{pmatrix} \in M_4(\mathbb{R}); {}^tgJg = J\},\ J = \begin{pmatrix} 0 & 1_2 \\ -1_2 & 0 \end{pmatrix}.$$

Now we define vector valued Siegel modular forms of degree two. We denote by $\rho_{k,j}$ the following irreducible representation of $GL_2(\mathbb{C})$

$$\rho_{k,j} : GL_2(\mathbb{C}) \to \rho_{k,j}(g) = \det(g)^k Sym_j(g) \in GL_{j+1}(\mathbb{C}),$$

where Sym_j is the symmetric tensor representation of $GL_2(\mathbb{C})$ of degree j. Let $\Gamma \subset Sp(2,\mathbb{R})$ be a discrete subgroup with $\mathrm{vol}(\Gamma\backslash H_2) < \infty$.

A holomorphic Siegel modular form of degree two of weight $\rho_{k,j}$ belonging to Γ is a holomorphic vector valued function $F : H_2 \to \mathbb{C}^{j+1}$ such that $F(\gamma Z) = \rho_{k,j}(CZ + D)F(Z)$ for any $\gamma = \begin{pmatrix} A & B \\ C & D \end{pmatrix} \in \Gamma$ and $Z \in H_2$. The space of these functions is denoted by $A_{k,j}(\Gamma)$. The functions which vanish on cusps of Γ (i.e. the boundaries of the Satake compactification of $\Gamma\backslash H_2$) are called cusp forms. The space of cusp forms is denoted by $S_{k,j}(\Gamma)$. When $j = 0$, we say that the weight is k and we write $S_k(\Gamma) = S_{k,0}(\Gamma)$.

1.2 Dimension formula of Siegel modular forms

To calculate the dimension of Siegel cusp forms, we can use Selberg trace formula or Riemann-Roch theorem when $k > 2n$ for weight k and degree n. If $n = 2$, we can manage to calculate explicit values of dimensions in many cases. As for general degree n, there is an explicit conjectural formula for torsion free principal congruence groups of any degree n (cf. [12], [13]). But this is a conjecture only for $k > 2n$. Main obstruction for the case $k \leq 2n$ is non-convergence of trace formula with some representation theoretic problems, or non-vanishing of cohomology. If $k < n/2$, all Siegel modular forms are singular modular forms, and there are no cusp forms.

When k satisfies $n/2 \leq k < 2n$, we have no general way to calculate the dimension. When $n = 2$, mysterious weights are $k = 1, 2, 3$. When $k = 1$, there exist no Siegel cusp forms for $\Gamma_0(N)$ for any natural number N (cf. [15]). When $k = 2$, the dimension is really mysterious, and we have no conjecture at all. The case $k = 3$ is the subject of this article.

1.3 Discrete subgroups

We prepare discrete subgroups we shall treat. We define
(1) The full modular group $Sp(2,\mathbb{Z}) := Sp(2,\mathbb{R}) \cap M_4(\mathbb{Z})$.
(2) The paramodular group for a prime p.

$$K(p) := Sp(2,\mathbb{Q}) \cap \begin{pmatrix} \mathbb{Z} & p\mathbb{Z} & \mathbb{Z} & \mathbb{Z} \\ \mathbb{Z} & \mathbb{Z} & \mathbb{Z} & p^{-1}\mathbb{Z} \\ \mathbb{Z} & p\mathbb{Z} & \mathbb{Z} & \mathbb{Z} \\ p\mathbb{Z} & p\mathbb{Z} & p\mathbb{Z} & \mathbb{Z} \end{pmatrix}.$$

This is isomorphic to the group defining the moduli of polarized abelian varieties having $\{1, p\}$ as elementary divisors of the polarization.
(3) The Iwahori subgroup for a prime p.

$$B(p) := Sp(2,\mathbb{Z}) \cap \begin{pmatrix} \mathbb{Z} & p\mathbb{Z} & \mathbb{Z} & \mathbb{Z} \\ \mathbb{Z} & \mathbb{Z} & \mathbb{Z} & \mathbb{Z} \\ p\mathbb{Z} & p\mathbb{Z} & \mathbb{Z} & \mathbb{Z} \\ p\mathbb{Z} & p\mathbb{Z} & p\mathbb{Z} & \mathbb{Z} \end{pmatrix}.$$

(4) A congruence subgroup

$$\Gamma_0(4) := \left\{ \gamma = \begin{pmatrix} A & B \\ C & D \end{pmatrix} \in Sp(2,\mathbb{Z});\ C \equiv 0 \bmod 4. \right\}.$$

2 Conjectures on dimensions of weight 3

We first give conjectural dimensions of vector valued Siegel modular forms of weight $\det^3 Sym_{2j}$ of $Sp(2,\mathbb{Z})$.

Conjecture 2.1. *We should have the following generating function of dimensions*

$$\sum_{j=0}^{\infty} \dim S_{3,2j}(Sp(2,\mathbb{Z}))s^j = \frac{s^{18}}{(1-s^3)(1-s^4)(1-s^5)(1-s^6)}$$

where s is a variable.

Numerical Examples of conjectured dimensions:
$\dim S_{3,j}(Sp(2,\mathbb{Z}))$ should be 0 for any $j < 36$ and we should have

j	36	38	40	42	44	46	48	50	52	54	56
dim	1?	0?	0?	1?	1?	1?	2?	1?	2?	3?	3?

This conjecture comes from much deeper conjecture on Shimura type correspondence, which will be explained later.

Next we fix a prime p and we give conjectures on dimensions related with the paramodular group $K(p)$ and the Iwahori subgroup $B(p)$.
We can show that $\dim S_3(K(p)) = 0$ for $p \le 5$.

Conjecture 2.2. *Assume that p is a prime with $p > 5$. Then the dimension of Siegel paramodular forms of weight 3 should be given as follows.*

$$\begin{aligned}
&\dim S_3(K(p)) = \\
&-1 + \frac{1}{2880}(p^2-1) + \frac{1}{64}(p+1)\left(1-\left(\frac{-1}{p}\right)\right) \\
&+\frac{5}{192}(p-1)\left(1+\left(\frac{-1}{p}\right)\right) + \frac{1}{72}(p+1)\left(1-\left(\frac{-3}{p}\right)\right) \\
&+\frac{1}{36}(p-1)\left(1+\left(\frac{-3}{p}\right)\right) \\
&+\begin{cases} 2/5 & \text{if } p \equiv 2,3 \bmod 5, \\ 0 & \text{otherwise}, \end{cases} \\
&+\frac{1}{8}\left(1-\left(\frac{2}{p}\right)\right) + \begin{cases} 1/6 & \text{if } p \equiv 5 \bmod 12, \\ 0 & \text{otherwise.} \end{cases}
\end{aligned}$$

Numerical examples.

p	2	3	5	7	11	13	17	19	23
dim	0	0	0	0	0	1	1	1	1

p	29	31	37	41	43	47	53	59
dim	2?	2?	4?	3?	4?	3?	5?	4?

Here the numbers without question marks are true values checked by several ad hoc methods. The numbers with question marks are conjectured values.

We denote by $S_k^{new}(B(p))$ the space of Siegel cusp forms of weight k with respect to $B(p)$ whose local admissible representation at p is the Steinberg representation. Equivalently, $S_k^{new}(B(p))$ is the orthogonal complement of the space spanned by Siegel cusp forms belonging to some discrete subgroup $\Gamma \neq B$ such that $B(p) \subset \Gamma \subset Sp(2, \mathbb{Q})$. We can show that $\dim S_3^{new}(B(p)) = 0$ for $p = 2$.

Conjecture 2.3. *Conjectural dimension of the space $S_3^{new}(B(p))$ of Siegel "new" cusp forms of weight 3 belonging to the Iwahori subroup $B(p)$ is 0 for $p = 3$ and given for $p \geq 5$ as follows.*

$$\dim S_3^{new}(B(p)) =$$

$$\begin{aligned}
&1 + \frac{1}{2880}(p-1)(p^3-1) - \frac{7}{576}(p-1)^2 \\
&- \frac{1}{32}(p-1)\left(1 - \left(\frac{-1}{p}\right)\right) - \frac{1}{24}(p-1)\left(1 - \left(\frac{-3}{p}\right)\right) \\
&- \frac{1}{12}\left(1 - \left(\frac{-1}{p}\right)\right)\left(1 - \left(\frac{-3}{p}\right)\right) \\
&- \frac{1}{9}\left(1 - \left(\frac{-3}{p}\right)\right)^2 \\
&- \begin{cases} 1/5 & \text{if } p = 5\,, \\ 2/5 & \text{if } p \equiv 2, 3 \bmod 5\,, \\ 4/5 & \text{if } p \equiv 4 \bmod 5\,, \\ 0 & \text{otherwise} \end{cases} \\
&- \begin{cases} 1/2 & \text{if } p \equiv 7 \bmod 8\,, \\ 0 & \text{otherwise}\,, \end{cases} \\
&- \begin{cases} 1/6 & \text{if } p \equiv 11 \bmod 12\,, \\ 0 & \text{otherwise.} \end{cases}
\end{aligned}$$

Numerical examples.

p	2	3	5	7	11	13	17	19	23
dim	0	0?	0?	0?	2?	8?	23?	38?	83?

p	29	31	37	41	43	47
dim	225?	298?	618?	935?	1136?	1625?

3 Conjecture on Eichler type correspondence

Conjecture 2.2 or 2.3 comes from an Eichler-Jacquet-Langlands type (conjectural) correspondence between $Sp(2,\mathbb{R})$ and its compact form $USp(2)$ which the author proposed in the early eighties.

The prototype of Jacquet-Langlands correspondence is due to Eichler. So we review this first. Let D be the definite quaternion algebra over $\mathbb{Q}$ with discriminant p. Let O be a maximal order of D and for any prime q, put $O_q = O \otimes_{\mathbb{Z}} \mathbb{Z}_p$. Let $D_A^\times$ be the adelization of $D^\times$ and let $D_\infty = \mathbb{H}$ be the division quaternion algebra over $\mathbb{R}$. We denote by $A_\nu(O_A^\times)$ the space of automorphic forms of $D_A^\times$ with respect to $O_A^\times = D_\infty^\times \prod_q O_q^\times$ of "weight" ν. Here "weight" ν means that the representation at the real place is the representation of $D_\infty^\times$ induced from the ν-th symmetric tensor representation of $SU(2)$ trivial on the center (i.e. ν is even) through the isomorphism $SU(2)/\{\pm 1\} \cong D_\infty^\times/center$. When $\nu > 0$, we put $S_\nu(O_A^\times) = A_\nu(O_A^\times)$. When $\nu = 0$, we denote by $S_0(O_A^\times)$ the orthogonal complement in $A_0(O_A^\times)$ of the one-dimensional subspace of $A_0(O_A^\times)$ spanned by constant functions.

Denote by $\Gamma_0^{(1)}(p)$ the usual congruence subgroup of $SL_2(\mathbb{Z})$ consisting of elements $\gamma = \begin{pmatrix} a & b \\ c & d \end{pmatrix} \in SL_2(\mathbb{Z})$ such that $c \equiv 0 \bmod p$. We denote by $S_k^{new}(\Gamma_0^{(1)}(p))$ the space of new forms of weight k belonging to $\Gamma_0^{(1)}(p)$ in the usual sense of Atkin-Lehner. This is equivalent to say that the local representation at p is the Steinberg representation. We note that the algebras generated by Hecke operators $T(q)$ (q is a prime such that $q \neq p$) for $GL(2)$ and $D_A^\times$ are isomorphic with each other.

Theorem 3.1 (Eichler). *For any integer $k \geq 2$, there exists an isomorphism*

$$S_k^{new}(\Gamma_0^{(1)}(p)) \cong S_{k-2}(O_A^\times)$$

as Hecke algebra modules. In particular, $\dim S_2(\Gamma_0(p)) + 1$ *is equal to the class number of D.*

Now we would like to consider the same sort of precise statement for the degree two case. This kind of problem was raised around in 1963 by

Y. Ihara [16] and now included in the general problem of Langlands. We gave several precise conjectures of this kind of correspondence before (cf. [6], [7]). We shall explain these here.

We first explain $\mathbb{Q}$-form of the compact real form of $Sp(2,\mathbb{R})$. We put

$$USp(2) = \{g \in M_2(\mathbb{H}); g\,{}^t\overline{g} = 1_2\}$$

where $\mathbb{H}$ is the division quaternion algebra over $\mathbb{R}$ and the bar means to take the main involution of each component. This is the compact real form of $Sp(2,\mathbb{R})$. Let D be the definite division quaternion algebra D with discriminant p as before. We introduce the following $\mathbb{Q}$-form of $GSp(2,\mathbb{C})$ whose semi-simple part is the $\mathbb{Q}$ form of $USp(2)$.

$$G = \{g \in M_2(D);\ g\,{}^t\overline{g} = n(g)1_2,\ n(g) \in \mathbb{Q}_+^\times\}.$$

We denote by G_A the adelization of G and by G_∞ the real component of G_A. The p-adic completion G_p of G is isomorphic to G_p^* given by

$$G_p \cong G_p^* := \left\{g \in M_2(D_p);\ {}^t\overline{g}\begin{pmatrix}0 & 1\\ 1 & 0\end{pmatrix}g = \begin{pmatrix}0 & 1\\ 1 & 0\end{pmatrix}\right\}.$$

In case of the split symplectic group, for any prime p, $Sp(2,\mathbb{Q}_p)$ has 7 standard parahoric subgroups, excluding $Sp(2,\mathbb{Q}_p)$ itself. On the other hand, the group G_p^* has three standard parahoric subgroups, excluding G_p^* itself. Namely, there are two maximal compact subgroups $U_{1,p}$ and $U_{2,p}$ of G_p^* and the minimal parahoric subgroup $U_{0,p} = U_{1,p} \cap U_{2,p}$, where

$$\begin{aligned} U_{1,p} &= G_p^* \cap GL_2(O_p)\,, \\ U_{2,p} &= G_p^* \cap \begin{pmatrix} O_p & \pi^{-1}O_p \\ \pi O_p & O_p \end{pmatrix}^\times, \\ U_{0,p} &= G_p^* \cap \begin{pmatrix} O_p & O_p \\ \pi O_p & O_p \end{pmatrix}^\times. \end{aligned}$$

Here π is the prime element of O_p. We note that $U_{1,p}$ and $U_{2,p}$ are very different and not conjugate with each other. For any $q \neq p$, we have $G_q \cong$

$GSp(2, Q_q)$ where $GSp(2, \mathbb{Q}_q)$ is the group of symplectic similitudes and we put $U_q = GSp(2, Z_q)$.

Put $U_i = G_\infty U_{i,p} \prod_{q \neq p} U_q \subset G_A$. A weight of automorphic forms of G_A means an irreducible representation of G_∞. For any integer ν, we denote by ρ_ν the irreducible representation of $USp(2)$ corresponding to the Young diagram parameter (ν, ν), namely to the Young diagram

1	2	$\cdots$	ν
1	2	$\cdots$	ν

.

This representation ρ_ν is trivial on the center $\{\pm 1_2\}$ of $USp(2)$ and we regard ρ_ν as a representation of G_∞ through the isomorphism $G_\infty/center \cong USp(2)/\{\pm 1_2\}$. We denote by V_ν the representation space of ρ_ν. A V_ν-valued function f on G_A is said to be an automorphic form of G_A of weight ν with respect to U_i when it satisfies

$$f(axu) = \rho_\nu(u_\infty) f(x)$$

for any $a \in G$, $x \in G_A$, and $u \in U_i$, where u_∞ is the component at the real place of u. We denote by $A_\nu(U_i)$ = the space of automorphic forms of weight ν with respect to U_i. By the way, when $\nu = 0$, $\dim A_0(U_1)$ or $\dim A_0(U_2)$ is nothing but the class number of the principal genus, or non-principal genus of maximal O-lattices in D^2, respectively. When $\nu = 0$, $A_0(U_i)$ contains constant functions, so it is reasonable to consider the subspace $S_0(U_i)$ of $A_0(U_i))$ which is the orthogonal complement of the one dimensional space spanned by constant functions in $A_0(U_i)$. When $\nu > 0$, we put $S_\nu(U_i) = A_\nu(U_i)$. The dimensions of $A_\nu(U_i)$ is known for any ν and i and p except for the case $(i, p) = (0, 3)$ (cf. [1], [2]). We denote by $S_\nu^{new}(U_0)$ the orthogonal complement of the space spanned by $S_\nu(U_1)$ and $S_\nu(U_2)$ in $S_\nu(U_0)$.

Conjecture 3.2 (cf. [6]). *For $k \geq 3$, there should exist an linear isomorphism*

$$S_k^{new}(B(p)) \cong S_{k-3}^{new}(U_0)$$

which preserves L functions.

As evidence of this conjecture, we have a dimensional equality for $k > 4$ (cf. [2]) and a lot of numerical examples of Euler factors supporting the conjecture (cf. [6]). If we believe that the same isomorphism holds even for $k = 3$, we get our precise conjectural dimension which we gave before, since the dimension of the right-hand side is rigorously known for any ν (including the case where $p = 3$ and $i = 0$).

Conjecture 3.3 ([7]). *For $k \geq 3$, there should exist an isomorphism*

$$S_k^{new}(K(p)) \cong S_{k-3}^{new}(U_2)$$

which preserves L functions.

Here the definition of "new forms" is not representation theoretic at moment. We shall give the definition later. As evidence for this conjecture, we have the following dimensional equality for $k > 4$

$$\begin{aligned}&\dim S_k(K(p)) - 2\dim S_k(Sp(2,\mathbb{Z})) = \\ &\dim S_{k-3}(U_2) - \dim A_2(\Gamma_0^1(p)) \times \dim S_{2k-2}(SL_2(\mathbb{Z})),\end{aligned}$$

where $A_2(\Gamma_0^{(1)}(p))$ is the space of holomorphic modular forms of one variable of weight 2 of $\Gamma_0^{(1)}(p))$ and $S_{2k-2}(SL_2(\mathbb{Z}))$ is the space of holomorphic cusp forms of one variable of weight $2k-2$ of $SL_2(\mathbb{Z})$ (cf [7]).

If we believe that the same equality should also hold for $k = 3$, then since it is well known that $S_4(SL_2(\mathbb{Z})) = 0$ and $S_3(Sp(2,\mathbb{Z})) = 0$, the above equality means that $\dim S_3(K(p)) = \dim S_0(U_2)$. Since the dimension of the right-hand side is equal to $\dim A_0(U_2) - 1$, and $\dim A_0(U_2)$ is already known (cf. [1] (II)), this leads to a precise dimension formula for $S_3(K(p))$. This is our conjecture on the dimension.

Now for general k, we define new forms. Put

$$\rho = \begin{pmatrix} 1_2 & 0 \\ 0 & p1_2 \end{pmatrix}.$$

The groups $K(p)$, $Sp(2,\mathbb{Z})$ and $\rho Sp(2,\mathbb{Z})\rho^{-1}$ contains $B(p)$ as a subgroup and we can define mappings from $S_k(Sp(2,\mathbb{Z}))$ and from $S_k(\rho Sp(2,\mathbb{Z})\rho^{-1})$ to $S_k(K(p))$ by taking the "trace" from $B(p)$ to $K(p)$. Call the subspace of $S_k(K(p))$ spanned by the images of $S_k(Sp(2,\mathbb{Z}))$ and $S_k(\rho Sp(2,\mathbb{Z})\rho^{-1})$ the space of old forms. We denote by $S_k^{new}(K(p))$ the orthogonal compliment of the space of old forms. On the other hand, for U_1 or U_2, there exists an analogue of the Saito-Kurokawa lifting (actually discovered much earlier than Saito-Kurokawa lifting by Ihara [16] and later generalized by Ibukiyama and Ihara[9]). As for the group U_2, there exists a lifting from the pair of forms in $A_2(\Gamma_0(p))$ and $S_{2k-2}(SL_2(\mathbb{Z}))$ to $S_{k-3}(U_2)$, which might vanish in general. We denote by $S_{k-3}^{new}(U_2)$ the orthogonal complement of the space obtained by this lifting in $S_{k-3}(U_2)$.

The actual interpretation of the dimension of new forms is a little more subtle. If k is even, we should read

$$\begin{aligned}\dim S_k^{new}(K(p)) &= \dim S_k(K(p)) - 2\dim S_k(Sp(2,\mathbb{Z})) \\ &\qquad + \dim S_{2k-2}(SL_2(\mathbb{Z}))\,, \\ \dim S_{k-3}^{new}(U_2) &= S_{k-3}(U_2) - \dim S_2(\Gamma_0(p)) \times \dim S_{2k-2}(SL_2(\mathbb{Z}))\,.\end{aligned}$$

If k is odd, we should read

$$\begin{aligned}\dim S_k^{new}(K(p)) &= \dim S_k(K(p)) - 2\dim S_k(Sp(2,\mathbb{Z}))\,, \\ \dim S_{k-3}^{new}(U_2) &= \dim S_{k-3}(U_2) - \dim A_2(\Gamma_0(p)) \times S_{2k-2}(SL_2(\mathbb{Z}))\,.\end{aligned}$$

The reason is as follows. If k is even, there is the Saito-Kurokawa lifting from $S_{2k-2}(SL_2(\mathbb{Z}))$ to $Sp(2,\mathbb{Z})$ and for these, the two trace maps from $Sp(2,\mathbb{Z})$ and $\rho Sp(2,\mathbb{Z})\rho^{-1}$ are not linearly independent. So, it is very plausible that the dimension of old forms in $S_k(K(p))$ is $2\dim S_k(Sp(2,\mathbb{Z})) - S_{2k-2}(SL_2(\mathbb{Z}))$ if k is even and $2\dim S_k(Sp(2,\mathbb{Z})$ if k is odd. Also for the compact twist, this formulation fits examples well.

4 Geometric interpretation

The lowest weight automorphic forms has good relation with some geometry. We review this in this section. Let K be an algebraically closed field of characteristic p. An elliptic curve E defined over K is called super-singular if it has no p-torsion point. An abelian variety A is called super-singular if A is isogenous to E^n for some super-singluar elliptic curve E. Abelian surface is super-singular if and only if it has no p-torsion point. (For general dimension, this is false.)

The prototype of the relation we shall explain later is due to Deuring and Eichler.

Theorem 4.1 (Deuring and Eichler). *The class number of the definite quaternion algebra D with $disc(D) = p$ is equal to the number of* *isomorphism classes of super-singular elliptic curves. This is also equal to* $\dim A_2(\Gamma_0^{(1)}(p)) = \dim S_2(\Gamma_0^{(1)}(p)) + 1$.

For abelian surfaces, we shall suggest the following conjectures.

Conjecture 4.2. $\dim S_3(K(p)) + 1$ *should be equal to the number of irreducible components of the locus of principally polarized super-singular abelian surfaces in the moduli space.*

Conjecture 4.3. $\dim S_3^{new}(B(p))$ *should be equal to the (generalized) arithmetic genus of the super-singular locus of principally polarized abelian surfaces in the moduli space.*

Here the locus is in general a reducible singular curve, but we have a notion of an arithmetic genus of such object defined by J. P. Serre (I learned this from Ekedahl). The reason for these conjectures is as follows. Each dimension which appears in each of the above conjectures should be equal to the dimension of automorphic forms of weight 0 of U_2 or U_0, respectively by the conjectures we gave before. But it was rigorously proved that if we replace the dimension by the one for U_2 or U_0, these conjectures are true. (As for Conjecture 4.2, this is due to Katsura and Oort [17], and as for 4.3, due to the author [10]).

5 Conjecture on Shimura type correspondence

Here we briefly explain about conjectures on Shimura type correspondence of degree two. More precise details will be published elsewhere [14]. We first review the definition of Siegel modular forms of half-integral weight. To define an automorphic factor of weight $1/2$, for $Z \in H_2$ we put

$$\theta(Z) := \sum_{p \in \mathbb{Z}^2} e^{2\pi i\, {}^t pZp} .$$

For $\gamma = \begin{pmatrix} A & B \\ C & D \end{pmatrix} \in \Gamma_0(4)$, we define $j_{1/2}(\gamma, Z) = \theta(\gamma Z)/\theta(Z)$. We also define a character ψ of $\Gamma_0(4)$ by $\psi(\gamma) = \left(\frac{-4}{\det(D)}\right)$. It is well known that

$$j_{1/2}(\gamma, Z)^2 = \psi(\gamma) \det(CZ + D) .$$

A Siegel modular form of degree two of weight $\det^{k-1/2} Sym_j$ with character ψ with respect to $\Gamma_0(4)$ is defined to be a holomorphic vector valued function $F : H_2 \to \mathbb{C}^{j+1}$ such that

$$F(\gamma Z) = \psi(\gamma) j_{1/2}(\gamma, Z)^{2k-1} Sym_j(CZ + D)F(Z)$$

for all $\gamma \in \Gamma_0(4)$. Such F automatically vanishes on all cusps of $\Gamma_0(4)$, so they are cusp forms. (Without character, this is false.) We denote by $S_{k-1/2,j}(\Gamma_0(4), \psi)$ the vector space of such vector valued cusp forms. Now, in order to extract the "level one" part, i.e. a kind of new forms from this space, we introduced the plus space $S^+_{k-1/2,j}(\Gamma_0(4), \psi)$ of Kohnen type which is characterized by a certain condition on Fourier coefficients. For $F \in S_{k-1/2}(\Gamma_0(4), \psi)$, we have the Fourier expansion

$$F(Z) = \sum_T a(T) e^{2\pi i \mathrm{Tr}}(TZ)$$

where $Z \in H_2$ and T runs over half-integral symmetric matrices $T = \begin{pmatrix} t_1 & t_{12}/2 \\ t_{12} & t_2 \end{pmatrix}$ $(t_1, t_2, t_{12} \in \mathbb{Z})$. We say that F is in the plus space if $a(T) = 0$ unless T satisfies the condition that

$$t_1 \equiv (-1)^{k+1} \nu^2 \bmod 4, t_2 \equiv (-1)^{k+1} \mu^2 \bmod 4$$

and $t_{12} \equiv 2\nu\mu(-1)^{k+1} \bmod 4$ for some $\nu, \mu \in \mathbb{Z}$.

Note that the plus space is exactly the subspace of $S_{k-1/2}(\Gamma_0(4), \psi)$ which correspond to the holomorphic, or skew-holomorphic Jacobi forms of index 1 of $Sp(2,\mathbb{Z})^J = Sp(2,\mathbb{Z}) \cdot \mathbb{Z}^4$ for odd k, or for even k, respectively (cf. [11], [3], [4]). But we omit the details here.

Now we can state our conjecture.

Conjecture 5.1. *For any natural number $k \geq 3$ and even integer $j \geq 0$, we should have a linear isomorphism σ.*

$$\sigma : S^+_{k-1/2,j}(\Gamma_0(4), \psi) \cong S_{j+3,2k-6}(Sp(2,\mathbb{Z}))$$

such that for a common Hecke eigen function $F \in S^+_{k-1/2,j}(\Gamma_0(4), \psi)$, we have $L(s, F) = L(s, \sigma(F))$. Here $L(s, F)$ is the L function defined by Zhuravlev [20] and $L(s, \sigma(F))$ is the usual spinor L function.

In particular, if $j = 0$, then we should have

$$S^+_{k-1/2}(\Gamma_0(4), \psi) \cong S_{3,2k-6}(Sp(2,\mathbb{Z})) .$$

Since the dimension of the left-hand side is known in Tsushima [19], this leads to the conjecture on dimension we gave before.

Remarks.

(1) The above conjecture is false for odd j.

(2) The weight correspondence is reasonable (we omit the explanation here).

(3) For $k > 4$, we have dimensional equality under standard conjecture on vanishing of cohomology. (We used conjectural results by Tsushima).

(4) There are a lot of numerical examples supporting this conjecture.

6 Concluding remarks

If $k > 4$, all $\dim S_{k,j}(Sp(2,\mathbb{Z}))$, $\dim S_k(K(p))$ and $\dim S_k^{new}(B(p))$ are known and these are elementary functions of k. So if we put $k = 3$ formally in these formulas for big $k > 4$, we get some numbers. Now if our conjecture on dimensions for $k = 3$ is true, then each of such numbers we mentioned above is equal to $\dim S_{3,j}(Sp(2,\mathbb{Z}))$, $\dim S_3(K(p)) - 1$, or $\dim S_3^{new}(B(p)) + 1$, respectively. This means that the usual trace formula should almost work in these cases too. This seems very interesting. (Such easy description cannot be expected in general setting as is noticed in Hiraga [5])

We also note that the appearance of super-singular abelian surfaces might suggest that dimensions of Siegel cusp forms of weight 3 might have some relation with geometry of the reduction of Shimura varieties mod p.

References

[1] K. Hashimoto and T. Ibukiyama, On class numbers of positive definite binary quaternion harmitian forms (I) *J.Fac.Sci.Univ.Tokyo Sect.IA Math.*, 27(1980), 549–601, (II) ibid. 28(1982), 695–699, (III) ibid. 30(1983), 393–401.

[2] K. Hashimoto and T. Ibukiyama, On relations of dimensions of automorphic forms of $Sp(2,\mathbf{R})$ and its compact twist $Sp(2)$ (II). *Advanced Studies in Pure Math.*, 7(1985), 30–102.

[3] S. Hayashida, Skew-holomorphic Jacobi forms of index 1 and Siegel modular forms of half-integral weight. J. Number Theory 106 (2004), no. 2, 200–218.

[4] S. Hayashida and T. Ibukiyama, Siegel modular forms of half integral weight and a lifting conjecture. J. Math. Kyoto Univ. 45 (2005), no. 3, 489–530.

[5] K. Hiraga, On the multiplicities of the discrete series of semisimple Lie groups. Duke Math. J. 85 (1996), no. 1, 167–181.

[6] T. Ibukiyama, On symplectic Euler factors of genus two. *J.Fac.Sci.Univ.Tokyo Sec.IA Math.* 30(1984), 587–614.

[7] T. Ibukiyama, On relations of dimensions of automorphic forms of $Sp(2, \mathbf{R})$ and its compact twist $Sp(2)$ (I). *Advanced Studies in Pure Math.*, 7(1985), 7–29.

[8] T. Ibukiyama, T. Katsura, and F. Oort. Supersingular curves of genus two and class numbers. *Compositio Math.* 57(1986), 127–152.

[9] T. Ibukiyama and Y. Ihara. On automorphic forms on the unitary symplectic group $Sp(n)$ and $SL_2(R)$. *Math. Ann.*, 278(1987), 307–327.

[10] T. Ibukiyama, Arithmetic theory of supersingular abelian varieties, RIMS K = oky = uroku No. 775, *in Japanese Einstein metric and Yang-Mills connection* Ed. By T. Ochiai (1992), 154-166.

[11] T. Ibukiyama, On Jacobi forms and Siegel modular forms of half integral weights. Comment. Math. Univ. St. Paul. 41 no. 2(1992), 109–124.

[12] T. Ibukiyama and H. Saito, On zeta functions associated to symmetric matrices and an explicit conjecture on dimensions of Siegel modular forms of general degree, Internt. Math. Res. Notices 1992, no.8, 161-169.

[13] T. Ibukiyama and H. Saito, On zeta functions associated to symmetric matrices II, Functional equations and special values MPI preprint series 97–37 (1997).

[14] T. Ibukiyama, Conjecture on Shimura correspondence on Siegel modular forms of degree two, in preparation.

[15] T. Ibukiyama and N-P. Skoruppa, Appendix to the article by C. Poor and D. S. Yuen, "Dimensions of cusp forms of $\Gamma_0(p)$ in degree two and small weights", preprint.

[16] Y. Ihara, On certain arithmetical Dirichlet series. J. Math. Soc. Japan 16(1964), 214–225.

[17] T. Katsura and F. Oort, Supersingular abelian varieties of dimension two or three and class numbers. Algebraic geometry, Sendai, 1985, Adv. Stud. Pure Math.10, Kinokuniya, Japan and North-Holland, Amsterdam, (1987), 253–281.

[18] R. Tsushima, An explicit dimension formula for the spaces of generalized automorphic forms with respect to Sp(2, $\mathbb{Z}$). Proc. Japan Acad. Ser. A Math. Sci. 59 (1983), no. 4, 139–142.

[19] R. Tsushima, On the dimension formula for the spaces of Jacobi forms of degree two. Automorphic forms and *L*-functions (Kyoto, 1999). RIMS K = oky = uroku No. 1103 (1999), 96–110. 11F50.

[20] V. G. Zhuravlev, Euler expansions of theta transforms of Siegel modular forms of half-integral weight and their analytic properties, *Math, sbornik* **123 (165)** (1984), 174-194.

Tomoyoshi Ibukiyama
Department of Mathematics,
Osaka University,
Machikaneyama 1-16, Toyonaka,
Osaka, 560-0043 Japan
Email: ibukiyam@math.wani.osaka-u.ac.jp

Convolutions of Fourier Coefficients of Cusp Forms and the Circle Method

M. JUTILA[1]

0 Introduction

The underlying idea of the classical circle method is to express a coefficient of a Fourier series by use of a Farey dissection of the interval of integration in the analytic formula for the coefficient. In particular, the constant term of the Fourier series

$$\sum_{n=-\infty}^{\infty} b_n e(nx) \tag{0.1}$$

of a periodic function ψ with period 1 is

$$b_0 = \int_0^1 \psi(x)\,dx, \tag{0.2}$$

and b_n can be reduced to this on replacing $\psi(x)$ by $\psi(x)e(-nx)$. Now a decomposition of the unit interval gives rise to a linear decomposition of its characteristic function $\chi(x)$, and thus to a decomposition of the integral (0.2). In the special case $\psi(x) = e(Nx)$ with N an integer, the integral (0.2) amounts to the *δ-function* $\delta(N)$ which equals 1 or 0 for $N = 0$ or $N \neq 0$, respectively. Thus the circle method gives a decomposition of the δ-function. There are various other formulae for the δ-function in literature, for instance one due to W. Duke, J. Friedlander and H. Iwaniec [1]. We are going to discuss an *approximate* version $\delta^*(N)$ of $\delta(N)$ such that $\delta^*(0) = 1$ and $\delta^*(N)$ is "small" for $N \neq 0$.

A notorious obstacle in applications of the circle method is the irregularity of the lengths of the Farey intervals. In his lecture at ICM 1962

[1]The author was supported by the grant 8205966 from the Academy of Finland.

in Stockholm, Yu. V. Linnik [14] considered what he called the *levelling problem*, namely replacing the Farey intervals by intervals of uniform length. A well-known device for this purpose is the Kloosterman refinement of the circle method; its idea is to introduce a finer subdivision of the Farey intervals related to rational numbers with a given denominator. In his lecture, Linnik suggested the following argument: to work with intervals of the same length with an attempt to cope with the error term. Interestingly, he pointed out that *overlapping* of the intervals is not excluded. We presented in [4] a variant of the circle method based actually on *ample* overlapping of the intervals. In this context, it is essential that the multiplicity of the overlapping turns out to be mostly close to its statistical expectation. This means that if $\chi^*(x)$ is a suitable linear combination of the characteristic functions (possibly with a smooth weight) of neighborhoods of a system of rational numbers, then the mean square of $\chi(x) - \chi^*(x)$ over the unit interval is small. In [7], we applied this method to convolutions of Fourier coefficients of Maass forms. In the present paper, we give a variant of the argument of [7] on the basis of another version of the method of overlapping intervals developed in [10]. Its main novelty is an explicit formula (stated in (1.5) below) for the error when b_0 in (0.2) is approximated by

$$b_0^* = \int_0^1 \chi^*(x)\psi(x)\,dx. \tag{0.3}$$

In particular, the above mentioned approximation to the δ-function is

$$\delta^*(N) = \int_0^1 \chi^*(x)e(NX)\,dx. \tag{0.4}$$

The general scheme of the argument is expounded in Sec. 2, and applications to convolutions of Fourier coefficients of Maass forms are considered in Sec. 3.

Notation. As to the spectral theory of automorphic functions, we follow the notation of [17]. We let ε stand for a small positive constant not necessarily the same at each occurrence. The constants implied by the symbols $\ll$ or $O(\cdots)$ may depend on ε without mentioning. Further, we write $A \asymp B$ to mean that A and B are positive and $A \ll B \ll A$. The standard notation is used for the Bessel functions $J_\nu(x)$ and $K_\nu(x)$, as well as for the arithmetic functions $\varphi(n)$ and $\mu(n)$. In sums $\sum^*$, the asterisk denotes the coprimality condition $(a, q) = 1$.

1 An arithmetic formula for Fourier coefficients

Following [10], we first describe the construction of $\chi^*(x)$ as an approximation of the characteristic function $\chi(x)$ of the interval $I = [0, 1]$.

Let Q be a large parameter and let $w(q)$ be a function satisfying $0 \leq w(q) \leq 1$ for natural numbers q and $w(q) = 0$ for $q \notin [Q, 2Q]$. Further, let $\Delta \in (0, 1/3]$ be another parameter and let $v(x)$ be a non-negative continuous function with support in $[-\Delta, \Delta]$ which is bounded by an absolute constant. Also, for normalization, we suppose that

$$\frac{1}{2\Delta}\int_{-\Delta}^{\Delta} v(x)\,dx = 1.$$

The function $v(x)$ will play the role of a continuous analogue of the characteristic function of the interval $[-\Delta, \Delta]$. In [10] we supposed that $v(x)$ be even, but this assumption, made just for convenience, is redundant. Putting

$$L = \sum_{q=1}^{\infty} w(q)\varphi(q) \tag{1.1}$$

and $\lambda = 2\Delta L$, we define

$$\chi^*(x) = \frac{1}{\lambda}\sum_{q=1}^{\infty} w(q) \sum_{a=-\infty}^{\infty}{}^{*} v\left(\frac{a}{q} - x\right)$$

as a tentative approximation to $\chi(x)$ in the interval I. Alternatively, this can be written as

$$\chi^*(x) = \frac{1}{\lambda}\sum_{q=1}^{\infty} w(q) \sum_{a=1}^{q}{}^{*} v^*\left(\frac{a}{q} - x\right), \tag{1.2}$$

where $v^*(x)$ is the extension of $v(x)$ to a periodic function with period 1; to verify this, consider separately the ranges $0 \leq x < \Delta$, $\Delta \leq x \leq 1 - \Delta$, and

$1 - \Delta < x \leq 1$. The function $\chi^*(x)$ is periodic with period 1 and its integral over I equals 1. We proved in [10] (Theorem 1) that

$$\int_0^1 (\chi^*(x) - \chi(x))^2 \, dx \ll \Delta^{-1} L^{-2} Q^2 \log^3(1/\Delta) \tag{1.3}$$

as a slight sharpening of an estimate in [4]. Note that typically $L \asymp Q^2$, and then the right hand side is $\ll (\Delta Q^2)^{-1} \log^3(1/\Delta)$. In particular, if $\Delta \asymp Q^{-1}$ (as we are going to choose in Sec. 3), this is $\ll Q^{-1} \log^3 Q$, which is small if Q is large.

Let now ψ be a continuous function with period 1 and Fourier coefficients b_n. Denote the Fourier coefficients of the function $\nu^*(x)$ by a_n. As an approximation of b_0, we now have b_0^* defined by (0.3). Thus, by (1.2),

$$\begin{aligned} b_0^* &= \frac{1}{\lambda} \sum_{q=1}^{\infty} w(q) \sum_{a=1}^{q}{}^{*} \int_0^1 \nu^*\left(\frac{a}{q} - x\right) \psi(x) \, dx \\ &= \frac{1}{\lambda} \sum_{q=1}^{\infty} w(q) \sum_{a=1}^{q}{}^{*} \int_{-\infty}^{\infty} \nu\left(\frac{a}{q} - x\right) \psi(x) \, dx; \end{aligned} \tag{1.4}$$

to verify the latter expression, note that the range of integration in the first integral can be replaced by $[a/q - \Delta, a/q - \Delta + 1]$. Now the formula (3.3) in [10] gives the following connection between b_0 and b_0^*:

$$b_0 = b_0^* - \frac{1}{\lambda} \sum_{d=1}^{\infty} d \sum_{m \neq 0} a_{dm} b_{dm} \sum_{r=1}^{\infty} w(dr) \mu(r), \tag{1.5}$$

where $\mu(r)$ stands for the Möbius function. This is apparently a relation of arithmetic nature. In typical applications, the difference $b_0 - b_0^*$ is estimated as an error term, and b_0^* is treated as in the circle method, with conveniently "levelled" intervals.

If we interpret $b_0 - b_0^*$ as an integral by (0.2) and (0.3), then by Cauchy's inequality and (1.3) we get

$$b_0 - b_0^* \ll L^{-1} Q \Delta^{-1/2} \|\psi\|_2 \log^{3/2}(1/\Delta); \tag{1.6}$$

this is (3.4) in [10] and represents the scope of the method of [4]. There should be more potential in the method of [10], at least in principle, since

(1.5) reveals also the *structure* of $b_0 - b_0^*$, not only its order of magnitude. Moreover, the occurrence of the b_n - coefficients on both sides of the relation (1.5) is a new phenomenon indicating possibilities of recursions or iterative arguments.

If we are interested in the other coefficients in addition to b_0, we simply interpret b_n as the constant term for $\psi(x)e(-nx)$, and then (1.5) takes the form

$$b_n = b_n^* - \frac{1}{\lambda} \sum_{d=1}^{\infty} d \sum_{m \neq 0} a_{dm} b_{n+dm} \sum_{r=1}^{\infty} w(dr)\mu(r), \tag{1.7}$$

where b_n^* is defined as b_0^* in (1.4) with $\psi(x)e(-nx)$ standing in place of $\psi(x)$. Also, we get a formula for linear combinations

$$B_n = \sum_{f=0}^{F} \beta_f b_{n+f} \tag{1.8}$$

and for analogous expressions B_n^* involving b_{n+f}^*, with arbitrary complex coefficients β_j. In fact, it follows immediately from (1.7) that

$$B_n = B_n^* - \frac{1}{\lambda} \sum_{d=1}^{\infty} d \sum_{m \neq 0} a_{dm} B_{n+dm} \sum_{r=1}^{\infty} w(dr)\mu(r). \tag{1.9}$$

The subsequent argument will run, in a nutshell, as follows. Our final goal is the estimation of the coefficients b_n, either individually or in mean. Since b_n^* is more easily tractable, we are interested in the approximation $b_n \approx b_n^*$. Of course, an approximation like this is feasible only under more specific assumptions, and such a case will be discussed in the next section. However, by the following theorem, we may estimate B_0 in terms of a suitable sum B_n^* under fairly general conditions. Now, in the application to convolutions we have in mind, the latter sum can be expressed in terms of Kloosterman sums, and an estimate for it will be obtained by spectral methods. Then get an estimate for B_0 as well, and this, in turn, implies an estimate for $b_n - b_n^*$.

Theorem 1. *Let ψ be a continuous function with period 1 and Fourier coefficients b_n. Suppose that*

$$|\psi(x)| \leq b \tag{1.10}$$

for all x. Let Q and F be integers with Q large and

$$0 \le F \ll Q^c \tag{1.11}$$

for some constant $c > 0$. Let $w(q)$ be a weight function such that $L \gg Q^2$ for the quantity L defined in (1.1). Further, let $\Delta = D/Q$, where D is a sufficiently large constant, and let $v(x)$ be a suitable smooth weight function satisfying the conditions imposed in the construction of the function χ^* in (1.2). Suppose $|\beta_f| \le 1$ in (1.8). Then we have

$$B_0 \ll bQ^{-A} + \max_{|n| \le Q^{1+\varepsilon}} |B_n^*| \tag{1.12}$$

for any fixed numbers $A > 0$ and $\varepsilon > 0$.

Proof. We may suppose that $b > 0$ and $|B_0| \gg bQ^{-A}$, for otherwise there is nothing to prove. Note that $|b_n| \le b$ by (1.10). The function $v(x)$ can be chosen in such a way that its Fourier coefficients a_n are of rapid decay, say

$$a_n \ll \min(\Delta, n^{-1}(n\Delta)^{-A}) \ll (n^{A+1}\Delta^A + \Delta^{-1})^{-1} \tag{1.13}$$

for a large constant A. For instance, a suitable "Vinogradov function" (see [12], Ch. I, §2) gives an example of such a function.

By our choices of Δ and $w(q)$, we have $\lambda = 2\Delta L \gg DQ$, where the implied constant is independent of D. By (1.9), we now have

$$B_0 = B_0^* - \frac{1}{\lambda}\sum_{d=1}^{\infty} d \sum_{m \ne 0} a_{dm} B_{dm} \sum_{r=1}^{\infty} w(dr)\mu(r). \tag{1.14}$$

Consider the following possibilities: either

$$|B_n| \le 2|B_0| \text{ for all } |n| \le Q^{1+\varepsilon/2}, \tag{1.15}$$

or else there exists an index n_1 with $|n_1| \le Q^{1+\varepsilon/2}$ such that

$$|B_{n_1}| > 2|B_0|. \tag{1.16}$$

Here ε is the same as in (1.12). In the case (1.15), we estimate in (1.14) the contribution of the terms with $|dm| > Q^{1+\varepsilon/2}$ as an error term which is at most bQ^{-A} in absolute value, in view of (1.10), (1.11), and (1.13). Then $|B_{dm}| \leq 2|B_0|$ in the remaining terms. Further, the relevant values of d lie in the interval $1 \leq d \leq 2Q$, and for given d, we have

$$\left|\sum_{r=1}^{\infty} w(dr)\mu(r)\right| \ll \frac{Q/d}{\log^2(3Q/d)}$$

by the known estimate for the sum function of the Möbius function; we may suppose that the weight function w is smooth enough to allow effective summation by parts. Also, by (1.13) with $A = 1$,

$$\sum_{m\neq 0} |a_{dm}| \ll \sum_{m\neq 0} ((dm)^2\Delta + \Delta^{-1})^{-1} \ll d^{-1}.$$

Thus the second term on the right of (1.14) is at most $|B_0|/2 + bQ^{-A}$ in absolute value if D is chosen sufficiently large. Therefore we get the estimate

$$|B_0| \leq 2|B_0^*| + 2bQ^{-A}$$

proving (1.12) in the case (1.15).

On the other hand, if (1.16) holds, then we repeat the same argument for B_{n_1} in place of B_0, that is we apply (1.9) to B_{n_1}. Then either

$$|B_n| \leq 2|B_{n_1}| \text{ for all } |n| \leq 2Q^{1+\varepsilon/2},$$

or else there exists n_2 with $|n_2| \leq 2Q^{1+\varepsilon/2}$ such that

$$|B_{n_2}| > 2|B_{n_1}| > 4|B_0|.$$

In the former case, we get

$$|B_0| < |B_{n_1}|/2 \leq |B_{n_1}^*| + bQ^{-A}$$

as above, and (1.12) follows; otherwise we repeat the procedure. After k steps, the latter of the two alternatives means that there exists n_k with $|n_k| \leq kQ^{1+\varepsilon/2}$ such that $|B_{n_k}| > 2^k|B_0|$. This is clearly impossible if $k/\log Q$ is sufficiently large, by our initial assumption on B_0, and then $kQ^{1+\varepsilon/2} \ll Q^{1+\varepsilon}$. Thus, in any case, we end up with the inequality (1.12). □

Remark. Returning to the remarks we made before stating Theorem 1, we see now that if we manage to bound the sums B_n^* nontrivially in some sense, Theorem 1 implies a good estimate for a weighted average of b_n as n ranges over an interval. This information can be fed back to the identity (1.7) to show eventually that b_n and b_n^* lie close to each other.

2 Applications to convolutions

In [7] we studied sums of the form

$$T(N;f) = \sum_{n=1}^{N} t(n)t(n+f),$$
$$T(N) = \sum_{n=1}^{N-1} t(n)t(N-n),$$

where the coefficients $t(n)$ are the Hecke eigenvalues of a Maass form u for the full modular group. Then u is an eigenfunction of the hyperbolic Laplacian for an eigenvalue $1/4+\kappa^2$ with $\kappa > 0$. The Fourier series of u is of the form

$$u(x+yi) = y^{1/2}\sum_{n\neq 0}\rho(n)K_{i\kappa}(2\pi|n|y)e(nx) \quad (y>0),$$

where $\rho(n)=\rho(1)t(n)$ for $n\geq 1$. Further, let ϵ be the parity sign of u, that is $u(-x+yi)=\epsilon u(x+yi)$. We refer to [2] or [17] for basic properties of Maass forms. Let us consider the sum $T(N;f)$ as an example of our modification of the argument of [7].

It is convenient to insert a smooth weight function to our sum. Let $W(x)$ be a function supported in the interval $[N-M,N]$, where M and N are positive integers such that

$$N^{1/2+\varepsilon} \leq M \leq N/4. \tag{2.1}$$

We suppose that

$$W^{(j)}(x) \ll_j M^{-j} \tag{2.2}$$

for sufficiently many derivatives. We are going to study the sum

$$T(W;f) = \sum_{n=1}^{\infty} W(n)t(n)t(n+f) \tag{2.3}$$

by use of the exponential sum

$$S(W;\alpha) = \sum_{n=1}^{\infty} W(n)t(n)e(n\alpha). \tag{2.4}$$

One may consider the sum (2.3) either for an individual shift $f \geq 1$ or in a mean value sense for f varying over an interval; these problems are in fact closely related in view of the identity (1.7). To clarify this point, we first introduce the relevant periodic function ψ in this context, in the sense of the preceding section. Let $W_0(x)$ be an auxiliary smooth weight function of support in the interval $[N/3, cN]$ with c a sufficiently large constant. Suppose that $W_0(x) = 1$ for $N/2 \leq x \leq cN/2$ and define

$$\psi(\alpha) = S(W_0; \alpha) S(W; -\alpha). \tag{2.5}$$

If this is written as a Fourier series with coefficients b_f, then

$$b_f = \sum_{n=1}^{\infty} W(n) W_0(n+f) t(n) t(n+f),$$

thus

$$b_f = T(W; f) \text{ for } -N/4 \leq f \leq \left(\frac{c}{2} - 1\right) N; \tag{2.6}$$

note that here f is not restricted to positive values. Or if we prefer to interpret $T(W; f)$ as the constant term b_0, then the factor $e(-f\alpha)$ should be inserted to our function (2.5). Now (1.4), for suitable weight functions w and v and with this modified function playing the role of ψ, gives an approximation, say $T^*(W; f)$, to $T(W; f)$. That is, by definition, $T^*(W; f)$ equals b_0^* in the present case. The following lemma, concerning the average of $|T^*(W; f)|$ over f, is the main step in our approach to the sum $T(W; f)$.

Lemma 1. *Let M and N be positive integers as in (2.1), and*

$$Q = MN^{-\delta}, \quad \Delta = D/Q, \tag{2.7}$$

where D and δ are positive constants with δ small and D large. Let the weight function v be as in Sec. 2, and suppose moreover that its support lies in the interval $[-\Delta, -\Delta/2]$. Also, suppose that the weight function w is such that $L \gg Q^2$ in (1.1). Then, for $1 \leq F \ll M$, we have

$$\sum_{1 \leq |f| \leq F} |T^*(W; f)| \ll (FN^{1/2} + F^{1/2} M^{-1/2} N) N^{\varepsilon}. \tag{2.8}$$

Proof. By definition,

$$T^*(W;f) \tag{2.9}$$
$$= \frac{1}{\lambda}\sum_{q=1}^{\infty} w(q) \sum_{a=1}^{q}{}^{*} \times \int v(-\eta)S(W_0;\frac{a}{q}+\eta)S(W;-\frac{a}{q}-\eta)e(-f(\frac{a}{q}+\eta))\,d\eta,$$

where $\lambda = 2\Delta L \gg Q$.

Next we transform the exponential sums here by the following summation formula of the Voronoi type (see [15], Theorem 2): if F is a smooth function with compact support in $(0,\infty)$, then

$$\sum_{n=1}^{\infty} t(n)e\left(\frac{na}{q}\right)F(n) = \frac{\pi i}{q\sinh\pi\kappa}\sum_{n=1}^{\infty} t(n)e\left(-\frac{n\bar{a}}{q}\right) \tag{2.10}$$
$$\times \int_0^{\infty}\left\{J_{2i\kappa}\left(\frac{4\pi}{q}\sqrt{nx}\right) - J_{-2i\kappa}\left(\frac{4\pi}{q}\sqrt{nx}\right)\right\}F(x)\,dx$$
$$+\frac{4\epsilon\cosh\pi\kappa}{q}\sum_{n=1}^{\infty} t(n)e\left(\frac{n\bar{a}}{q}\right)\int_0^{\infty} K_{2i\kappa}\left(\frac{4\pi}{q}\sqrt{nx}\right)F(x)\,dx,$$

where $a\bar{a} \equiv 1 \pmod q$; recall that ϵ is the parity sign of the Maass form u and κ is the related spectral parameter. We apply here the asymptotic formula (see [13], eq. (5.11.6))

$$J_{2i\kappa}(x) \sim \left(\frac{2}{\pi x}\right)^{1/2}\cos(x - i\kappa\pi - \frac{\pi}{4}) \tag{2.11}$$

for $x \geq N^{\varepsilon}$, and otherwise we use the formulae (see [13], eqs. (5.10.4) and (5.10.23))

$$J_\nu(x) = \frac{(x/2)^\nu}{\Gamma(1/2)\Gamma(\nu+1/2)}\int_0^{\pi}\cos(x\cos\theta)(\sin\theta)^{2\nu}\,d\theta \tag{2.12}$$

and

$$K_\nu(x) = \int_0^{\infty} e^{-x\cosh u}\cosh(\nu u)\,du. \tag{2.13}$$

Now, since $\eta \asymp N^{\delta}M^{-1}$ by (2.7) and our assumptions on the support of the v-function, the factor $e(\pm x\eta)$ in the integrands in the formula (2.10), applied to our S-sums in (2.9), is a genuinely oscillating function in the interval $[N-M, N]$. Therefore, applying (2.13) and repeated integration by parts or simply direct estimations, we see that the terms involving the K-Bessel functions are negligibly small. Next, as to the terms involving the J-Bessel functions, we apply (2.11) for $n \geq Q^2N^{\delta-1}$ and (2.12) for $n < Q^2N^{\delta-1}$. Again, repeated integration by parts shows that the contribution of the last mentioned terms is negligible. Otherwise the integrand has a saddle point only if $n \asymp NQ^2\eta^2 \asymp N$, and the other terms can be neglected. The formula (2.11) contains just the leading term of the asymptotic expansion, the subsequent terms being similar but smaller, so it will suffice to work with the leading term.

Denote by x_1 and x_2 the variables of integration in the Voronoi formula just applied to our exponential sums. Then the η-integral in (2.9) is

$$\int v(-\eta)e((x_1 - x_2 - f)\eta)\,d\eta,$$

which is small unless $x_1 - x_2 \ll M$; recall that $1 \leq f \ll M$. Therefore we may suppose in the sequel that $|x_i - N| \ll M$; this is clear for $i = 2$ owing to the factor $W(x_2)$. We substitute the exponential sums in the transformed shape into (2.9), summing over the same values of f as in (2.8) with suitable coefficients β_f on the unit circle such that

$$\sum_f \beta_f T^*(W; f) = \sum_f |T^*(W; f)|.$$

Then it remains to prove the estimate (2.8) for the sum

$$\begin{aligned} &Q^{-1}\sum_{q=1}^{\infty} w(q)q^{-1} \\ &\times \sum_{n_1 \asymp N}\sum_{n_2 \asymp N} t(n_1)t(n_2)(n_1n_2)^{-1/4}\sum_f \beta_f S(n_2 - n_1, f; q) \\ &\times \iint W_0(x_1)W(x_2)e^{4\pi i(\sqrt{n_2x_2} - \sqrt{n_1x_1})/q}(x_1x_2)^{-1/4} \\ &\times \left(\int v(-\eta)e((x_1 - x_2 - f)\eta)\,d\eta\right) dx_1\,dx_2. \end{aligned}$$

There appear also integrals involving the sum of the square roots in the exponent, but these are seen to be small on integrating over x_1 for a given value of $x_1 - x_2$. The Kloosterman sums $S(n_2 - n_1, f; q)$ appeared after the summation over a. We fix x_1, x_2, and η for a moment, estimating the corresponding integrals trivially in the end of the argument.

The above expression should be compared with (6.4) and (6.6) in [7], putting $n_2 - n_1 = p$ and $n_2 = n$. Namely, if we replace x_1 and x_2 heuristically by N in the exponential, then the coincidence is perfect except that we now have a sum over f to deal with instead of an individual value of f. Moreover, replacing the numbers x_i by N does not make an essential change, for $\sqrt{n_i x_i}/q$ is then changed only by an amount $\ll M/Q \ll N^{\delta}$. Thus, if we consider the case of individual f, we get again the result of [7] in a somewhat simplified way (in [7], we used the automorphy of the Maass form rather than the Voronoi formula, so that we had to work out the transformation by an *ad hoc* argument). In particular, we have

$$T(N; f) \ll N^{2/3+\varepsilon} \text{ for } 1 \le f \ll N^{2/3},$$

which is (1.15) in [7].

The presence of the f-sum entails only a minor modification. Applying Kuznetsov's sum formula for sums of Kloosterman sums, we end up with a spectral expression involving factors $t_j(f)t_j(p)$, where the t_j-numbers are Hecke eigenvalues of the jth Maass form with a spectral parameter $\kappa_j \ll (N/M)N^{\varepsilon}$. Moreover, there is a factor depending on κ_j, n/p, and f. As in [7], we split up the double sum over n and p into subsums such that n/p is essentially constant in each. Then p is separated from κ_j and f. Moreover, the parameters κ_j and f can be separated (as in [8], or in [6], p. 454) if we restrict κ_j for a moment to a short interval $k \le \kappa_j < k+1$ to eliminate, essentially, the dependence on κ_j for a moment. Then we apply Cauchy's inequality to the j-sum and invoke the spectral large sieve inequality

$$\sum_{k \le \kappa_j < k+1} \alpha_j \left| \sum_{f=1}^{F} c_f t_j(f) \right|^2 \ll (k + F^{1+\varepsilon}) \sum_{f=1}^{F} |c_f|^2,$$

where $\alpha_j = |\rho_j(1)|^2 / \cosh(\pi\kappa_j)$. After that, we sum over k, again by Cauchy's inequality, and apply the "long range" version of the spectral large sieve to the p-sum. Now, in [7] we had the square root of the sum

$$\sum_{\kappa_j \le K} \alpha_j t_j^2(f) \ll K^2 + f^{1/2+\varepsilon}$$

with $K \ll (N/M)N^{\varepsilon}$ as factor in the final estimate, while in the present case this is to be replaced by $(K(K+F)F)^{1/2}N^{\varepsilon}$, and then the estimate (2.8) follows. □

In particular, it follows from (2.8) that

$$\sum_{|f| \ll M} |T^*(W; f)| \ll MN^{1/2+\varepsilon}; \tag{2.14}$$

note that the term for $f = 0$ can be included because $T^*(W; 0) \ll N^{1+\varepsilon}$.

In the following lemma, we show that (2.14) holds even for the sums $T(W; f)$ if we restrict the summation a bit.

Lemma 2. *Let M and N be as in (2.1). Then*

$$\sum_{-M \le f \ll M} |T(W; f)| \ll MN^{1/2+\varepsilon}. \tag{2.15}$$

Proof. We may interpret the sum (2.15) as B_0 defined in (1.8) with $F \asymp M$ for

$$\psi(\alpha) = S(W_0; \alpha)S(W; -\alpha)e(M\alpha)$$

and suitable coefficients β_f. with $F \asymp M$. Next we estimate B_0 by (1.12) noting that $Q^{1+\varepsilon} \ll M$ if ε is chosen sufficiently small. Then (2.15) follows from (2.14). □

We are now in a position to estimate the accuracy of the approximation $T(W; f) \approx T^*(W; f)$ for an *individual* value of f.

Theorem 2. *Let M and N be as in (2.1) and* $1 \le f \ll M$. *Then*

$$T(W; f) = T^*(W; f) + O(N^{1/2+\varepsilon}). \tag{2.16}$$

Proof. We interpret $T(W; f)$ and $T^*(W; f)$ as b_f and b_f^* in the notation of (1.7) with $\psi(\alpha)$ as in (2.5). By the estimate (1.13) for the coefficients a_n, we may restrict the summation over d and m in (1.7) to values such that $|dm| \le M$. Since the mean value of $|b_{f+n}|$ over $|n| \le M$ is $\ll N^{1/2+\varepsilon}$ by Lemma 2, we see that by straightforward estimations in (1.7) that $b_f - b_f^* \ll N^{1/2+\varepsilon}$. □

Finally, combining this theorem with Lemma 1, we obtain a statistical result on $T(W; f)$.

Theorem 3. *With M and N as in (2.1), and $F \ll M$, we have*

$$\sum_{1\le f\le F} |T(W;f)| \ll (FN^{1/2} + F^{1/2}M^{-1/2}N)N^{\varepsilon}. \tag{2.17}$$

Remark 1. We may apply (2.17) to the mean value of $|T(N;f)|$ over f. Suppose, for simplicity, that the Ramanujan -Petersson conjecture $t(n) \ll n^{\varepsilon}$ holds. Then $T(W;f) \ll MN^{\varepsilon}$ by a trivial estimation. We may approximate $T(N;f)$, up to an error $\ll M_0N^{\varepsilon}$, by at most $\ll \log N$ sums $T(W;f)$ with $M \ge M_0 \gg N^{1/2+\varepsilon}$. Hence, by (2.17),

$$\sum_{1\le f\le F} |T(N;f)| \ll (FM_0 + F^{1/2}M_0^{-1/2}N)N^{\varepsilon}.$$

We minimize the right hand side on choosing $M_0 = F^{-1/3}N^{2/3}$. Then, for $F \ll N^{1/2-\varepsilon}$, we have $M_0 \gg N^{1/2+\varepsilon}$ and

$$\sum_{1\le f\le F} |T(N;f)| \ll (FN)^{2/3}N^{\varepsilon}. \tag{2.18}$$

This holds unconditionally for the analogous sum in the case of holomorphic cusp forms, which can also be treated by a spectral identity (see [11]) for $T(W;f)$, an analogue for an identity of Y. Motohashi [16] in the additive divisor problem. The mean value estimate (2.18) should be compared with the following a mean value result due to A. Ivić and Y. Motohashi [3] for the error term $E(N;f)$ in the additive divisor problem: for $F \ll N^{1/2-\varepsilon}$,

$$\sum_{1\le f\le F} E(N;f)^2 \ll F^{1/3}N^{4/3+\varepsilon};$$

this implies an analogue for (2.18) by Cauchy's inequality.

Remark 2. In applications, the unconditional result (2.17) may serve as a substitute for (2.18). For instance, this is the case in [11], where a certain spectral mean square estimate for Rankin-Selberg zeta-functions is studied.

Remark 3. The main new ingredient in the proof of Theorem 2 is a *comparison principle* used in the proof of Theorem 1; that is, we compare different Fourier coefficients to draw conclusions on a single coefficient, and utilize the familiar phenomenon that an average is usually easier to deal with than a single term. The argument of [7] yields the same if we assume the Ramanujan-Petersson conjecture. Indeed, with $\psi(\alpha)$ as in (2.5), we deduce from (1.6) that

$$\begin{aligned} (2.19) \qquad & T(W; f) - T^*(W; f) \\ & \ll Q^{-1/2}(\log Q)^{3/2}\left(\int_0^1 |S(W_0; \alpha)S(W; -\alpha)|^2\, d\alpha\right)^{1/2}. \end{aligned}$$

We have here $S(W_0; \alpha) \ll N^{1/2}\log N$ (see [2], Theorem 8.1) and

$$\int_0^1 |S(W; -\alpha)|^2\, d\alpha \ll \sum_{N-M\le n\le N} |t(n)|^2 \ll \min\left(MN^{2\theta}, M + N^{3/5}\right),$$

where θ is a constant such that $t(n) \ll n^{\theta}$. The exponent $3/5$ here is the classical Rankin-Selberg exponent in the error term for the mean square of Fourier coefficients of a cusp form. Now (2.19) reduces to (2.16) if $\theta = \varepsilon$ is admissible.

Remark 4. An alternative approach to convolutions makes use of the generating function of $t(n)t(n+f)$ (see, e. g., [5], [6] or [9]). Then we encounter a non-trivial problem, namely the estimation of the inner products (u^2, u_j). In [5], [6], this was reduced, by a variant of the Rankin-Selberg unfolding method, to convolutions analogous to those which occurred in the proof of Lemma 1. Thus these inner products are hidden in a disguised form even in the present argument.

References

[1] W. Duke, J. Friedlander, and H. Iwaniec: *Bounds for automorphic L-functions*, Invent. Math. 112 (1993), 1–8.

[2] H. Iwaniec: *Introduction to the Spectral Theory of Automorphic Forms*, Bibl. de la Revista Mathématica Iberoamericana, Madrid 1995; 2nd ed. Amer. Math. Soc. 2002.

[3] A. Ivić and Y. Motohashi: *On some estimates involving the binary additive divisor problem*, Quart. J. Math. Oxford 46 (1995), 471–483.

[4] M. Jutila: *A Variant of the Circle Method*, Sieve Methods, Exponential Sums, and their Applications in Number Theory, Cambridge Univ. Press 1996, 245–254.

[5] M. Jutila: *The additive divisor problem and its analogs for Fourier coefficients of cusp forms. I*, Mat. Z. 223 (1996), 435–461.

[6] M. Jutila: *The additive divisor problem and its analogs for Fourier coefficients of cusp forms. II*, Math. Z. 225 (1997), 625–637.

[7] M. Jutila: *Convolutions of Fourier coefficients of cusp forms*, Publ. de l'Institut Math. Nouv. sér. 65 (79) (1999), 31–51.

[8] M. Jutila: *On spectral large sieve inequalities*, Funct. Approx. Comment. Math. 28 (2000), 7–18.

[9] M. Jutila: *Sums of the additive divisor problem type and the inner product method*, Zap. nauchn. sem. POMI 322 (2005), 239–250.

[10] M. Jutila: *Distribution of rational numbers in short intervals*, The Ramanujan J. (to appear).

[11] M. Jutila and Y. Motohashi: *Uniform bounds for Rankin-Selberg L-functions*, Proc. of the Workshop on Multiple Dirichlet Series, Amer. Math. Soc. (to appear).

[12] A. A. Karatsuba: *Basic analytic number theory*, Springer-Verlag , Berlin 1993.

[13] N. N. Lebedev: *Special Functions and their Applications*, Dover Publications, New York 1972.

[14] Yu. V. Linnik *Additive problems and eigenvalues of modular operators*, Proc. Internat. Congr. Mathematicians (Stockholm 1962), Inst. Mittag-Leffler, Djursholm 1963, 270–284.

[15] T. Meurman: *On exponential sums involving Fourier coefficients of Maass wave forms*, J. reine angew. Math. 384 (1988), 192–207.

[16] Y. Motohashi: *The Binary Additive Divisor Problem*, Ann. Sci. l'Ecole Norm. Sup. 27 (1994), 529–572.

[17] Y. Motohashi: *Spectral Theory of the Riemann Zeta-Function*, Cambridge Univ. Press 1997.

Matti Jutila
Department of Mathematics,
University of Turku,
FI-20014 Turku,
Finland
E-mail: jutila@utu.fi

On an Extension of the Derivation Relation for Multiple Zeta Values

M. KANEKO

In this paper, we propose a conjectural generalization of the derivation relation for multiple zeta values. This extension was inspired by works of Alain Connes and Henri Moscovici on a certain Hopf algebra of transverse geometry [1], [2], and is thought of as a first attempt to materialize the suggestion given at the end of Section 7 in [4].

The multiple zeta value (MZV) is a real number defined for each index set $\mathbf{k} = (k_1, k_2, \ldots, k_n)$ of positive integers with $k_1 > 1$, by the convergent series

$$\zeta(\mathbf{k}) = \zeta(k_1, k_2, \ldots, k_n) = \sum_{m_1 > m_2 > \cdots > m_n > 0} \frac{1}{m_1^{k_1} m_2^{k_2} \cdots m_n^{k_n}}.$$

One of the fascinating features of these numbers is that the structures of (linear or algebraic) relations over $\mathbb{Q}$ among them reflect various other structures in mathematics and mathematical physics.

To describe the derivation relation of MZV's, we use the algebraic setup introduced by Hoffman [3]. Let $\mathfrak{H} = \mathbb{Q}\langle x, y\rangle$ be the non-commutative polynomial algebra over the rationals in two indeterminates x and y, and $\mathfrak{H}^1$ and $\mathfrak{H}^0$ its subalgebras $\mathbb{Q} + \mathfrak{H}y$ and $\mathbb{Q} + x\mathfrak{H}y$, respectively. Let $Z : \mathfrak{H}^0 \to \mathbb{R}$ be the $\mathbb{Q}$-linear map defined by $Z(1) = 1$ and

$$Z(x^{k_1-1}yx^{k_2-1}y \cdots x^{k_n-1}y) = \zeta(k_1, k_2, \ldots, k_n).$$

This "evaluation map" arises naturally from the iterated integral representation of MZV's. To find a linear relation among MZV's is to find an element in the kernel of the map Z.

For each integer $n \geq 1$, define the derivation ∂_n on $\mathfrak{H}$ by

$$\partial_n(x) = x(x+y)^{n-1}y, \quad \partial_n(y) = -x(x+y)^{n-1}y.$$

Since $\mathfrak{H}$ is the polynomial algebra on x and y, this uniquely determines the derivation ∂_n. (A derivation on $\mathfrak{H}$ is a linear map $\partial : \mathfrak{H} \to \mathfrak{H}$ satisfying Leibniz's rule $\partial(uv) = \partial(u)v + u\partial(v)$ for any $u, v \in \mathfrak{H}$.) Note that the derivation ∂_n preserves $\mathfrak{H}^1$ and $\mathfrak{H}^0$. The derivation relation of MZV's is then the following.

Theorem. ([4, Corollary 6]) *One has $Z(\partial_n(w)) = 0$ for all $n \geq 1$ and all $w \in \mathfrak{H}^0$.*

This relation supplies a large class of linear relations of MZV's. To add the derivation relation to (finite) double shuffle relation is equivalent to extend the double shuffle relation by using divergent MZV's. Also, the derivation relation is equivalent to Ohno's relation (minus duality). See [4] for the details.

We now proceed to generalize this. In [4, p.328], it is noticed that the derivation ∂_n has an alternative description: Let θ be the derivation on $\mathfrak{H}$ defined by

$$\theta(x) = x^2 + (xy + yx)/2, \quad \theta(y) = y^2 + (xy + yx)/2.$$

Then we have

$$\partial_n = \mathrm{ad}(\theta)^{n-1}(\partial_1)/(n-1)!,$$

where $\mathrm{ad}(\theta)(\partial) := [\theta, \partial] = \theta\partial - \partial\theta$. Let H be the "Euler operator", namely, the linear endomorphism of $\mathfrak{H}$ which sends each homogeneous element w to $\deg(w)w$.

Definition. *Let c be a rational number. For any $n \geq 1$, we define a linear endomorphism $\partial_n^{(c)}$ of $\mathfrak{H}$ by*

$$\partial_n^{(c)} = \mathrm{ad}(\theta^{(c)})^{n-1}(\partial_1)/(n-1)!,$$

where $\theta^{(c)}$ is the linear endomorphism defined by the rules

$$\theta^{(c)}(x) = \theta(x), \quad \theta^{(c)}(y) = \theta(y)$$

and

$$\theta^{(c)}(uv) = \theta^{(c)}(u)v + u\theta^{(c)}(v) + c\partial_1(u)H(v)$$

for any $u, v \in \mathfrak{H}$.

Remark. This definition is modeled after the "twisted" Hopf algebra of Connes-Moscovici (see [1], [2], [4]).

Note that $\theta^{(0)} = \theta$ and $\partial_n^{(0)} = \partial_n$, and, if $c \neq 0$, $\theta^{(c)}$ and $\partial_n^{(c)}$ are not derivations.

Conjecture 1. *For all $n \geq 1$, $c \in \mathbb{Q}$, and $w \in \mathfrak{H}^0$, we have*

$$Z(\partial_n^{(c)}(w)) = 0.$$

When we view c as a parameter, we see that $\partial_n^{(c)}(w)$ is a polynomial in c of degree $n-1$. Hence, the above conjecture is equivalent to the following.

Conjecture 1′. *For all $n \geq 1$ and i with $0 \leq i \leq n-1$ and $w \in \mathfrak{H}^0$, we have*

$$Z(\text{coefficient of } c^i \text{ in } \partial_n^{(c)}(w)) = 0.$$

We have checked this by using Mathematica for all n, i and w with $n + \deg(w) \leq 9$. Tatsushi Tanaka has extended the experiments by using Risa/Asir and verified the conjecture up to weight 12. Below we give a table of various numbers of independent relations of MZV's obtained by $\partial_n^{(c)}$, detailed descriptions of the table being given in the remarks after the table. Recall the weight of the MZV $\zeta(k_1, k_2, \ldots, k_n)$ is $k_1 + k_2 + \cdots + k_n$.

Table 1: (Conjectural) Numbers of Independent Relations

weight	3	4	5	6	7	8	9	10	11	12
$\partial_n(w)$	1	2	5	10	22	44	90	181	363	727
$\partial_n^{(c)}(w), c \neq -2$ fixed	1	2	5	10	22	44	90	181	363	727
$\partial_{n_1}^{(c_1)}(w_1) \cup \partial_{n_2}^{(c_2)}(w_2)$	1	2	5	10	23	46	97	196	396	
$\partial_n^{(c)}(w)$ for all c $\cup$ duality	1	2	5	10	23	46	98	200	413	838

Remarks. i) The first line of the data gives the number of linearly independent elements in $\mathfrak{H}$ among all $\partial_n(w)$ with $n \in \mathbb{N}$ and $w \in \mathfrak{H}^0$ varying under the condition $n + \deg(w) =$ weight. (The weight of each term of $Z(\partial_n(w))$ is $n + \deg(w)$.)

ii) The second line is the same experiment with various fixed $c \in \mathbb{Q}$. The result says that, at least for tens of randomly chosen values of $c \neq -2$, the elements $\partial_n^{(c)}(w)$ as n and w vary give the *same* number of independent relations as $\partial_n(w)$, regardless of the choice of c. We have no proof of this (nor of $Z(\partial_n^{(c)}(w)) = 0$, except in low weights). The reason why the value $c = -2$ is exceptional is explained by the identity $\partial_n^{(-2)} = (-1)^{n-1}\partial_1^n$, which, as Seidai Yasuda and Kentaro Ihara pointed out, is easily proved.

iii) The experiments for various single c as in ii) may suggest that $\partial_n^{(c)}$ only gives the same set of relations as the original derivation ∂_n does. However, we do have new relations, as the third line of the above table show. The third line of the table gives the number of linearly independent elements supplied by the union of $\partial_n^{(c_1)}(w)$ and $\partial_n^{(c_2)}(w)$ for randomly chosen different values of c_1 and c_2 ($\neq -2$). It is remarkable here again that any choice of c_1 and c_2 seems to give the *same* number of independent relations. This too is only experimental and is not yet proved.

iv) The final line is the result of taking all c into account and moreover adding the duality relations. Up to weight 9, we used the form of Conjecture 1′ to ensure the "for all c" part, but from weight 10 on we only used the data for several rational values of c for computational reason. Thus the number given may increase if we add more values of c, but we believe that does not occur, in light of Conjecture 2 below. The reason why we add the duality is also explained by Conjecture 2. If the duality is not added, the numbers obtained are strictly less at weights 11 and 12 (and probably for higher weights).

v) Seidai Yasuda pointed out the identity

$$\theta^{(c)} = \theta^{(0)} + \frac{c}{2}(\theta^{(0)} - \theta^{(-2)}),$$

or in general

$$\theta^{(c)} = \theta^{(0)} - \frac{c}{c'}(\theta^{(0)} - \theta^{(c')})$$

for any $c' \neq 0$. Hence, to consider $\partial_n^{(c)}$ for all c is also equivalent to consider $\mathrm{ad}(\theta^{(c_1)})\mathrm{ad}(\theta^{(c_2)})\cdots\mathrm{ad}(\theta^{(c_{n-1})})(\partial_1)$ with each $c_i = 0$ and 1 (say).

Now let a_k be the entry for weight k of the last line of the above table (e.g., $a_8 = 46$, $a_9 = 98$, $a_{10} = 200, \ldots$). Then, we have observed the following remarkable (conjectural) identity.

Conjecture 2. *For $k \geq 3$, we have*

$$2^{k-2} - a_k = \frac{1}{k-1} \sum_{d|k-1} \mu(\frac{k-1}{d}) 2^d.$$

Here, μ is the Möbius function.

Remark. Let L_2 be the free graded Lie algebra of] rank 2 over $\mathbb{Q}$ generated by x and y of degree 1, and L_∞ the free graded Lie algebra over $\mathbb{Q}$ with infinitely many generators z_i $(i = 1, 2, 3, \ldots)$ whose degrees are given by $\deg(z_i) = i$. The number on the right-hand side of the conjecture is equal to the dimension of degree $k-1$ part of both L_2 and L_∞, as Witt's formula shows. We note that the universal enveloping algebra of L_2 is identified with $\mathfrak{H}$, whereas that of L_∞ is identified with $\mathfrak{H}^1$ under the identification $z_i = x^{i-1}y$. (The algebra $\mathfrak{H}^1$ is freely generated by $z_i = x^{i-1}y$, $i = 1, 2, 3, \ldots$.) This observation suggests intimate connections between these Lie algebras and the relations of MZV's. Note that the number 2^{k-2} on the left is the total number of MZV's of weight k. Thus the number on the left-hand side of the conjecture is the maximal number of independent MZV's, under the conjectural relations $Z(\partial_n^{(c)}(w)) = 0$ and the duality.

The most curious feature of Conjecture 2 is that the weights concerned on both sides differ by 1. The conjecture implies the possibility of constructing linearly independent (under the conjectural relations and the duality) elements of MZV's of weight k from a basis of homogeneous part of degree $k-1$ of L_2 or L_∞. But how to construct such elements? We tried several ways at random and found one way (using the derivation δ_y in [4, §4] and the regularization) which bears the numerical test up to weight 13, but there is no theoretical support to believe that is the right way. It is most desirable to prove (or disprove) this conjecture and clarify the suggested connection of the MZV to the Lie algebras L_2 and L_∞.

Acknowledgment. The author would like to thank Seidai Yasuda at RIMS, Kyoto, for valuable comments and suggestions.

References

[1] Connes, A. and Moscovici, H., *Modular Hecke algebras and their Hopf symmetry*, Moscow Mathematical Journal **4** (2004), 67–109.

[2] Connes, A. and Moscovici, H., *Rankin–Cohen brackets and the Hopf algebra of transverse geometry*, Moscow Mathematical Journal **4** (2004), 111–130.

[3] Hoffman, M., *The algebra of multiple harmonic series*, J. Algebra **194** (1997), 477–495.

[4] Ihara, K., Kaneko, M., and Zagier, D., *Derivation and double shuffle relations for multiple zeta values.* Compositio Math. **142-02** (2006), 307–338.

Masanobu Kaneko
Graduate School of Mathematics,
Kyushu University,
Fukuoka 812-8581,
Japan
Email: mkaneko@math.kyushu-u.ac.jp

On Symmetric Powers of Cusp Forms on GL_2

Henry H. KIM

Introduction

Let $\mathbb{A}$ be the ring of adeles of a number field F. Let $\pi = \otimes_v \pi_v$ be a cuspidal automorphic representation of $GL_2(\mathbb{A})$ with central character ω_π. Let $Sym^m: GL_2(\mathbb{C}) \longrightarrow GL_{m+1}(\mathbb{C})$ be the mth symmetric power representation of $GL_2(\mathbb{C})$ on symmetric tensors of rank m. By the local Langlands correspondence, $Sym^m(\pi_v)$ is a well-defined irreducible admissible representation of $GL_{m+1}(F_v)$ for every v (see [Ki1] for the details). Let $Sym^m(\pi) = \otimes_v Sym^m(\pi_v)$. It is an irreducible, admissible representation of $GL_{m+1}(\mathbb{A})$. Langlands functoriality conjecture is that $Sym^m(\pi)$ is an automorphic representation of $GL_{m+1}(\mathbb{A})$. It is one of the most important problems in the Langlands program, and its solution would imply the Ramanujan conjecture and Sato-Tate conjecture for π [Ki-Sh1]. So far it is proved only up to $m = 4$, even though we can prove the functoriality for all m for cuspidal representations attached to Galois representations [Ki3]. In this note, we give a new proof of the cuspidality criterion of the symmetric fourth [Ki-Sh2], and supplement various results on $Sym^m(\pi)$ in [Ki1, Ki-Sh1, Ki-Sh2]. For convenience, let $A^m(\pi) = Sym^m(\pi) \otimes \omega_\pi^{-1}$. Suppose Π is an isobaric automorphic representation of $GL_n(\mathbb{A})$ of the form $\Pi = \pi_1 \boxplus \cdots \boxplus \pi_k$, $n_1 + \cdots + n_k = n$, where π_i is a cuspidal representation of $GL_{n_i}(\mathbb{A})$. Then we say Π is of type $(n_1, ..., n_k)$. We refer [Ki1] for unexplained notations.

1 Symmetric fourth

In this section, we assume that $Sym^3(\pi)$ is a cuspidal representation of $GL_4(\mathbb{A})$. Then by the result in [Ki-Sh1], $Sym^2(\pi) \not\simeq Sym^2(\pi) \otimes \chi$ for any $\chi \neq 1$. Note that $Sym^2(\pi) = Ad(\pi) \otimes \omega_\pi$. In [Ki1], we proved that

$Sym^4(\pi)$ is an automorphic representation of $GL_5(\mathbb{A})$, either cuspidal or of type (3,2). We gave the cuspidality criterion of $Sym^4(\pi)$ in [Ki-Sh2]. Namely,

Theorem 1.1. *$A^4(\pi)$ is not cuspidal if and only if there exists a non-trivial quadratic character η of F such that $A^3(\pi) \simeq A^3(\pi) \otimes \eta$. It is equivalent to: $Ad(\pi_E) \simeq Ad(\pi_E) \otimes \chi$ for a grössencharacter χ of E, where E/F be the quadratic extension, determined by η, and π_E is the base change of π to $GL_2(\mathbb{A}_E)$. In that case, $A^4(\pi) = \sigma_1 \boxplus \sigma_2$, where $\sigma_1 = Ad(\pi) \otimes (\omega_\pi \eta)$, and $\sigma_2 = \pi(\chi^{-1}) \otimes \omega_\pi$.*

We had to use the following unpublished result on the descent of cuspidal representations of GL_4 to GSp_4 of Jacquet, Piatetski-Shapiro and Shalika:

Theorem 1.2 (Jacquet, Piatetski-Shapiro and Shalika). *Let π be a cuspidal automorphic representation of $GL_4(\mathbb{A})$ such that there exist a grössen–character χ and a finite set S of places as above for which $L_S(s, \pi, \wedge^2 \otimes \chi^{-1})$ has a pole at $s = 1$. Then there exists a globally generic cuspidal automorphic representation τ of $GSp_4(\mathbb{A})$ with central character χ such that π is the Langlands' functorial lift of τ under the natural embedding ${}^LGSp_4 = GSp_4(\mathbb{C}) \hookrightarrow GL_4(\mathbb{C})$.*

Here we prove Theorem 1.1 without using the above theorem. Throughout this note, S always denotes a set of places of F such that for $v \notin S$, every representation is unramified.

First we note that for any cuspidal representation σ of $GL_k(\mathbb{A})$,

$$L^S(s, \sigma \times \pi \times A^3(\pi)) = L^S(s, \sigma \times A^4(\pi)) L^S(s, \sigma \times Sym^2(\pi)). \tag{1.1}$$

Note also that $A^4(\pi)$ is not cuspidal if and only if $L^S(s, \sigma \times A^4(\pi))$ has a pole at $s = 1$ for some cuspidal representation σ of $GL_2(\mathbb{A})$. Since $L^S(s, \sigma \times Sym^2(\pi))$ is invertible at $s = 1$, by (1.1), $A^4(\pi)$ is not cuspidal if and only if $L^S(s, \sigma \times \pi \times A^3(\pi))$ has a pole at $s = 1$. However,

$$L^S(s, \sigma \times \pi \times A^3(\pi)) = L^S(s, (\sigma \boxtimes \pi) \times A^3(\pi)), \tag{1.2}$$

where $\sigma \boxtimes \pi$ is the functorial product of σ, π which is an automorphic representation of $GL_4(\mathbb{A})$ [R, Theorem M]. Since $A^3(\pi)$ is cuspidal, $L^S(s, (\sigma \boxtimes$

$\pi) \times A^3(\pi))$ has a pole at $s = 1$ if and only if $\sigma \boxtimes \pi$ is cuspidal, and $\sigma \boxtimes \pi \simeq \widetilde{A^3(\pi)} \simeq A^3(\pi) \otimes \omega_\pi^{-1}$. Consider

$$\wedge^2(\sigma \boxtimes \pi) \simeq \wedge^2(A^3(\pi) \otimes \omega_\pi^{-1}). \tag{1.3}$$

Both are automorphic representations of $GL_6(\mathbb{A})$ [Ki1]. By direct calculation, we see that the left hand side of (1.3) is $Sym^2(\sigma)\otimes\omega_\pi \boxplus Sym^2(\pi)\otimes\omega_\sigma$. (We see that for almost all v, $\wedge^2(\sigma_v \boxtimes \pi_v) \simeq Sym^2(\sigma_v)\otimes\omega_{\pi_v} \boxplus Sym^2(\pi_v)\otimes\omega_{\sigma_v}$. Since both $\wedge^2(\sigma \boxtimes \pi)$ and $Sym^2(\sigma) \otimes \omega_\pi \boxplus Sym^2(\pi) \otimes \omega_\sigma$ are isobaric automorphic representations of $GL_6(\mathbb{A})$, by the strong multiplicity one theorem, $\wedge^2(\sigma \boxtimes \pi) \simeq Sym^2(\sigma) \otimes \omega_\pi \boxplus Sym^2(\pi) \otimes \omega_\sigma$.) By the main result in [Ki1], the right hand side of (1.3) is $(A^4(\pi) \otimes \omega_\pi^{-2}) \boxplus \omega_\pi^{-1}$. Hence $Sym^2(\sigma)$ cannot be cuspidal, i.e., σ is monomial. Let $\sigma = I_E^F \chi \otimes \omega_\pi^{-1}$ for some quadratic extension E/F, and χ is a grössencharacter of E. Let η be the quadratic character attached to E/F by class field theory. Then $Ad(\sigma) = \pi(\chi\chi'^{-1}) \boxplus \eta$, and $\omega_\sigma = (\chi|_{\mathbb{A}^*})\eta$. Hence we see that $\chi|_{\mathbb{A}^*} = 1$, and $A^4(\pi) = (Ad(\pi) \otimes \omega_\pi \eta) \boxplus (I_E^F \chi \otimes \omega_\pi^{-1})$.

Note that since $\sigma = I_E^F \chi \otimes \omega_\pi^{-1}$, by [R, Lemma 3.1.1], $\sigma \boxtimes \pi = I_E^F(\pi_E \otimes \chi) \otimes \omega_\pi^{-1}$. So by (1.2),

$$L^S(s, \sigma \times \pi \times A^3(\pi)) = L^S(s, I_E^F(\pi_E \otimes \chi) \times A^3(\pi) \otimes \omega_\pi^{-1}).$$

By the adjointness property of the base change [R, Proposition 2.3.1],

$$L^S(s, I_E^F(\pi_E \otimes \chi) \times A^3(\pi) \otimes \omega_\pi^{-1}) = L^S(s, (\pi_E \otimes \chi) \times A^3(\pi)_E \otimes \omega_E^{-1}),$$

where $\omega_E = \omega_\pi \circ N_{E/F} = \omega_{\pi_E}$. By (1.1), the right hand side has a pole at $s = 1$. Hence $A^3(\pi)_E$ is not cuspidal. By the cuspidality criterion of the base change (see [R, Proposition 2.3.1].), $A^3(\pi) \simeq A^3(\pi) \otimes \eta$.

On the other hand, $A^3(\pi)_E = A^3(\pi_E)$. Since $Ad(\pi)$ is cuspidal and E/F is a quadratic extension, $Ad(\pi)_E$ is cuspidal. Hence by the cuspidality criterion of $A^3(\pi_E)$ in [Ki-Sh1], $Ad(\pi_E) \simeq Ad(\pi_E) \otimes \mu$ for a grössencharacter μ of E. Here $\mu^3 = 1$ and $\mu \neq 1$. Then $A^3(\pi_E) = (\pi_E \otimes \mu) \boxplus (\pi_E \otimes \mu^2)$. We have

$$\begin{aligned} &L^S(s, (\pi_E \otimes \chi) \times A^3(\pi)_E \otimes \omega_E^{-1}) \\ &= L^S(s, \pi_E \times \widetilde{\pi_E} \otimes \chi\mu) L^S(s, \pi_E \times \widetilde{\pi_E} \otimes \chi\mu^2). \end{aligned}$$

Since it has a pole at $s = 1$, $\chi = \mu$ or μ^{-1}. This finishes the proof of Theorem 1.1.

2 Symmetric mth powers

In this section, we assume that $Sym^m(\pi) = \otimes_v Sym^m(\pi_v)$ is an automorphic representation of $GL_{m+1}(\mathbb{A})$ for $m \geq 5$. In [Ki3], we proved that it is the case if π is of icosahedral type, attached to a holomorphic cusp form of weight 1. It is also true if π is of dihedral, tetrahedral or octahedral type, attached to a holomorphic cusp form of weight 1. Let us briefly explain them. We assume $F = \mathbb{Q}$. By Deligne-Serre theorem [DS], π is attached to an odd irreducible 2-dimensional representation $\rho : W_{\mathbb{Q}} \longrightarrow GL_2(\mathbb{C})$ ("ρ is odd" means that $det(\rho)(c) = -1$ where c is the complex conjugate. Conjecturally Maass forms give rise to even irreducible 2-dimensional representations.) By twisting by a character of $W_{\mathbb{Q}}^*$, we can assume that ρ factors through $G_{\mathbb{Q}} = Gal(\bar{\mathbb{Q}}/\mathbb{Q})$, denoted again by ρ. Let $\bar{\rho} : G_{\mathbb{Q}} \longrightarrow PGL_2(\mathbb{C})$ be the composition of ρ and the projection $GL_2(\mathbb{C}) \longrightarrow PGL_2(\mathbb{C})$. Then the image of $\bar{\rho}$ is D_{2n}, A_4, S_4, or A_5. We say π is of dihedral, tetrahedral, octahedral, or icosahedral type if the image of $\bar{\rho}$ is $G = D_{2n}, A_4, S_4, A_5$, resp. Then ρ factors through G', which is a central extension of G.

Note that $H^2(G, \mathbb{C}^*) = 1$ if $G = D_{2n}$ for n odd. Otherwise, $H^2(G, \mathbb{C}^*) = \mathbb{Z}/2\mathbb{Z}$ [Ka]. (We used the fact that $H^2(A_5, \mathbb{C}^*) = \mathbb{Z}/2\mathbb{Z}$ in [Ki3].) Hence ρ factors through G if $G = D_{2n}$, n odd, and otherwise through $\tilde{G}$, the unique central extension of G by $\{\pm 1\}$. Then G' is an amalgamated product $\tilde{G} \times_{\{\pm 1\}} C_{2n}$, where C_{2n} is the cyclic group of order $2n$ [Ki3]. Hence irreducible representations of G' are equivalent to a twist of irreducible representations of $\tilde{G}$. (In the case of D_{2n}, n odd, irreducible representations of G' are equivalent to a twist of those of D_{2n}.) Hence ρ is equivalent to a twist of an irreducible 2-dimensional representation of $\tilde{G}$.

More explicitly, $\tilde{G} = SL_2(\mathbb{F}_3), GL_2(\mathbb{F}_3)$ for $G = A_4, S_4$, resp. When $G = D_{2n}$, n even, $\tilde{G} = Q_{4n}$, the dicyclic group of order $4n$, defined by $Q_{4n} =< a, b \,|\, a^{2n} = 1, b^2 = a^n, b^{-1}ab = a^{-1} >$. Note that $Q_{4n}/ < a^n > \simeq D_{2n}$. It has four 1-dimensional characters, and $n-1$ two-dimensional representations. On the other hand, the irreducible representations of the dihedral group, D_{2n}, n odd, are: two 1-dimensional characters, and $\frac{n-1}{2}$ two-dimensional representations. Hence by Langlands-Tunnell theorem [La, Tu], all irreducible representations of D_{2n}, Q_{4n} are automorphic.

Next consider $\tilde{G} = GL_2(\mathbb{F}_3)$. It has order 48. The character table is given as follows. (See [FH] for character tables of $SL_2(\mathbb{F}_3)$ and $GL_2(\mathbb{F}_3)$.)

	I	$-I$	$\begin{pmatrix}1&1\\0&1\end{pmatrix}$	$\begin{pmatrix}2&1\\0&2\end{pmatrix}$	$\begin{pmatrix}1&0\\0&-1\end{pmatrix}$	$\begin{pmatrix}0&-1\\1&0\end{pmatrix}$	$\begin{pmatrix}1&2\\1&1\end{pmatrix}$	$\begin{pmatrix}2&2\\1&2\end{pmatrix}$
size	1	1	8	8	12	6	6	6
χ_1	1	1	1	1	1	1	1	1
χ_1'	1	1	1	1	-1	1	-1	-1
χ_2	2	-2	-1	1	0	0	$-\sqrt{-2}$	$\sqrt{-2}$
χ_2'	2	2	-1	-1	0	2	0	0
χ_2''	2	-2	-1	1	0	0	$\sqrt{-2}$	$-\sqrt{-2}$
χ_3	3	3	0	0	1	-1	-1	-1
χ_3'	3	3	0	0	-1	-1	1	1
χ_4	4	-4	1	-1	0	0	0	0

Here $I = \begin{pmatrix}1&0\\0&1\end{pmatrix}$. Note the relations: $\chi_3 = Sym^2(\chi_2)$, $Sym^2(\chi_2') = \chi_2' + \chi_1$, $\chi_2'' = \chi_2 \otimes \chi_1'$, $\chi_3' = \chi_3 \otimes \chi_1'$, and $\chi_4 = \chi_2 \otimes \chi_2'$.

Also χ_2 is the 2-dimensional representation which comes from the natural map $GL_2(\mathbb{F}_3) \longrightarrow GL_2(\mathbb{C})$, and has been used in the proof of Fermat's last theorem by Andrew Wiles (see [Ge, page 160]): More explicitly, χ_2 is the character of the representation $\psi : GL_2(\mathbb{F}_3) \longrightarrow GL_2(\mathbb{Z}[\sqrt{-2}]) \subset GL_2(\mathbb{C})$, defined by

$$\psi\begin{pmatrix}-1&1\\-1&0\end{pmatrix} = \begin{pmatrix}-1&1\\-1&0\end{pmatrix}, \quad \psi\begin{pmatrix}1&-1\\1&1\end{pmatrix} = \begin{pmatrix}1&-1\\-\sqrt{-2}&-1+\sqrt{-2}\end{pmatrix}.$$

Here $\begin{pmatrix}-1&1\\-1&0\end{pmatrix}$ and $\begin{pmatrix}1&-1\\1&1\end{pmatrix}$ are two generators of $GL_2(\mathbb{F}_3)$.

Note that $GL_2(\mathbb{F}_3)/\{\pm I\} \simeq S_4$, and hence the representations that factor through representations of S_4 are $\chi_1, \chi_1', \chi_2, \chi_3, \chi_3'$.

Consider $\tilde{G} = SL_2(\mathbb{F}_3)$. It has order 24. The character table is

	I	$-I$	$\begin{pmatrix}1&1\\0&1\end{pmatrix}$	$\begin{pmatrix}1&2\\0&1\end{pmatrix}$	$\begin{pmatrix}2&1\\0&2\end{pmatrix}$	$\begin{pmatrix}2&2\\0&2\end{pmatrix}$	$\begin{pmatrix}0&-1\\1&0\end{pmatrix}$
size	1	1	4	4	4	4	6
χ_1	1	1	1	1	1	1	1
μ_1	1	1	u	u^2	$-u$	$-u^2$	1
μ_1'	1	1	u^2	u	$-u^2$	$-u$	1
χ_2	2	-2	-1	-1	1	1	0
η_2	2	-2	s	$-s^2$	$-s$	s^2	0
η_2'	2	-2	$-s^2$	s	s^2	$-s$	0
χ_3	3	3	0	0	0	0	-1

Here u is a primitive cube root of 1, i.e., $u = -\frac{1}{2} + \frac{1}{2}\sqrt{-3}$, and $s = \frac{1}{2} + \frac{1}{2}\sqrt{-3} = u + 1$. Note that $\chi_4|_{SL_2(\mathbb{F}_3)} = \eta_2 + \eta_2'$, and $\chi_2'|_{SL_2(\mathbb{F}_3)} = \mu_1 + \mu_1'$.

Note that $SL_2(\mathbb{F}_3)/\{\pm I\} \simeq A_4$, and hence the representations that factor through representations of A_4 are those taking positive values at $-I$, namely, $\chi_1, \mu_1, \mu_1', \chi_3$.

Hence by Langlands-Tunnell [La, Tu], Gelbart-Jacquet [Ge-J], and Ramakrishnan [R], all irreducible representations of $SL_2(\mathbb{F}_3)$, $GL_2(\mathbb{F}_3)$ are automorphic: If χ_2, χ_2' give rise to cuspidal representations π, π', resp., then χ_3, χ_4 give rise to $Sym^2(\pi), \pi \boxtimes \pi'$, resp.

Any representation of a finite group is a direct sum of irreducible representations of the group. Hence $Sym^m(\rho)$ is equivalent to a twist of a direct sum of irreducible representations of $\tilde{G}$. Since all irreducible representations of $\tilde{G}$ are automorphic, and so is $Sym^m(\rho)$. Namely, $Sym^m(\pi)$ is an automorphic representation of $GL_{m+1}(\mathbb{A})$ for all m. It is clear from the above calculation that if π is of dihedral type, $Sym^m(\pi)$ is not cuspidal for all $m \geq 2$; if π is of tetrahedral type, $Sym^m(\pi)$ is not cuspidal for all $m \geq 3$; if π is of octahedral type, $Sym^m(\pi)$ is not cuspidal for all $m \geq 4$.

We prove here that if $Sym^4(\pi)$ is cuspidal, then $Sym^5(\pi)$ is always cuspidal. We can assume that F is an arbitrary number field.

Theorem 2.1. *Suppose $Sym^4(\pi)$ is a cuspidal representation of $GL_5(\mathbb{A})$. Then $Sym^5(\pi)$ is always a cuspidal representation of $GL_6(\mathbb{A})$*

Proof. First note

$$L^S(s, \sigma \times \pi \times A^4(\pi)) = L^S(s, \sigma \times A^5(\pi)) L^S(s, \sigma \times Sym^3(\pi)). \tag{2.1}$$

Since $A^4(\pi)$ is cuspidal, $L^S(s, \sigma \times \pi \times A^4(\pi)) = L^S(s, (\sigma \boxtimes \pi) \times A^4(\pi))$ cannot have a pole at $s = 1$ if σ is a cuspidal representation of $GL_2(\mathbb{A})$. Let σ be a cuspidal representation of $GL_3(\mathbb{A})$. We use

Proposition 2.2. *[R-W] Let σ be a cuspidal representation of $GL_3(\mathbb{A})$ and π is a non-monomial cuspidal representation of $GL_2(\mathbb{A})$. Then $\sigma \boxtimes \pi$ is not cuspidal if and only if $\sigma \simeq Ad(\pi) \otimes \chi$.*

In order that $L^S(s, (\sigma \boxtimes \pi) \times A^4(\pi))$ has a pole at $s = 1$, $\sigma \boxtimes \pi$ is not cuspidal. Hence $\sigma = Ad(\pi) \otimes \chi$ for some χ. Then $\sigma \boxtimes \pi = (A^3(\pi) \otimes \chi) \boxplus (\pi \otimes \chi)$. Hence

$$L^S(s, \sigma \times \pi \times A^4(\pi)) = L^S(s, (A^3(\pi) \otimes \chi) \times A^4(\pi)) L^S(s, (\pi \otimes \chi) \times A^4(\pi)),$$

has no pole at $s = 1$. □

By Theorem 2.1, we can recover

Corollary 2.3. *[W] Let π be a cuspidal representation of $GL_2(\mathbb{A})$ of icosahedral type. Then $Sym^5(\pi)$ is a cuspidal representation of $GL_6(\mathbb{A})$.*

Proof. By [Ki3], $Sym^5(\pi)$ is an automorphic representation of $GL_6(\mathbb{A})$, and $Sym^4(\pi)$ is a cuspidal representation of $GL_5(\mathbb{A})$. We apply Theorem 2.1. □

Corollary 2.4. *Suppose $Sym^4(\pi)$ is a cuspidal representation of $GL_5(\mathbb{A})$. Then there does not exist $\eta \neq 1$ such that $Sym^4(\pi) \simeq Sym^4(\pi) \otimes \eta$.*

Proof. Consider the following: By setting $\sigma = A^3(\pi) \otimes \omega_\pi^{-2}\eta$ in (2.1), we have

$$\begin{aligned} & L^S(s, (A^3(\pi) \otimes \omega_\pi^{-2}\eta) \times \pi \times A^4(\pi)) \\ = \; & L^S(s, A^4(\pi) \otimes \omega_\pi^{-2}\eta \times A^4(\pi)) L^S(s, Ad(\pi) \otimes \omega_\pi^{-1}\eta \times A^4(\pi)) \\ = \; & L^S(s, A^3(\pi) \otimes \omega_\pi^{-2}\eta \times A^5(\pi)) L^S(s, A^3(\pi) \otimes \omega_\pi^{-1}\eta \times A^3(\pi)). \end{aligned}$$

Here $A^4(\pi) \otimes \omega_\pi^{-2} \simeq \widetilde{A^4(\pi)}$. Hence we can write

$$\begin{aligned} & L^S(s, \widetilde{A^4(\pi)} \times A^4(\pi) \otimes \eta) L^S(s, (Ad(\pi) \otimes \omega_\pi^{-1}\eta) \times A^4(\pi)) \\ = \; & L^S(s, (A^3(\pi) \otimes \omega_\pi^{-2}\eta) \times A^5(\pi)) L^S(s, (A^3(\pi) \otimes \omega_\pi^{-1}\eta) \times A^3(\pi)). \end{aligned}$$

Since $L^S(s,(A^3(\pi)\otimes\omega_\pi^{-2}\eta)\times A^5(\pi))$ has no pole at $s=1$, there does not exist $\eta\neq 1$ such that $A^4(\pi)\otimes\eta\simeq A^4(\pi)$.

We can prove this fact directly as follows: Suppose that $A^4(\pi)\otimes\eta\simeq A^4(\pi)$ for $\eta\neq 1$. Then $\eta^5=1$, and η determines a cyclic quintic extension K/F, and $A^4(\pi)=I_K^F\chi$ (automorphic induction) for a grössencharacter χ of K. Let $Gal(K/F)=\langle\theta\rangle$. Hence $A^4(\pi)_K=A^4(\pi_K)=\chi\boxplus\chi^\theta\boxplus\chi^{\theta^2}\boxplus\chi^{\theta^3}\boxplus\chi^{\theta^4}$, where $\chi^\theta=\chi\circ\theta$. Since $A^4(\pi)$ is cuspidal, $\chi\neq\chi^{\theta^i}$. However, according to our calculation in [Ki-Sh2], since $A^4(\pi_K)$ is of type (1,1,1,1,1), one constituent occurs with multiplicity. Contradiction. □

Remark 2.1. It is remarkable that there is no cuspidal representation π of $GL_2(\mathbb{A})$ such that $Sym^4(\pi)$ is cuspidal and $Sym^4(\pi)\simeq Sym^4(\pi)\otimes\eta$ for $\eta\neq 1$. Consider

$$L^S(s,Sym^4(\pi)\times Sym^4(\pi))=L^S(s,\pi,Sym^8)L^S(s,\pi,Sym^6\otimes\omega_\pi)\\ \times L^S(s,Sym^4(\pi)\otimes\omega_\pi^2)L^S(s,Sym^2(\pi)\otimes\omega_\pi^3)L^S(s,\omega_\pi^4)$$

and

$$L^S(s,Sym^3(\pi)\times Sym^5(\pi))=L^S(s,\pi,Sym^8)L^S(s,\pi,Sym^6\otimes\omega_\pi)\\ \times L^S(s,Sym^4(\pi)\otimes\omega_\pi^2)L^S(s,Sym^2(\pi)\otimes\omega_\pi^3).$$

Hence if $Sym^5(\pi)$ is cuspidal, $L^S(s,\pi,Sym^8)$ is always holomorphic at $s=1$. On the other hand, if $\omega_\pi^4\neq 1$ and $Sym^4(\pi)$ is self-contragredient, i.e., $Sym^4(\pi)\simeq Sym^4(\pi)\otimes\omega_\pi^{-4}$, then $L^S(s,\pi,Sym^8)$ has a pole at $s=1$. However, we showed above that there does not exist such π.

More generally, suppose that $Sym^m(\pi)$ is an automorphic representation of $GL_{m+1}(\mathbb{A})$ whose irreducible constituents are cuspidal representations of $GL_k(\mathbb{A})$ for $k>4$. Consider, for any cuspidal representation σ of $GL_k(\mathbb{A})$,

$$L^S(s,\sigma\times\pi\times A^m(\pi))=L^S(s,\sigma\times A^{m+1}(\pi))L^S(s,\sigma\times Sym^{m-1}(\pi)). \tag{2.2}$$

Suppose σ is a cuspidal representation of GL_2. Consider $\sigma\boxtimes\pi$. Then by our assumption, $L^S(s,\sigma\times\pi\times A^m(\pi))$ cannot have a pole at $s=1$. Therefore, $A^{m+1}(\pi)$ cannot have a cuspidal representation of $GL_2(\mathbb{A})$ as an irreducible constituent. This implies that $A^{m+1}(\pi)$ cannot have a cuspidal representation of $GL_m(\mathbb{A})$ as an irreducible constituent, either. So $L^S(s,\sigma\times$

$\pi \times A^m(\pi))$ does not have a pole at $s = 1$ for $\sigma = A^{m-1}(\pi) \otimes \omega_\pi^{-(m-2)}\eta$ in (2.2). Then we have

$$\begin{aligned}
&L^S(s,(A^{m-1}(\pi) \otimes \omega_\pi^{-(m-2)}\eta) \times \pi \times A^m(\pi)) \\
&\quad = L^S(s, A^m(\pi) \otimes \omega_\pi^{-(m-2)}\eta \times A^m(\pi)) \\
&\qquad \times L^S(s, A^{m-2}(\pi) \otimes \omega_\pi^{-(m-3)}\eta \times A^m(\pi)) \\
&\quad = L^S(s, A^{m-1}(\pi) \otimes \omega_\pi^{-(m-2)}\eta \times A^{m+1}(\pi)) \\
&\qquad \times L^S(s, A^{m-1}(\pi) \otimes \omega_\pi^{-(m-3)}\eta \times A^{m-1}(\pi)).
\end{aligned}$$

Here $A^m(\pi) \otimes \omega_\pi^{-(m-2)} \simeq \widetilde{A^m(\pi)}$. Hence we can write

$$\begin{aligned}
&L^S(s, \widetilde{A^m(\pi)} \times A^m(\pi) \otimes \eta)\, L^S(s, (A^{m-2}(\pi) \otimes \omega_\pi^{-(m-3)}\eta) \times A^m(\pi)) \\
&= L^S(s, (A^{m-1}(\pi) \otimes \omega_\pi^{-(m-2)}\eta) \times A^{m+1}(\pi)) \\
&\qquad \times L^S(s, (A^{m-1}(\pi) \otimes \omega_\pi^{-(m-3)}\eta) \times A^{m-1}(\pi)).
\end{aligned}$$

Hence if $\eta \neq 1$, $L^S(s, \widetilde{A^m(\pi)} \times (A^m(\pi) \otimes \eta))$ has no pole at $s = 1$, i.e., $A^m(\pi) \otimes \eta \not\simeq A^m(\pi)$.

We have proved the following:

Proposition 2.5. *Suppose $m \geq 5$. Suppose $Sym^m(\pi)$ is of type $(n_1, ..., n_k)$, where $n_i > 4$ for each i. Then there does not exist $\eta \neq 1$ such that $Sym^m(\pi) \otimes \eta \simeq Sym^m(\pi)$.*

Since $Sym^{k-1}(\pi) \boxtimes \pi = Sym^k(\pi) \boxplus Sym^{k-2}(\pi) \otimes \omega_\pi$, if $\sigma = Sym^l(\pi) \boxtimes \tau$, then $\sigma \boxtimes \pi$ is not cuspidal. One wonders if the converse is true. It would be a natural generalization of Proposition 2.2.

3 First occurences of poles of symmetric power L-functions

We summarize results on poles of symmetric power L-functions [Ki-Sh2]. Let π be a cuspidal representation of GL_2 with the central character ω.

First, π is of dihedral type. Then $L(s, \pi, Sym^2)$ has a pole at $s = 1$ if $\omega \neq 1$, and $\omega^2 = 1$. If $\omega = 1$, $L(s, \pi, Sym^4)$ has a pole at $s = 1$ for the first time.

Second, π is of tetrahedral type. Then $L(s,\pi,Sym^4\otimes\omega^{-1})$ has a pole at $s=1$ if $\omega\neq 1$. Here $\omega=\eta$ or η^2, where η is a cubic character such that $Ad(\pi)\simeq Ad(\pi)\otimes\eta$. If $\omega=1$, $L(s,\pi,Sym^6)$ has a pole at $s=1$ for the first time: Consider

$$L^S(s,Sym^2(\pi)\times Sym^3(\pi))=L^S(s,\pi,Sym^5)L^S(s,Sym^3(\pi))L^S(s,\pi).$$

Since $Sym^3(\pi)=(\pi\otimes\mu)\boxplus(\pi\otimes\mu^2)$, where μ is a cubic character, the left hand side is holomorphic at $s=1$, and $L^S(s,\pi,Sym^5)$ is holomorphic at $s=1$. Consider

$$\begin{aligned}&L^S(s,Sym^3(\pi)\times Sym^3(\pi))\\&=L^S(s,\pi,Sym^6)L^S(s,Sym^4(\pi))L^S(s,Sym^2(\pi))L^S(s,1).\end{aligned}$$

The left hand side has a double pole at $s=1$, hence $L^S(s,\pi,Sym^6)$ has a pole at $s=1$.

Third, π is of octahedral type. Then by [Ki-Sh2, Theorem 3.3.7], $L^S(s,\pi,Sym^6\otimes\omega^{-3}\eta)$ has a pole at $s=1$. Here η is a quadratic character such that $A^3(\pi)\simeq A^3(\pi)\otimes\eta$. If $\omega=1$, $L^S(s,\pi,Sym^8)$ has a pole at $s=1$ for the first time: We can show as above that $L^S(s,\pi,Sym^m)$ does not have a pole at $s=1$ for $m\leq 7$. Consider

$$\begin{aligned}L^S(s,Sym^4(\pi)\times Sym^4(\pi))&=L^S(s,\pi,Sym^8)\\&\quad\times L^S(s,\pi,Sym^6)L^S(s,Sym^4(\pi))L^S(s,Sym^2(\pi))L^S(s,1).\end{aligned}$$

Since $Sym^4(\pi)=\pi(\chi^{-1})\boxplus Sym^2(\pi)\otimes\eta$ (cf. Theorem 1.1), the left hand side has a double pole at $s=1$. Hence $L^S(s,\pi,Sym^8)$ has a pole at $s=1$.

Fourth, π is of icosahedral type. Note that $L^S(s,\pi,Sym^m\otimes\chi)$ has no pole at $s=1$ for all $m\leq 11$ for any χ. For example, for $m=11$, consider $L^S(s,Sym^5(\pi)\times Sym^6(\pi)\otimes\chi)$

$$=L^S(s,\pi,Sym^{11}\otimes\chi)\prod_{i=1}^{5}L^S(s,\pi,Sym^{11-2i}\otimes\chi\omega^i).$$

Since $Sym^5(\pi)$ is cuspidal, and $Sym^6(\pi)$ is equivalent to a twist of $Sym^2(\pi^\tau)\boxplus(\pi\boxtimes\pi^\tau)$, where π^τ is a certain cuspidal representation of $GL_2(\mathbb{A})$ [Ki3, Theorem 6.4], the left hand side is holomorphic at $s=1$. Hence $L^S(s,\pi,Sym^{11}\otimes\chi)$ is holomorphic at $s=1$.

If $\omega = 1$, $L^S(s,\pi,Sym^{12})$ has a pole at $s = 1$ for the first time: Consider $L^S(s, Sym^6(\pi) \times Sym^6(\pi))$

$$= L^S(s,\pi,Sym^{12}) \prod_{i=1}^{5} L^S(s,\pi,Sym^{12-2i}) L^S(s,1).$$

If $\omega = 1$, $Sym^6(\pi) = Sym^2(\pi^\tau) \boxplus (\pi \boxtimes \pi^\tau)$, and the left hand side has a double pole at $s = 1$. So $L^S(s,\pi,Sym^{12})$ has a pole at $s = 1$.

If π is not of Weil type, i.e., π is not attached to any of the homomorphisms $\phi : W_F \longrightarrow GL_2(\mathbb{C})$, it is expected that $L(s,\pi,Sym^m \otimes \chi)$ is entire for all m, and for all χ. At the moment, we know that it is the case only for $m \leq 4$ [Ki1]. For $5 \leq m \leq 9$, we only know that they are meromorphic in all of $\mathbb{C}$, and invertible for $Re(s) \geq 1$ [Ki-Sh2]. For $m \geq 10$, no such results are available.

4 Descent to cuspidal representations on classical groups

For $g \in GL_2(\mathbb{C})$, $Sym^m(g) \in GSp_{m+1}(\mathbb{C})$ if m is odd, and $Sym^m(g) \in GSO_{m+1}(\mathbb{C})$ if m is even. (See [G-W, page 244].) For $m = 1$, we need to identify $GSp_2(\mathbb{C}) \simeq GL_2(\mathbb{C})$.

Moreover, if m is even, $Sym^m(g)(det\, g)^{-\frac{m}{2}} \in SO_{m+1}(\mathbb{C})$. Let $m = 2k$ and $B(\pi) = Sym^m(\pi) \otimes \omega_\pi^{-k}$. Then $B(\pi)$ is a self-dual representation of $GL_{2k+1}(\mathbb{A})$ with the trivial central character. Then $L^S(s, B(\pi), Sym^2)$ always has a pole at $s = 1$, since $L^S(s, B(\pi), \wedge^2)$ is holomorphic at $s = 1$ [Ki2]. However, we can see directly as follows:

$$L^S(s, B(\pi), Sym^2) = L^S(s,\pi,Sym^{2m} \otimes \omega_\pi^{-m})$$
$$L^S(s,\pi,Sym^{2m-4} \otimes \omega_\pi^{-m+2}) \cdots L^S(s,\pi,Sym^4 \otimes \omega_\pi^{-2}) L^S(s,1).$$

Since all the L-functions on the right side have no zeros at $s = 1$, our result follows. Hence by the result of Ginzburg-Rallis-Soudry [GRS], $B(\pi)$ descends to a generic cuspidal representation τ of $Sp_{2k}(\mathbb{A})$.

Let $m = 2k-1$, and $B(\pi) = Sym^m(\pi) \otimes \omega_\pi^{-k+1}$. Then $B(\pi)$ is a representation of $GL_{2k}(\mathbb{A})$ with the central character ω_π^k and $\widetilde{B(\pi)} \otimes \omega_\pi \simeq B(\pi)$. We claim that $L^S(s, B(\pi), \wedge^2 \otimes \omega_\pi^{-1})$ has a pole at $s = 1$: Consider

$$L^S(s, B(\pi), \wedge^2 \otimes \omega_\pi^{-1}) = L^S(s, \pi, Sym^{2m-2} \otimes \omega_\pi^{-m+1})$$
$$\times L^S(s, \pi, Sym^{2m-6} \otimes \omega_\pi^{-m+3}) \cdots L^S(s, \pi, Sym^4 \otimes \omega_\pi^{-2}) L^S(s, 1).$$

Since all the L-functions on the right side have no zeros at $s = 1$, our result follows.

It is conjectured that $B(\pi)$ descends to a generic cuspidal representation of $GSpin_{2k+1}(\mathbb{A})$. When $k = 2$, $GSpin_5 \simeq GSp_4$, and it is the unpublished result of Jacquet, Piatetski-Shapiro, and Shalika (Theorem 1.2). If $\omega_\pi = 1$, then it is expected that $B(\pi)$ descends to a generic cuspidal representation of $SO_{2k+1}(\mathbb{A})$.

We state the result for $m = 4$ as

Theorem 4.1. *Suppose $Sym^4(\pi)$ is cuspidal. Then $Sym^4(\pi) \otimes \omega_\pi^{-2}$ descends to a generic cuspidal representation τ of $Sp_4(\mathbb{A})$. Hence $L(s, \tau) = L(s, Sym^4(\pi) \otimes \omega_\pi^{-2})$.*

We examine the relationship between the infinity components π_∞ and τ_∞. Let π be a cuspidal representation of $GL_2(\mathbb{A}_\mathbb{Q})$ attached to a holomorphic cusp form of weight l. Then π_∞ is a discrete series D_l of lowest weight l. Recall that the discrete series D_l of $GL_2(\mathbb{R})$ is the subrepresentation of $\pi(|\ |^{\frac{l-1}{2}}, \epsilon|\ |^{-\frac{l-1}{2}})$, where l is an integer, $l \geq 2$, and $\epsilon = \begin{cases} 1, & \text{if } l \text{ is even} \\ sgn, & \text{if } l \text{ is odd} \end{cases}$.

When $l = 1$, it is called the limit of discrete series. In particular, D_l is parametrized by

$$\phi_l : W_\mathbb{R} = \mathbb{C}^\times \cup j\mathbb{C}^\times \longrightarrow GL_2(\mathbb{C}),$$

where $\phi_l(re^{i\theta}) = \begin{pmatrix} e^{i\theta(l-1)} & 0 \\ 0 & e^{-i\theta(l-1)} \end{pmatrix}$, $\phi_l(j) = \begin{pmatrix} 0 & (-1)^{l-1} \\ 1 & 0 \end{pmatrix}$.

Here $\phi_l = Ind_{W_\mathbb{R}}^{W_\mathbb{C}}\, \omega$, where $\omega(z) = (z\bar{z})^{-\frac{l-1}{2}} z^{l-1}$, and $L(s, D_l)$

$$= L(s, \phi_l) = L(s, \omega) = \Gamma_\mathbb{C}(s + \frac{l-1}{2}) = 2(2\pi)^{-(s+\frac{l-1}{2})}\Gamma(s + \frac{l-1}{2}).$$

Then $Sym^4(\phi_l) = diag(e^{4i\theta(l-1)}, e^{2i\theta(l-1)}, 1, e^{-2i\theta(l-1)}, e^{-4i\theta(l-1)})$, and

$$L(s, Sym^4(D_l)) = \Gamma_\mathbb{R}(s)\Gamma_\mathbb{C}(s + l - 1)\Gamma_\mathbb{C}(s + 2(l-1)),$$

where $\Gamma_{\mathbb{R}}(s) = \pi^{-\frac{s}{2}}\Gamma(\frac{s}{2})$. Hence $Sym^4(D_l) = 1 \boxplus D_{2(l-1)+1} \boxplus D_{4(l-1)+1}$.

Discrete series of $Sp_4(\mathbb{R})$ are parametrized by the infinitesimal character $\chi(k,l)$, $k > l > 0$, $k,l \in \mathbb{Z}$ [Ad]. The lowest weight is $(k+1, l+2)$. It is the infinity component of a vector-valued Siegel cusp form of weight $(k+1, l+2)$. The Langlands parameter is

$$\phi_{k,l} \ : \ W_{\mathbb{R}} \longrightarrow SO_5(\mathbb{C}),$$

$$\phi_{k,l}(re^{i\theta}) \ = \ diag(t(2k\theta), t(2l\theta), 1), \quad \phi_{k,l}(j) = J = \begin{pmatrix} 0 & 0 & I \\ 0 & 1 & 0 \\ -I & 0 & 0 \end{pmatrix}$$

where $t(\theta) = \begin{pmatrix} \cos\theta & \sin\theta \\ -\sin\theta & \cos\theta \end{pmatrix}$. Since $t(\theta)$ is conjugate to $\begin{pmatrix} e^{i\theta} & 0 \\ 0 & e^{-i\theta} \end{pmatrix}$, $\phi_{k,l}$ can be written in a form

$$\phi_{k,l}(z) = diag(z^k\bar{z}^{-k}, z^l\bar{z}^{-l}, 1, z^{-l}\bar{z}^l, z^{-k}\bar{z}^k), \quad z \in \mathbb{C}.$$

If $l = 0$, it is the limit of discrete series. By comparing the parameters, we can see that $Sym^4(\phi_l)$ parametrizes an L-packet of representations of $Sp_4(\mathbb{R})$. The L-packet consists of four representations [Ad]. One of them is a generic representation, and it is the infinity component τ_∞. Another one is a discrete series, denoted by $Sym^4(D_l)$. Its infinitesimal character is $\chi(2(l-1), l-1)$. The lowest weight is $(2l-1, l+1)$. According to Arthur' conjecture, it is expected that there exists a vector-valued Siegel cusp form of weight $(2l-1, l+1)$ whose functorial lift to $GL_5(\mathbb{A}_\mathbb{Q})$ is $Sym^4(\pi)\otimes\omega_\pi^{-2}$.

More explicitly, let us consider Δ, the cusp form of weight 12 with respect to $SL_2(\mathbb{Z})$. Let $\pi = \pi_\Delta$ be the cuspidal representation of $GL_2(\mathbb{A}_\mathbb{Q})$ attached to Δ. Then $\pi_\infty = D_{12}$, and $Sym^4(\pi)$ descends to a generic cuspidal representation $\tau = \otimes\tau_p$ of $Sp_4(\mathbb{A}_\mathbb{Q})$. Here the infinitesimal character of τ_∞ is $\chi(22, 11)$. T. Ibukiyama [Ib] has a conjectural construction of the vector-valued Siegel cusp form whose infinity component is $Sym^4(D_{12})$.

Suppose π is a cuspidal representation of $GL_2(\mathbb{A}_\mathbb{Q})$ of icosahedral type. Then π_∞ is either $\pi(1,1)$ or $\pi(1, sgn)$. If $\pi_\infty = \pi(1, sgn)$, by [DS], π gives rise to an icosahedral representation $\phi : W_\mathbb{Q} \longrightarrow GL_2(\mathbb{C})$. We assume the same if $\pi_\infty = \pi(1,1)$. Then $Sym^4(\pi)\otimes\omega_\pi^{-2}$ descends to a generic cuspidal representation τ of $Sp_4(\mathbb{A}_\mathbb{Q})$, which is parametrized by $Sym^4(\phi)\otimes\omega^{-2}$: $W_\mathbb{Q} \longrightarrow SO_5(\mathbb{C})$.

Let π be a cuspidal representation of $GL_2(\mathbb{A})$ with the central character ω_π. Consider $Sym^3(\pi), Sym^4(\pi)$. Assume that they are cuspidal. Then by Theorem 4.1, $Sym^4(\pi)\otimes\omega_\pi^{-2}$ descends to a generic cuspidal representation τ of $Sp_4(\mathbb{A})$. On the other hand, by the unpublished result of Jacquet, Piatetski-Shapiro and Shalika (Theorem 1.2), $Sym^3(\pi)\otimes\omega_\pi^{-1}$ descends to a generic cuspidal representation Π on $GSp_4(\mathbb{A})$ with the central character ω_π. Let τ' be any irreducible constituent of $\Pi|_{Sp_4(\mathbb{A})}$.

Proposition 4.2. *τ is nearly equivalent to τ', i.e., $\tau_v \simeq \tau'_v$ for almost all v.*

Proof. By construction, $L^S(s,\sigma\times\tau) = L^S(s,\sigma\times A^4(\pi)\otimes\omega_\pi^{-1})$ for any cuspidal representation σ of $GL_k(\mathbb{A})$. Since $\wedge^2(A^3(\pi)) = A^4(\pi)\boxplus\omega_\pi$ [Ki1], $L^S(s,\sigma\times A^4(\pi)) = L^S(s,(\sigma\otimes\omega_\pi)\times\tau')$ [Ki-Sh2, (4.2)]. Hence $L^S(s,\sigma\times\tau) = L^S(s,\sigma\times\tau')$, and our result follows. □

Remark 4.1. It is expected that $\tau\simeq\tau'$. It would follow from the strong multiplicity one result for generic cuspidal representations of $Sp_4(\mathbb{A})$.

5 Remark on the images of functorial lift

We are interested in the image of $Sym^3(\pi)$ and other functorial lifts to $GL_4(\mathbb{A})$. For $Sym^2(\pi)$, we have

Proposition 5.1. *(Flicker, Ramakrishnan) Let Π be a cuspidal representation of $GL_3(\mathbb{A})$ which is essentially self-dual, i.e., $\widetilde{\Pi\otimes\chi} = \Pi\otimes\chi$ for some χ. Then $\Pi = Ad(\pi)\otimes\chi^{-1}$, for some cuspidal representation π of $GL_2(\mathbb{A})$.*

There are six ways to generate a cuspidal representation of $GL_4(\mathbb{A})$ in the literature.

1. The symmetric cube $Sym^3(\pi)$, where π is a cuspidal representation of GL_2 [Ki-Sh1]. In this case, $\wedge^2(Sym^3(\pi)) = Sym^4(\pi)\otimes\omega_\pi\boxplus\omega_\pi^3$ [Ki1];

2. The tensor product $\pi\boxtimes\pi'$, where π,π' are cuspidal representations of GL_2 [R]. This corresponds to the tensor product map $GL_2(\mathbb{C})\times GL_2(\mathbb{C})\longrightarrow GL_4(\mathbb{C})$; $(g,g')\longmapsto g\otimes g'$. In this case, $\wedge^2(\pi\boxtimes\pi') = Ad(\pi)\otimes\omega\boxplus Ad(\pi')\otimes\omega$, where $\omega=\omega_\pi\omega_{\pi'}$;

3. The Asai lift $As(\pi)$, where π is a cuspidal representation of $GL_2(\mathbb{A}_E)$, and E/F is a quadratic extension [R2]. This corresponds to the L-group homomorphism

$$r : {}^L R_{E/F} GL_2 = (GL_2(\mathbb{C}) \times GL_2(\mathbb{C})) \rtimes Gal(E/F) \longrightarrow GL(\mathbb{C}^2 \otimes \mathbb{C}^2) \simeq GL_4(\mathbb{C}),$$
$$r(g, g'; 1)(x \otimes y) = g(x) \otimes g'(y); \quad r(1, 1; \theta)(x \otimes y) = y \otimes x.$$

In this case, $As(\pi)_E = \pi \boxtimes (\pi \circ \theta)$, and $\wedge^2(As(\pi)) = I_E^F(Ad(\pi)) \otimes (\omega_\pi|_{\mathbb{A}_F^*})$.

4. The automorphic induction $I_E^F \pi$, where π is a cuspidal representation of $GL_2(\mathbb{A}_E)$, and E/F is a quadratic extension [AC]. This corresponds to the L-group homomorphism

$$r : {}^L R_{E/F} GL_2 = (GL_2(\mathbb{C}) \times GL_2(\mathbb{C})) \rtimes Gal(E/F) \longrightarrow GL(\mathbb{C}^2 \oplus \mathbb{C}^2) \simeq GL_4(\mathbb{C}),$$
$$r(g, g'; 1)(x \oplus y) = g(x) \oplus g'(y), \quad r(1, 1; \theta)(x \oplus y) = y \oplus x.$$

In this case, $(I_E^F \pi)_E = \pi \boxplus (\pi \circ \theta)$, and $\wedge^2(I_E^F \pi) = As(\pi) \otimes \omega_{E/F} \boxplus I_E^F(\omega_\pi)$ [Ki4], where $\omega_{E/F}$ is the quadratic character attached to E/F by class field theory. This includes $I_K^F \chi$, where χ is a grössencharacter of K, and K/F is a quartic extension (normal or non-normal). In [Ki4], we assumed that $I_E^F(\omega_\pi)$ is cuspidal. However it is not necessary: $I_E^F(\omega_\pi)$ is not cuspidal if and only if $\omega_\pi = \omega_\pi \circ \theta$. It is the case if and only if $\omega_\pi = \mu \circ N_{E/F}$ for a grössencharacter μ of F. Then $I_E^F(\omega_\pi) = \mu \boxplus \mu\omega_{E/F}$;

5. The functorial lift from GSp_4, corresponding to the L-group homomorphism $\phi : GSp_4(\mathbb{C}) \longrightarrow GL_4(\mathbb{C})$ [As-Sh], where ϕ is the embedding. Namely, given a generic cuspidal representation τ of $GSp_4(\mathbb{A})$, there exists an automorphic representation $\phi(\tau)$ of $GL_4(\mathbb{A})$ corresponding to ϕ. This includes the functorial lift from SO_5 to GL_4, corresponding to the embedding ${}^L SO_5 = Sp_4(\mathbb{C}) \hookrightarrow GL_4(\mathbb{C})$, since $SO_5 \simeq PGSp_4$. In this case, $\wedge^2(\phi(\tau)) = \Pi \boxplus \omega_\tau$, where Π is an automorphic representation of $GL_5(\mathbb{A})$. By the unpublished result of Jacquet, Piatetski-Shapiro and Shalika (Theorem 1.2), we know the image of this lift completely, namely, π comes from GSp_4 if and only if $L(s, \pi, \wedge^2 \otimes \chi)$ has a pole at $s = 1$ for some χ. If σ is a cuspidal representation of $GL_2(\mathbb{A})$, then $Sym^3(\sigma)$ is contained in this image, since $L(s, Sym^3(\sigma), \wedge^2 \otimes \omega_\sigma^{-3})$ has a pole at $s = 1$.

6. The stable base change from $U(2,2)_{F/F_0}$ [Ki-Kr], where F/F_0 is a quadratic extension. This corresponds to the L-group homomorphism

$$BC : GL_4(\mathbb{C}) \rtimes Gal(F/F_0) \longrightarrow (GL_4(\mathbb{C}) \times GL_4(\mathbb{C})) \rtimes Gal(F/F_0),$$
$$(g, 1) \longmapsto (g, \theta(g), 1); \quad (g, \theta) \longmapsto (g, \theta(g), \theta),$$

where $\theta(g) = J_4^{-1} {}^t g^{-1} J_4$, J_4 is the matrix that defines $U(2,2)_{F/F_0}$. Note that $GL_4(\mathbb{C}) \rtimes Gal(F/F_0)$, $(GL_4(\mathbb{C}) \times GL_4(\mathbb{C})) \rtimes Gal(F/F_0)$ are L-groups of $U(2,2)_{F/F_0}$, $R_{F/F_0}GL_4$, resp. and $R_{F/F_0}GL_4(\mathbb{A}_{F_0}) = GL_4(\mathbb{A})$. Let π be a generic cuspidal representation of $U(2,2)(\mathbb{A}_{F_0})$. Then the stable base change lift $BC(\pi)$ is an automorphic representation of $GL_4(\mathbb{A})$ of the form $\sigma_1 \boxplus \cdots \boxplus \sigma_k$, where $n_1 + \cdots + n_k = 4$ and for each i, σ_i is a cuspidal representation of $GL_{n_i}(\mathbb{A})$, $\tilde{\sigma}_i \simeq \sigma_i \circ \theta$, $\sigma_i \not\simeq \sigma_j$ for $i \neq j$, and $L(s, \sigma_i, As \otimes \omega_{F/F_0})$ has a pole at $s = 1$. Here ω_{F/F_0} is the quadratic character attached to F/F_0 by class field theory. Note that if χ is a character of $\mathbb{A}_F^*$, then $L(s, \chi, As) = L(s, \chi|_{\mathbb{A}_{F_0}^*})$. Hence "$L(s, \chi, As)$ has a pole at $s = 1$" means simply that $\chi|_{\mathbb{A}_{F_0}^*} = 1$. If σ is a cuspidal representation of $GL_2(\mathbb{A})$, $L(s, \sigma, As \otimes \omega_{F/F_0})$ has a pole at $s = 1$ if and only if σ is the stable base change lift from $U(1,1)_{F/F_0}$. In that case, $As(\sigma) = \Pi \boxplus \omega_{F/F_0}$ for an automorphic representation Π of $GL_3(\mathbb{A}_{F_0})$. So we know the image of the stable base change lift completely. A cuspidal representation π of $GL_4(\mathbb{A})$ comes from $U(2,2)_{F/F_0}$ if and only if $\tilde{\pi} \simeq \pi \circ \theta$ and $L(s, \pi, As \otimes \omega_{F/F_0})$ has a pole at $s = 1$. In this case, it is most likely that $\wedge^2(\pi)$ is cuspidal.

Let π be a cuspidal representation of $GL_4(\mathbb{A})$. We want to know when it is in the image of the above lifts. Let's summarize our discussions above as follows:

Suppose $L(s, \pi, \wedge^2 \otimes \chi)$ has a pole at $s = 1$ for a grössencharacter χ of F. Then π comes from GSp_4. Now π may be either $Sym^3(\sigma)$ for a cuspidal representation σ of $GL_2(\mathbb{A})$, or $I_E^F \sigma$ for a cuspidal representation σ of $GL_2(\mathbb{A}_E)$, or neither. (If $\omega_\pi = \mu \circ N_{E/F}$ for a grössencharacter μ of F, then $L(s, I_E^F \pi, \wedge^2 \otimes \mu^{-1})$ has a pole at $s = 1$, and $I_E^F \pi$ descends to a generic cuspidal representation of $GSp_4(\mathbb{A})$.) We need to consider the base change π_E. If π_E is not cuspidal for a quadratic extension E/F, then $\pi = I_E^F \sigma$ for some cuspidal representation σ of $GL_2(\mathbb{A}_E)$. We do not know when π is of the form $Sym^3(\sigma)$.

Suppose $\wedge^2 \pi$ is not cuspidal of type (3,3), then π is of the form $\sigma \boxtimes \sigma'$, where σ, σ' are cuspidal representations of $GL_2(\mathbb{A})$.

Suppose $\wedge^2 \pi$ is cuspidal. If there exists a quadratic extension E/F such that $(\wedge^2 \pi)_E$ is not cuspidal, then π is of the form $As(\sigma)$ for some cuspidal representation σ of $GL_2(\mathbb{A}_E)$. If there exists a quadratic extension F/F_0 such that $L(s, \pi, As \otimes \omega_{F/F_0})$ has a pole at $s = 1$, π is the stable base change lift from $U(2,2)_{F/F_0}$. Otherwise, π is not in the image of any lifts.

References

[AC] J. Arthur and L. Clozel, Simple Algebras, Base Change, and the Advanced Theory of the Trace Formula, Annals of Math. Studies **120**, Princeton, 1989.

[Ad] J. Adams, $THETA-10$ in Contributions to Automorphic Forms, Geometry and Number Theory, 39-56, Johns Hopkins Press, 2004.

[As-Sh] M. Asgari and F. Shahidi, *Generic Transfer from $GSp(4)$ to $GL(4)$*, Compos. Math. **142** (2006), 541-550.

[DS] P. Deligne and J.P. Serre, *Formes modulaires de poids 1*, Ann. scient. Ec. Norm. Sup. **7** (1974), 507-530.

[FH] W. Fulton and J. Harris, Representation Theory, A First Course, Springer Verlag, 2000.

[Ib] T. Ibukiyama, *Numerical example of a Siegel cusp form having the cubic zeta function*, preprint (2006).

[Ge] S. Gelbart, *Three lectures on the modularity of $\bar{\rho}_{E,3}$ and the Langlands reciprocity conjecture*, in Modular forms and Fermat's last theorem, 155-207, Springer, 1997.

[Ge-J] S. Gelbart and H. Jacquet, *A relation between automorphic representations of $GL(2)$ and $GL(3)$*, Ann. Sci. École Norm. Sup. **11** (1978), 471–552.

[G-W] R. Goodman and N. Wallach, Representations and invariants of the classical groups, Cambridge University Press, 1998.

[GRS] D. Ginzburg, S. Rallis and D. Soudry, *On explicit lifts of cusp forms from GL_m to classical groups*, Ann. of Math. **150** (1999), 807-866.

[Ka] G. Karpilovsky, The Schur Multiplier, Oxford University Press, 1987.

[Ki1] H. Kim, *Functoriality for the exterior square of GL_4 and symmetric fourth of GL_2*, J. Amer. Math. Soc. **16** (2003), 139-183.

[Ki2] H. Kim, *Langlands-Shahidi method and poles of automorphic L-functions: application to exterior square L-functions*, Can. J. Math. **51** (1999), 835-849.

[Ki3] H. Kim, *An example of non-normal quintic automorphic induction and modularity of symmetric powers of cusp forms of icosahedral type*, Inv. Math. **156** (2004), 495-502.

[Ki4] H. Kim, *An application of exterior square functoriality of GL_4; Asai lift*, in Number theory, CRM Proc. Lecture Notes, 36 (2004), 197–202.

[Ki-Kr] H. Kim and M. Krishnamurthy, *Stable base change lift from unitary groups to GL_N*, IMRP **1** (2005), 1-52.

[Ki-Sh1] H. Kim and F. Shahidi, *Functorial products for $GL_2 \times GL_3$ and the symmetric cube for GL_2*, Ann. of Math. **155** (2002), 837-893.

[Ki-Sh2] H. Kim and F. Shahidi, *Cuspidality of symmetric powers with applications*, Duke Math. J. **112** (2002), 177-197.

[La] R.P. Langlands, Base Change for $GL(2)$, Annals of Math. Studies **96**, Princeton, 1980.

[R] D. Ramakrishnan, *Modularity of the Rankin-Selberg L-series, and multiplicity one for $SL(2)$*, Ann. of Math. **152** (2000), 45-111.

[R2] D. Ramakrishnan, *Modularity of solvable Artin representations of $GO(4)$-type*, IMRN, No. 1 (2002), 1-54.

[R-W] D. Ramakrishnan and S. Wang, *A cuspidality criterion for the functorial product on $GL(2) \times GL(3)$ with a cohomological application*, IMRN (2004), 1355-1394.

[Tu] J. Tunnell, *Artin's conjecture for representations of octahedral type*, Bulletin AMS **5** (1981), 173-175.

[W] S. Wang, *On the symmetric powers of cusp forms on GL(2) of icosahedral type*, IMRN (2003), 2373-2390.

Henry H. Kim
Department of Mathematics,
University of Toronto,
Toronto, ON M5S 2E4, Canada
Email: henrykim@math.toronto.edu
and
Korea Institute for Advanced Study,
Seoul, Korea

Zeta Functions of Root Systems

Y. KOMORI, K. MATSUMOTO & H. TSUMURA

In this paper, we introduce multi-variable zeta-functions of roots, and prove the analytic continuation of them. For the root systems associated with Lie algebras, these functions are also called Witten zeta-functions associated with Lie algebras which can be regarded as several variable generalizations of Witten zeta-functions defined by Zagier. In the case of type A_r, we have already studied some analytic properties in our previous paper. In the present paper, we prove certain functional relations among these functions of types A_r ($r = 1, 2, 3$) which include what is called Witten's volume formulas. Moreover we mention some structural background of the theory of functional relations in terms of Weyl groups.

0 Introduction

Let $\mathbb{N}$ be the set of natural numbers, $\mathbb{N}_0 = \mathbb{N} \cup \{0\}$, $\mathbb{Z}$ the ring of rational integers, $\mathbb{Q}$ the field of rational numbers, $\mathbb{R}$ the field of real numbers, and $\mathbb{C}$ the field of complex numbers.

Let $\mathfrak{g}$ be any semisimple Lie algebra. In [25], Zagier defined the Witten zeta-function associated with $\mathfrak{g}$ by

$$\zeta_W(s;\mathfrak{g}) = \sum_{\varphi} (\dim \varphi)^{-s} \quad (s \in \mathbb{C}), \tag{0.1}$$

where φ runs over all finite dimensional irreducible representations of $\mathfrak{g}$. Using Witten's result [24], Zagier noted that

$$\zeta_W(2k;\mathfrak{g}) \in \mathbb{Q}\pi^{2kl} \tag{0.2}$$

for $k \in \mathbb{N}$, where l is the number of positive roots of $\mathfrak{g}$, which is called *Witten's volume formula* (see [25, Section 7]). In particular, $\zeta_W(s;\mathfrak{sl}(3))$

was already studied by Tornheim [19] and Mordell [15] in the 1950's. The above result for $\zeta_W(s;\mathfrak{sl}(3))$ was showed by Mordell and its explicit formula was given by Subbarao and Sitaramachandrarao in [18]. Recently Gunnells and Sczech [3] evaluated $\zeta_W(2k;\mathfrak{g})$ for $k\in\mathbb{N}$ by means of the generalized higher-dimensional Dedekind sums, and gave certain evaluation formulas for $\zeta_W(2k;\mathfrak{sl}(3))$ and $\zeta_W(2k;\mathfrak{sl}(4))$ for $k\in\mathbb{N}$ in terms of $\zeta(2j)$ for $j\in\mathbb{N}$, where $\zeta(s)$ is the Riemann zeta-function.

In [10], the second-named author defined multi-variable Witten zeta-functions by

$$\zeta_{\mathfrak{sl}(3)}(s_1,s_2,s_3)=\sum_{m,n=1}^{\infty} m^{-s_1}n^{-s_2}(m+n)^{-s_3}, \tag{0.3}$$

$$\zeta_{\mathfrak{so}(5)}(s_1,s_2,s_3,s_4)=\sum_{m,n=1}^{\infty} m^{-s_1}n^{-s_2}(m+n)^{-s_3}(m+2n)^{-s_4}. \tag{0.4}$$

Note that $\zeta_W(s;\mathfrak{sl}(3))=2^s\zeta_{\mathfrak{sl}(3)}(s,s,s)$ and $\zeta_W(s;\mathfrak{so}(5))=6^s\zeta_{\mathfrak{so}(5)}(s,s,s,s)$. The second-named author proved the analytic continuation of them by the method using the Mellin-Barnes integral formula (see [7, 8, 9]) and calculated their possible singularities. The function $\zeta_{\mathfrak{sl}(3)}(s_1,s_2,s_3)$ is also called the Tornheim double sum or the Mordell-Tornheim double zeta-function denoted by $T(s_1,s_2,s_3)$ or $\zeta_{MT,2}(s_1,s_2,s_3)$, because special values $\zeta_{\mathfrak{sl}(3)}(a_1,a_2,a_3)$ for $a_j\in\mathbb{N}$ $(j=1,2,3)$ were studied by Tornheim in [19].

Recently the third-named author [23] has proved certain functional relations between $\zeta_{\mathfrak{sl}(3)}(s_1,s_2,s_3)$ and $\zeta(s)$. These can be regarded as certain continuous relations including the known formulas between Tornheim's double sums and values of $\zeta(s)$ at positive integers obtained in [15, 19]. For example, we proved that

$$2\zeta_{\mathfrak{sl}(3)}(2,s,2)+\zeta_{\mathfrak{sl}(3)}(2,2,s)=4\zeta(2)\zeta(s+2)-6\zeta(s+4) \tag{0.5}$$

for any $s\in\mathbb{C}$ except for singularities of each function on the both sides. In particular, putting $s=2$ in (0.5), we obtain Witten's volume formula $\zeta_{\mathfrak{sl}(3)}(2,2,2)=\pi^6/2835$ given by Mordell (see [15]).

More recently, in [12], the second-named and the third-named authors defined multi-variable Witten zeta-functions $\zeta_{\mathfrak{sl}(r+1)}(\mathbf{s})$ $(r\in\mathbb{N})$ associated with $\mathfrak{sl}(r+1)$ and proved the analytic continuation of them. Furthermore, for $r\leq 3$, true singularities of them were determined, certain functional relations for them were given and new evaluation formulas for their special values were introduced. However, in that paper, we were unable to prove functional relations which can be regarded as continuous relations including Witten's volume formulas (0.2).

It seems possible to generalize these results for $\mathfrak{sl}(r+1)$ in [12] to those for any general semisimple Lie algebra $\mathfrak{g}$. Indeed, in [4], we define the Witten zeta-function associated with $\mathfrak{g}$ denoted by $\zeta_r(\mathbf{s};\mathfrak{g})$ or $\zeta_r(\mathbf{s};X_r)$ if $\mathfrak{g}$ is of type X_r ($X = A,B,C,D,E,F,G$), for $\mathbf{s} \in \mathbb{C}^n$, where n is the number of positive roots of $\mathfrak{g}$. For example, $\zeta_2(\mathbf{s};A_2)$ and $\zeta_2(\mathbf{s};B_2)$ coincide with $\zeta_{\mathfrak{sl}(3)}(s_1,s_2,s_3)$ defined by (0.3) and $\zeta_{\mathfrak{so}(5)}(s_1,s_2,s_3,s_4)$ defined by (0.4), respectively, except for the order of variables. More generally we further define zeta-functions of roots. Considering their properties, we study the analytic continuation of $\zeta_r(\mathbf{s};\mathfrak{g})$, and certain recursive structures among them. In [5], we prove certain functional relations among them, in particular, of types B_r and C_r ($r \leq 3$), which include Witten's volume formulas. Furthermore, in [6], we will study $\zeta_2(\mathbf{s};G_2)$ associated with the exceptional Lie algebra of type G_2. Indeed, we will construct a certain functional relation for $\zeta_2(\mathbf{s};G_2)$ which includes Witten's volume formula.

In the present paper, we give an overview of our previous result [12] and our recent results [4, 5], and prove some related facts. In Section 1, we introduce the definition of zeta-functions of roots which include the multi-variable Witten zeta-functions. We further study certain recursive relations for them. In Section 2, we study some structural background of functional relations for these functions. As its concrete examples, in Section 3, we explicitly give certain functional relations for $\zeta_r(\mathbf{s};A_r)$ ($r \leq 3$) which can be regarded as continuous relations including Witten's volume formula. For example, we can give a functional relation including

$$\text{(0.6)}\quad \zeta_3(2,2,2,2,2,2;A_3) = \sum_{l,m,n=1}^{\infty} \frac{1}{l^2m^2n^2(l+m)^2(m+n)^2(l+m+n)^2} = \frac{23}{2554051500}\pi^{12}$$

(see [3]) as a special value-relation. As mentioned above, we were unable to treat Witten's volume formulas in our previous paper ([12]), because we were only able to consider some limited types of relations for $\zeta_3(s_1,s_2,s_3,s_4,0,s_5;A_3)$ and $\zeta_3(s_1,s_2,s_3,0,s_4,s_5;A_3)$, that is, the case one-variable is equal to 0. In order to remove this limitation, we will study certain multiple polylogarithms of type A_3 (see (3.18) and (3.19)). By making use of them, we will be able to construct some functional relations including Witten's volume formula. This method can be applied to zeta-functions of root systems of other types (see [5, 6]).

1 Zeta functions of root systems

In this section, we give an overview of [12] and [4]. First, following [4], we introduce the definition of zeta-functions of root systems associated with semisimple Lie algebras, and more generally, zeta-functions of roots.

We quote some notation and results about semisimple Lie algebras from [2, 17]. Let $\mathfrak{g}$ be a complex semisimple Lie algebra of rank r. We denote by $\Delta = \Delta(\mathfrak{g})$ the set of all roots of $\mathfrak{g}$, by $\Delta_+ = \Delta_+(\mathfrak{g})$ (resp. $\Delta_- = \Delta_-(\mathfrak{g})$) the set of all positive roots (resp. negative roots) of $\mathfrak{g}$, and by $\Psi = \Psi(\mathfrak{g}) = \{\alpha_1, \ldots, \alpha_r\}$ the fundamental system of Δ. For any $\alpha \in \Delta$, we denote by $\alpha^\vee$ the associated coroot. Let $\lambda_1, \ldots, \lambda_r$ be the fundamental weights satisfying $\langle \alpha_i^\vee, \lambda_j \rangle = \lambda_j(\alpha_i^\vee) = \delta_{ij}$ (Kronecker's delta). Then we see that any dominant weight can be written as

$$\lambda = n_1\lambda_1 + \cdots + n_r\lambda_r \qquad (n_1, \ldots, n_r \in \mathbb{N}_0). \tag{1.1}$$

In particular, the lowest strongly dominant form is $\rho = \lambda_1 + \cdots + \lambda_r$.

Let d_λ be the dimension of the representation space corresponding to the dominant weight λ. Then Weyl's dimension formula (see [17, Section 3.8]) gives that

$$d_\lambda = \prod_{\alpha \in \Delta_+} \frac{\langle \alpha^\vee, \lambda + \rho \rangle}{\langle \alpha^\vee, \rho \rangle} = \prod_{\alpha \in \Delta_+} \frac{\langle \alpha^\vee, (n_1 + 1)\lambda_1 + \cdots + (n_r + 1)\lambda_r \rangle}{\langle \alpha^\vee, \lambda_1 + \cdots + \lambda_r \rangle}.$$

Putting $m_j = n_j + 1$, it follows from (0.1) that

$$\begin{aligned} \zeta_W(s; \mathfrak{g}) &= \sum_\lambda \prod_{\alpha \in \Delta_+} \left(\frac{\langle \alpha^\vee, m_1\lambda_1 + \cdots + m_r\lambda_r \rangle}{\langle \alpha^\vee, \lambda_1 + \cdots + \lambda_r \rangle} \right)^{-s} \\ &= K(\mathfrak{g})^s \sum_{m_1=1}^\infty \cdots \sum_{m_r=1}^\infty \prod_{\alpha \in \Delta_+} \langle \alpha^\vee, m_1\lambda_1 + \cdots + m_r\lambda_r \rangle^{-s}, \end{aligned} \tag{1.2}$$

where the sum on the second member of the above runs over all dominant weights of the form (1.1), and

$$K(\mathfrak{g}) = \prod_{\alpha \in \Delta_+} \langle \alpha^\vee, \lambda_1 + \cdots + \lambda_r \rangle. \tag{1.3}$$

Let Δ^* be a subset of $\Delta_+ = \Delta_+(\mathfrak{g})$ which satisfies the condition that for any λ_j $(1 \le j \le r)$, there exists an element $\alpha \in \Delta^*$ such that $\langle \alpha^\vee, \lambda_j \rangle \neq 0$. Then we define the zeta-function of Δ^* by

$$\zeta_r(\mathbf{s}; \Delta^*) = \sum_{m_1=1}^{\infty} \cdots \sum_{m_r=1}^{\infty} \prod_{\alpha \in \Delta^*} \langle \alpha^\vee, m_1\lambda_1 + \cdots + m_r\lambda_r \rangle^{-s_\alpha}, \tag{1.4}$$

where $\mathbf{s} = \mathbf{s}(\Delta^*) = (s_\alpha)_{\alpha \in \Delta^*} \in \mathbb{C}^{n^*}$ with $n^* = |\Delta^*|$. In particular when $\Delta^* = \Delta_+(\mathfrak{g})$, we denote $\zeta_r(\mathbf{s}; \Delta^+(\mathfrak{g}))$ by $\zeta_r(\mathbf{s}; \mathfrak{g})$ or $\zeta_r(\mathbf{s}; X_r)$ if $\mathfrak{g}$ is of type X_r, and call it the multi-variable Witten zeta-function associated with $\mathfrak{g}$ or X_r. Namely we have

$$\zeta_r(\mathbf{s}; \mathfrak{g}) = \zeta_r(\mathbf{s}; X_r) = \sum_{m_1=1}^{\infty} \cdots \sum_{m_r=1}^{\infty} \prod_{\alpha \in \Delta_+} \langle \alpha^\vee, m_1\lambda_1 + \cdots + m_r\lambda_r \rangle^{-s_\alpha}, \tag{1.5}$$

where $\mathbf{s} = (s_\alpha)_{\alpha \in \Delta_+} \in \mathbb{C}^n$ and $n = |\Delta_+|$ is the number of positive roots of $\mathfrak{g}$. Note that $K(\mathfrak{g})^s \zeta_r(s, \ldots, s; \mathfrak{g})$ coincides with the one-variable Witten zeta-function $\zeta_W(s; \mathfrak{g})$ defined by (0.1). The series (1.4) and (1.5) can be continued meromorphically to the whole space (see [10, Theorem 3]). In the case $X_r = A_r$, we see that

$$\zeta_r(\mathbf{s}; A_r) = \sum_{m_1=1}^{\infty} \cdots \sum_{m_r=1}^{\infty} \prod_{1 \le i < j \le r+1} (m_i + \cdots + m_{j-1})^{-s_{ij}}, \tag{1.6}$$

where $\mathbf{s} = \mathbf{s}(A_r) = (s_{ij}) \in \mathbb{C}^{r(r+1)/2}$ (see [4, (2.3)] or [12, (1.5)]). In Section 3, we prove some functional relations among $\zeta_r(\mathbf{s}; A_r)$ $(r \le 3)$ which can be regarded as continuous relations including Witten's volume formulas for $\mathfrak{sl}(4)$. Moreover, in [5], we give some functional relations of types B_r and C_r for $r \le 3$.

Now we consider $\mathfrak{g} = \mathfrak{sl}(r+1)$. Let ε_j be the j-th coordinate function. It is well-known that positive roots are

$$\Delta_+ = \Delta_+(A_r) = \left\{ \varepsilon_i - \varepsilon_j = \sum_{i \le k < j} \alpha_k \;\middle|\; 1 \le i < j \le r+1 \right\} \tag{1.7}$$

(see the list at the end of [2]). We further define

$$\Delta^*(A_r) = \{\varepsilon_i - \varepsilon_j \mid 2 \le i < j \le r+1\}; \tag{1.8}$$

$$\Delta_h^*(A_r) = \{\varepsilon_1 - \varepsilon_j \mid 2 \le j \le h\} \cup \Delta^*(A_r) \tag{1.9}$$

$(2 \le h \le r+1)$ which are subsets of $\Delta_+(A_r)$. For example, putting $r = 3$, we have the relation $\Delta^*(A_3) \subset \Delta_2^*(A_3) \subset \Delta_3^*(A_3) \subset \Delta_4^*(A_3) = \Delta_+(A_3)$. We can see that $\mathbf{s}(\Delta_h^*(A_3)) = (s_{12}, \ldots, s_{1h}, s_{23}, s_{24}, s_{34})$ $(h = 2, 3, 4)$.

We recall the Mellin-Barnes integral formula

$$(1+\lambda)^{-s} = \frac{1}{2\pi\sqrt{-1}} \int_{(c)} \frac{\Gamma(s+z)\Gamma(-z)}{\Gamma(s)} \lambda^z dz, \tag{1.10}$$

where s, λ are complex numbers with $\Re s > 0$, $|\arg \lambda| < \pi$, $\lambda \neq 0$, c is real with $-\Re s < c < 0$, and the path (c) of integration is the vertical line $\Re z = c$. Using this formula, we have

$$\begin{aligned} &\zeta_3(s_{12}, \ldots, s_{1h}, s_{23}, s_{24}, s_{34}; \Delta_h^*(A_3)) \\ &= \sum_{m_1=1}^{\infty} \sum_{m_2=1}^{\infty} \sum_{m_3=1}^{\infty} \prod_{2 \le i < j \le 4} (m_i + \cdots + m_{j-1})^{-s_{ij}} \\ &\quad \times \prod_{2 \le j < h} (m_1 + \cdots + m_{j-1})^{-s_{1j}} \frac{1}{2\pi\sqrt{-1}} \int_{(c_h)} \frac{\Gamma(s_{1h}+z_h)\Gamma(-z_h)}{\Gamma(s_{1h})} \\ &\quad \times (m_1 + \cdots + m_{h-2})^{-s_{1h}-z_h} m_{h-1}^{z_h} dz_h \\ &= \frac{1}{2\pi\sqrt{-1}} \int_{(c_h)} \frac{\Gamma(s_{1h}+z_h)\Gamma(-z_h)}{\Gamma(s_{1h})} \zeta_3(\mathbf{s}^*(A_3, z_h); \Delta_{h-1}^*(A_3)) dz_h \end{aligned} \tag{1.11}$$

for $h = 3, 4$, where

$$\mathbf{s}^*(A_3, z_3) = (s_{12} + s_{13} + z_3, s_{23} - z_3, s_{24}, s_{34});$$
$$\mathbf{s}^*(A_3, z_4) = (s_{12}, s_{13} + s_{14} + z_4, s_{23}, s_{24}, s_{34} - z_4).$$

Hence we find the recursive structure which we denote by

$$\zeta_3(\cdot\,; A_3) = \zeta_3(\cdot\,; \Delta_4^*(A_3)) \to \zeta_3(\cdot\,; \Delta_3^*(A_3)) \to \zeta_3(\cdot\,; \Delta_2^*(A_3)). \tag{1.12}$$

Note that

$$\zeta_3(\mathbf{s}^*(A_3, z_3); \Delta_2^*(A_3)) = \zeta_2(s_{23} - z_3, s_{24}, s_{34}; \Delta^*(A_3))\zeta(s_{12} + s_{13} + z_3). \tag{1.13}$$

Substituting (1.13) into (1.11), we have

(1.14)
$$\zeta_3(\mathbf{s};\Delta_3^*(A_3)) = \frac{1}{2\pi\sqrt{-1}}\int_{(c_3)} \frac{\Gamma(s_{13}+z_3)\Gamma(-z_3)}{\Gamma(s_{13})} \times \zeta_2(s_{23}-z_3, s_{24}, s_{34};\Delta^*(A_3))\zeta(s_{12}+s_{13}+z_3)dz_3.$$

Therefore we can rewrite (1.12) as

(1.15) $$\zeta_3(\cdot\,;A_3) = \zeta_3(\cdot\,;\Delta_4^*(A_3)) \to \zeta_3(\cdot\,;\Delta_3^*(A_3)) \to \zeta_2(\cdot\,;\Delta^*(A_3)),$$

by neglecting the Riemann zeta factor. Renaming ε_i as ε_{i-1} $(2 \le i \le r+1)$, we find that $\zeta_2(\cdot\,;\Delta^*(A_3))$ is equal to $\zeta_2(\cdot\,;A_2)$. Consequently we can write (1.15) as

(1.16) $$\zeta_3(\cdot\,;A_3) \to \zeta_3(\cdot\,;\Delta_3^*(A_3)) \to \zeta_2(\cdot\,;A_2).$$

The above argument can be applied to A_r and A_{r-1} for any $r \ge 2$. Hence we can obtain the following result which can be viewed as a refinement of [12, Theorem 2.2].

Theorem 1.1 ([4, Theorem 1]). *Between $\zeta_r(\cdot\,;A_r)$ and $\zeta_{r-1}(\cdot\,;A_{r-1})$ (for each $r \ge 2$) there is the recursive relation like (1.16), which can be summarizingly written as*

(1.17) $$\zeta_r(\cdot\,;A_r) \to \zeta_{r-1}(\cdot\,;A_{r-1}) \to \cdots \to \zeta_2(\cdot\,;A_2) \to \zeta_1(\cdot\,;A_1) = \zeta.$$

For zeta-functions of other types, we can also give certain recursive relations similar to (1.17). For the details, see [4, Sections 4 and 5].

2 Structural background of functional relations

Hereafter we discuss the topic on functional relations for zeta-functions of root systems. First, in this section, we study the structural background in a general framework. For the details, see [5].

As we mentioned in Section 0, we have already obtained some functional relations for $\zeta_2(s_1, s_2, s_3; A_2)$ like (0.5) (For the details, see [23], also

[13, 14]). Recently, in [16], Nakamura gave an alternative proof of functional relations given in [23]. Inspired by Nakamura's method, we aim to generalize these functional relations of type A_2 to those for zeta-functions of general root systems.

We use the same notation as in Section 1. Let $\mathfrak{g}$ be a complex simple Lie algebra of type X_r and V be the vector space generated by the roots of $\mathfrak{g}$. Let W be the Weyl group acting on V. We denote by P and P_+ the lattice of weights and the set of dominant weights, respectively. Let

$$C := \{v \in V \mid \langle v, \alpha_i^\vee \rangle \geq 0 \text{ for all } \alpha \in \Delta\} \tag{2.1}$$

be a fundamental domain of W, which we call the fundamental Weyl chamber. Let

$$H_\Delta = V \setminus \bigcup_{w \in W} w \mathring{C},$$

where by $\mathring{C}$ we denote the interior of C, and

$$\begin{aligned} S\left((s_\alpha)_{\alpha \in \Delta_+}; X_r\right) &:= \sum_{\lambda \in P \setminus H_\Delta} \prod_{\alpha \in \Delta_+} \langle \alpha^\vee, \lambda \rangle^{-s_\alpha} \\ &= \sum_{w \in W} \sum_{\lambda \in P_+} \prod_{\alpha \in \Delta_+} \langle \alpha^\vee, w(\lambda + \rho) \rangle^{-s_\alpha} \\ &= \sum_{w \in W} \sum_{\lambda \in P_+} \prod_{w\alpha \in \Delta_+} \langle \alpha^\vee, \lambda + \rho \rangle^{-s_{w\alpha}}, \end{aligned} \tag{2.2}$$

where s_α for $\alpha \in \Delta_-$ is defined by $s_\alpha = s_{-\alpha}$ and ρ is the lowest strongly dominant form. Let $\Delta_w = \Delta_+ \cap w^{-1} \Delta_-$. Divide the product on the right-hand side of (2.2) into two parts according to $\alpha \in \Delta_+$ or $\alpha \in \Delta_-$, and in the latter part replace α by $-\alpha$. Then we have

$$\begin{aligned} S\left((s_\alpha); X_r\right) &= \sum_{w \in W} \Big(\prod_{\alpha \in \Delta_w} (-1)^{-s_{w\alpha}} \Big) \sum_{\lambda \in P_+} \prod_{\alpha \in \Delta_+} \langle \alpha^\vee, \lambda + \rho \rangle^{-s_{w\alpha}} \\ &= \sum_{w \in W} \Big(\prod_{\alpha \in \Delta_w} (-1)^{-s_{w\alpha}} \Big) \zeta_r\left((s_{w\alpha}); X_r\right), \end{aligned} \tag{2.3}$$

which is a 'Weyl group symmetric' linear combination of our zeta-function.

Now we consider the Lerch-type series

$$L_s(x) = -\frac{\Gamma(s+1)}{(2\pi i)^s} \sum_{\substack{m\in\mathbb{Z} \\ m\neq 0}} \frac{e^{2\pi imx}}{m^s} \tag{2.4}$$

for $s \in \mathbb{C}$ with $\Re s > 1$, $x \in \mathbb{R}$ and $i = \sqrt{-1}$. Then we obtain the following general form of functional relations for zeta-functions of root systems (for the proof, see [5]).

Theorem 2.1.

(2.5)

$$\begin{aligned} S\left((s_\alpha); X_r\right) &= \sum_{w\in W}\Big(\prod_{\alpha\in\Delta_w}(-1)^{-s_{w\alpha}}\Big)\zeta_r\left((s_{w\alpha}); X_r\right) \\ &= (-1)^{|\Delta_+|}\sum_{\lambda\in P}\prod_{\alpha\in\Delta_+}\frac{(2\pi i)^{s_\alpha}}{\Gamma(s_\alpha+1)}\int_0^1 L_{s_\alpha}(x_\alpha)e^{-2\pi i\langle\alpha^\vee,\lambda\rangle x_\alpha}dx_\alpha \\ &= (-1)^{|\Delta_+|}\Big(\prod_{\alpha\in\Delta_+}\frac{(2\pi i)^{s_\alpha}}{\Gamma(s_\alpha+1)}\Big)\underbrace{\int_0^1\cdots\int_0^1}_{|\Delta_+\setminus\Psi|\ \text{times}}\Big(\prod_{\alpha\in\Delta_+\setminus\Psi}L_{s_\alpha}(x_\alpha)\Big) \\ &\quad\times\Big(\prod_{i=1}^{r}L_{s_{\alpha_i}}\Big(-\sum_{\alpha\in\Delta_+\setminus\Psi}x_\alpha\langle\alpha^\vee,\lambda_i\rangle\Big)\Big)\prod_{\alpha\in\Delta_+\setminus\Psi}dx_\alpha \end{aligned}$$

holds for any s_α in a certain region.

Remark 2.1. When $s_\alpha \in \mathbb{N}$, we see that $L_{s_\alpha}(x_\alpha) = B_{s_\alpha}(\{x_\alpha\})$, where $\{x\} = x-[x]$ denotes the fractional part of x and $B_n(\cdot)$ is the Bernoulli polynomial (see [1]). When $X_r = A_2$, the idea of Theorem 2.1 is essentially included in Nakamura's proof of certain functional relations for $\zeta_2(s_1, s_2, s_3; A_2)$ ([16, Theorem 1.2]) which coincide with those given by the third-named author in [23]. For example, see (0.5).

Example 2.1. In the case of type $B_2(= C_2)$, $\zeta_2(s_1, s_2, s_3, s_4; B_2)$ coincides with $\zeta_{\mathfrak{so}(5)}(s_2, s_3, s_4, s_1)$ (see (0.4) and (1.5)). As well as in [16], direct calculations on the right-hand side of (2.5) for $\mathfrak{g} = \mathfrak{so}(5)$ show that, for example,

(2.6)

$$\zeta_2(1,2,s,2;B_2)+\zeta_2(1,2,2,s;B_2)+\zeta_2(2,1,2,s;B_2)-\zeta_2(2,1,s,2;B_2) = 3\zeta(2)\zeta(s+3)-\frac{39}{8}\zeta(s+5);$$

(2.7)

$$\zeta_2(2,2,s,2;B_2)+\zeta_2(2,2,2,s;B_2)=\frac{3}{2}\zeta(2)\zeta(s+4)-\frac{39}{16}\zeta(s+6).$$

Putting $s=2$ in (2.6) and (2.7), we have

$$\zeta_2(1,2,2,2;B_2)=\frac{3}{2}\zeta(2)\zeta(5)-\frac{39}{16}\zeta(7); \tag{2.8}$$

$$\zeta_2(2,2,2,2;B_2)=\frac{3}{4}\zeta(2)\zeta(6)-\frac{39}{32}\zeta(8)=\frac{\pi^8}{302400}. \tag{2.9}$$

Note that (2.8) has been obtained in [22] by a different method, and (2.9) is just equal to Witten's volume formula (0.2) for $\mathfrak{g}=\mathfrak{so}(5)$. However it seems to be hard to calculate (2.5) for $\zeta_r(\cdot;X_r)$ directly if r is large. Hence we need another technique. In [5], we introduced a certain technique using multiple polylogarithms (which may be called the 'polylogathm technique') to prove general formulas for zeta-functions of types B_2, B_3 and C_3 including (2.8) and (2.9). In Section 3, using this technique, we treat $\zeta_r(\cdot;A_r)$ and prove certain functional relations for them.

At the end of this section, we consider a generating function of values of $S((s_\alpha);X_r)$ at positive integers. Let $\mathbf{k}=(k_\alpha)_{\alpha\in\Delta_+}\in\mathbb{N}_0^{|\Delta_+|}$ and $\mathbf{t}=(t_\alpha)_{\alpha\in\Delta_+}\in\mathbb{C}^{|\Delta_+|}$ with $|t_\alpha|<2\pi$ for each α. Then we define

$$F(\mathbf{t};X_r):=(-1)^{|\Delta_+|}\sum_{\mathbf{k}}S(\mathbf{k};X_r)\prod_{\alpha\in\Delta_+}\left(\frac{t_\alpha}{2\pi i}\right)^{k_\alpha}, \tag{2.10}$$

where

$$S(\mathbf{k};X_r)=(-1)^{|\Delta_+|}\Bigg(\prod_{\alpha\in\Delta_+}\frac{(2\pi i)^{k_\alpha}}{k_\alpha!}\Bigg)\underbrace{\int_0^1\cdots\int_0^1}_{|\Delta_+\setminus\Psi|\text{ times}}\Bigg(\prod_{\alpha\in\Delta_+\setminus\Psi}B_{k_\alpha}(x_\alpha)\Bigg)\times \Bigg(\prod_{i=1}^r B_{k_{\alpha_i}}\Big(\Big\{-\sum_{\alpha\in\Delta_+\setminus\Psi}x_\alpha\langle\alpha^\vee,\lambda_i\rangle\Big\}\Big)\Bigg)\prod_{\alpha\in\Delta_+\setminus\Psi}dx_\alpha$$

(see (2.5) and Remark 2.1). For example, we can explicitly calculate that

$$F(t_1,t_2,t_3,t_4;B_2)=\left(\prod_{j=1}^{4}\frac{t_j}{e^{t_j}-1}\right)(P(\mathbf{t})+Q(\mathbf{t})-R(\mathbf{t})), \tag{2.11}$$

where $t_1=t_{\alpha_1}, t_2=t_{\alpha_2}, t_3=t_{\alpha_1+\alpha_2}, t_4=t_{\alpha_1+2\alpha_2}$ and

$$P(\mathbf{t})=\frac{2(e^{t_1}-1)(1+e^{(t_2+t_3)/2})(e^{(t_2+t_3)/2}-e^{t_4})}{(2t_1+t_2-t_3)(t_2+t_3-2t_4)};$$

$$Q(\mathbf{t})=\frac{(e^{t_2}-1)(e^{t_1+t_4}-e^{t_3})}{(2t_1+t_2-t_3)(t_1-t_3+t_4)};\ R(\mathbf{t})=\frac{(e^{t_3}-1)(e^{t_1+t_2}-e^{t_4})}{(2t_1+t_2-t_3)(t_1+t_2-t_4)}.$$

Hence we can calculate the Taylor expansion as

$$\begin{aligned}F(t_1,t_2,t_3,t_4;B_2)&\\ =1+\frac{1}{2880}&(4t_1t_2t_3^2+2t_1t_2t_4^2-4t_1t_2^2t_3+t_1t_2^2t_4-2t_1t_3t_4^2-t_1t_3^2t_4\\ &+4t_1^2t_2t_3-2t_1^2t_2t_4-2t_1^2t_3t_4-4t_2t_3t_4^2-4t_2t_3^2t_4-4t_2^2t_3t_4)\\ &+\frac{1}{241920}(3t_1t_2t_3^2t_4^2-3t_1t_2^2t_3t_4^2-3t_1^2t_2t_3^2t_4-3t_1^2t_2^2t_3t_4\\ &+8t_1^2t_2^2t_3^2+2t_1^2t_2^2t_4^2+2t_1^2t_3^2t_4^2+8t_2^2t_3^2t_4^2)\\ &+\frac{1}{9676800}t_1^2t_2^2t_3^2t_4^2+\cdots.\end{aligned} \tag{2.12}$$

Comparing the coefficients of $t_1^2t_2^2t_3^2t_4^2$ in (2.10) and in (2.12), we can also obtain (2.9), namely

$$\zeta(2,2,2,2;B_2)=(-1)^4\frac{(2\pi i)^8}{2!\cdot 2^2}\frac{1}{9676800}=\frac{\pi^8}{302400}.$$

However it is also hard to evaluate the zeta-function of type X_r by this method when r is large.

3 Functional relations for $\zeta_3(\mathbf{s};A_3)$

In this section, we consider zeta-functions of type A_3. Though we have already considered certain special cases of this type in our previous paper

([12]), we will study this type more generally and in detail, by considering a certain multiple polylogarithm of this type.

From (1.5), we have

$$\zeta_2(s_{12}, s_{13}, s_{23}; A_2) = \sum_{m_1,m_2=1}^{\infty} \frac{1}{m_1^{s_{12}}(m_1+m_2)^{s_{13}} m_2^{s_{23}}};$$

$$\zeta_3(s_{12}, s_{13}, s_{14}, s_{23}, s_{24}, s_{34}; A_3)$$
$$= \sum_{m_1,m_2,m_3=1}^{\infty} \frac{1}{m_1^{s_{12}}(m_1+m_2)^{s_{13}}(m_1+m_2+m_3)^{s_{14}} m_2^{s_{23}}(m_2+m_3)^{s_{24}} m_3^{s_{34}}}.$$

In our previous paper [12], we proved a closed form of functional relation for

$$(-1)^a\zeta_3(s_1, s_3, b, s_2, 0, a; A_3) + (-1)^b\zeta_3(s_1, s_3, a, s_2, 0, b; A_3)$$
$$+ \zeta_3(a, s_1, s_3, 0, b, s_2; A_3) + \zeta_3(s_1, a, s_3, 0, s_2, b; A_3)$$

in terms of $\zeta_2(s_1, s_2, s_3; A_3)$ and $\zeta_1(s; A_2) = \zeta(s)$, where $a, b \in \mathbb{N}_0$ and $s_1, s_2, s_3 \in \mathbb{C}$. For example, we have

$$-\zeta_2(s_1, s_3, 2, s_2, 0, 1; A_3) + \zeta_2(s_1, s_3, 1, s_2, 0, 2; A_3)$$
$$+ \zeta_2(1, s_1, s_3, 0, 2, s_2; A_3) + \zeta_2(s_1, 1, s_3, 0, s_2, 2; A_3)$$
$$= -3\zeta_2(s_1, s_3+3, s_2; A_2) - \zeta_2(s_1+1, s_3, s_2+2; A_2)$$
$$+ 2\zeta(2)\zeta_2(s_1, s_3+1, s_2; A_2)$$

(see [12, Theorems 5.9 and 5.10]). However our previous result does not include Witten's volume formula (for example, (0.6)).

Therefore we aim to construct some functional relations for $\zeta_r(\mathbf{s}; A_r)$ ($r \le 3$) which include Witten's volume formulas as value-relations.

Let

$$\phi(s) = \sum_{m=1}^{\infty} \frac{(-1)^m}{m^s} = \left(2^{1-s} - 1\right)\zeta(s). \tag{3.1}$$

We recall that

$$\sum_{l=1}^{\infty} \frac{(-1)^l \cos(l\theta)}{l^{2p}} = \sum_{\nu=0}^{p} \phi(2p-2\nu)\,\frac{(-1)^\nu \theta^{2\nu}}{(2\nu)!}; \tag{3.2}$$

$$\sum_{l=1}^{\infty} \frac{(-1)^l \sin(l\theta)}{l^{2q+1}} = \sum_{\nu=0}^{q} \phi(2q-2\nu)\,\frac{(-1)^\nu \theta^{2\nu+1}}{(2\nu+1)!} \tag{3.3}$$

for $p \in \mathbb{N}$, $q \in \mathbb{N}_0$ and $\theta \in (-\pi, \pi) \subset \mathbb{R}$ (see, for example, [20, Lemma 2]). Note that $\phi(0) = \zeta(0) = -\frac{1}{2}$. We see that the both sides of (3.2) and of (3.3) are continuous for $\theta \in [-\pi, \pi]$ in the case $p, q \in \mathbb{N}$.

Fix $p \in \mathbb{N}$ and $s, t, u, x, y \in \mathbb{C}$ with $\Re s > 1$, $\Re t > 1$, $\Re u > 1$, $|x| \le 1$ and $|y| \le 1$. Put

(3.4)

$$F_0(\theta; 2p, s, t, u; x, y) = 2\left(\sum_{l=1}^{\infty} \frac{(-1)^l \cos(l\theta)}{l^{2p}} - \sum_{\nu=0}^{p} \phi(2p-2\nu) \frac{(-1)^\nu \theta^{2\nu}}{(2\nu)!}\right) \times \sum_{m,n=1}^{\infty} \frac{(-1)^{m+n} x^m y^n e^{i(m+n)\theta}}{m^s n^t (m+n)^u}.$$

By (3.2), we see that

$$F_0(\theta; 2p, s, t, u; x, y) = 0 \quad (-\pi \le \theta \le \pi). \tag{3.5}$$

By the assumptions on p, s, t, u, the right-hand side of (3.4) is absolutely convergent. Hence, by $\cos\theta = (e^{i\theta} + e^{-i\theta})/2$, we have

(3.6)

$$F_0(\theta; 2p, s, t, u; x, y) = \sum_{l,m,n=1}^{\infty} \frac{(-1)^{l+m+n} x^m y^n \left(e^{i(l+m+n)\theta} + e^{i(-l+m+n)\theta}\right)}{l^{2p} m^s n^t (m+n)^u} - 2\sum_{\nu=0}^{p} \phi(2p-2\nu) \frac{(-1)^\nu \theta^{2\nu}}{(2\nu)!} \sum_{m,n=1}^{\infty} \frac{(-1)^{m+n} x^m y^n e^{i(m+n)\theta}}{m^s n^t (m+n)^u}.$$

Note that in our previous paper ([12]), we only considered the case $(x, y) = (1, 1)$. Hence we were only able to treat some limited case of $\zeta_3(\cdot; A_3)$. In this paper, we consider $F_0(\theta; 2p, s, t, u; x, y)$ which is a certain triple polylogarithm. Considering the integration of F_0 with respect to the variable θ in the numerator of each summand, we can produce a certain factor in the denominator of each summand, which corresponds to some relevant root. This is the 'polylogarithm technique' mentioned in the preceding section. We will execute this procedure as follows.

Corresponding to $F_0(\theta; 2p, s, t, u; x, y)$, we define

(3.7)
$$
\begin{aligned}
G_d(\theta; 2p, s, t, u; x, y) = \frac{1}{i^d}\Bigg[& \sum_{l,m,n=1}^{\infty} \frac{(-1)^{l+m+n} x^m y^n e^{i(l+m+n)\theta}}{l^{2p} m^s n^t (m+n)^u (l+m+n)^d} \\
& + \sum_{\substack{l,m,n=1 \\ l \neq m+n}}^{\infty} \frac{(-1)^{l+m+n} x^m y^n e^{i(-l+m+n)\theta}}{l^{2p} m^s n^t (m+n)^u (-l+m+n)^d} \\
& - 2 \sum_{\nu=0}^{p} \phi(2p-2\nu) \sum_{\mu=0}^{2\nu} \binom{d-1+2\nu-\mu}{2\nu-\mu} \frac{(-i\theta)^\mu}{\mu!} \\
& \times \sum_{m,n=1}^{\infty} \frac{(-1)^{m+n} x^m y^n e^{i(m+n)\theta}}{m^s n^t (m+n)^{u+d+2\nu-\mu}} \Bigg]
\end{aligned}
$$

for $\theta \in [-\pi, \pi]$ and $d \in \mathbb{N}_0$, where

$$
\binom{X}{a} = \begin{cases} \frac{X(X-1)\cdots(X-a+1)}{a!} & (a \in \mathbb{N}); \\ 1 & (a = 0). \end{cases}
$$

In the case $d = 0$, it follows from (3.5) that

$$
G_0(\theta; 2p, s, t, u; x, y) = F_0(\theta; 2p, s, t, u; x, y) + C_0 = C_0, \tag{3.8}
$$

where

$$
C_0 = C_0(2p, s, t, u; x, y) = -\sum_{m,n=1}^{\infty} \frac{x^m y^n}{m^s n^t (m+n)^{u+2p}}. \tag{3.9}
$$

Note that the definition of $G_d(\theta; 2p, s, t, u; x, y)$ seems to have been given abruptly. However, by using the same consideration as in our previous papers [12, 23], we can presume the expression of $G_d(\theta; 2p, s, t, u; x, y)$ as (3.7).

Since we fix p, s, t, u, x, y, we regard $G_d(\theta; 2p, s, t, u; x, y)$ as a continuous function for $\theta \in [-\pi, \pi]$. We can show

$$
\frac{d}{d\theta} G_d(\theta; 2p, s, t, u; x, y) = G_{d-1}(\theta; 2p, s, t, u; x, y) \quad (d \in \mathbb{N}). \tag{3.10}
$$

Hence, by integrating the both sides of (3.8) and multiplying by i on the both sides, we have

$$iG_1(\theta; 2p, s, t, u; x, y) = C_0(i\theta) + C_1$$

for some $C_1 = C_1(2p, s, t, u; x, y)$. By repeating the integration, and by (3.10), we obtain

Lemma 3.1. *With the above notation, for $d \in \mathbb{N}_0$, we have*

$$i^d G_d(\theta; 2p, s, t, u; x, y) = \sum_{j=0}^{d} C_{d-j} \frac{(i\theta)^j}{j!}, \tag{3.11}$$

where $C_k = C_k(2p, s, t, u; x, y)$ $(0 \le k \le d)$ are certain constants.

We can determine $\{C_k\}$ as follows. Putting $\theta = \pm\pi$ in (3.11) with $d+1$, we have

$$\begin{aligned} &\frac{i^{d+1}}{2(i\pi)} \{G_{d+1}(\pi; 2p, s, t, u; x, y) - G_{d+1}(-\pi; 2p, s, t, u; x, y)\} \\ &\quad = \sum_{\mu=0}^{[d/2]} C_{d-2\mu} \frac{(i\pi)^{2\mu}}{(2\mu+1)!}. \end{aligned} \tag{3.12}$$

From (3.7), we see that the left-hand side of (3.12) is equal to

$$\begin{aligned} &2\sum_{\nu=0}^{p} \phi(2p-2\nu) \sum_{\sigma=0}^{\nu-1} \binom{d-1+2\nu-2\sigma}{2\nu-2\sigma-1} \frac{(i\pi)^{2\sigma}}{(2\sigma+1)!} \\ &\qquad \times \sum_{m,n=1}^{\infty} \frac{x^m y^n}{m^s n^t (m+n)^{u+d+2\nu-2\sigma}}. \end{aligned} \tag{3.13}$$

Applying [11, Lemma 2.1] to (3.13) with $a = 2p$ and

$$g(X) = 2\binom{d-1+X}{X-1} \sum_{m,n=1}^{\infty} \frac{x^m y^n}{m^s n^t (m+n)^{u+d+X}},$$

we have

$$-\binom{2p+d-1}{2p-1} \sum_{m,n=1}^{\infty} \frac{x^m y^n}{m^s n^t (m+n)^{u+2p+d}} = \sum_{\mu=0}^{[d/2]} C_{d-2\mu} \frac{(i\pi)^{2\mu}}{(2\mu+1)!} \tag{3.14}$$

for $d \in \mathbb{N}_0$. Applying [21, Lemma 3.3] to (3.14), we have

$$C_k = 2\sum_{\nu=0}^{k} \phi(k-\nu)\lambda_{k-\nu}\binom{2p+\nu-1}{2p-1}\sum_{m,n=1}^{\infty}\frac{x^m y^n}{m^s n^t (m+n)^{u+2p+\nu}} \tag{3.15}$$

for $k \in \mathbb{N}_0$, where $\lambda_j = \{1+(-1)^j\}/2$ ($j \in \mathbb{Z}$). Substitute (3.15) into (3.11). Then we have

(3.16)

$$\begin{aligned}
&i^d G_d(\theta; 2p, s, t, u; x, y)\\
&= 2\sum_{j=0}^{d}\sum_{\nu=0}^{d-j}\phi(d-j-\nu)\lambda_{d-j-\nu}\binom{2p+\nu-1}{2p-1}\sum_{m,n=1}^{\infty}\frac{x^m y^n}{m^s n^t (m+n)^{u+2p+\nu}}\frac{(i\theta)^j}{j!}\\
&= 2\sum_{\eta=0}^{d}\phi(d-\eta)\lambda_{d-\eta}\sum_{j=0}^{\eta}\binom{2p+\eta-j-1}{2p-1}\sum_{m,n=1}^{\infty}\frac{x^m y^n}{m^s n^t (m+n)^{u+2p+\eta-j}}\frac{(i\theta)^j}{j!},
\end{aligned}$$

by putting $\eta = \nu + j$. Combining (3.7) and (3.16) with $d = 2q$ for $q \in \mathbb{N}$, we obtain

$$\begin{aligned}
&\sum_{l,m,n=1}^{\infty}\frac{(-1)^{l+m+n}x^m y^n e^{i(l+m+n)\theta}}{l^{2p}m^s n^t (m+n)^u (l+m+n)^{2q}}\\
&\quad+\sum_{\substack{l,m,n=1\\ l\neq m+n}}^{\infty}\frac{(-1)^{l+m+n}x^m y^n e^{i(-l+m+n)\theta}}{l^{2p}m^s n^t (m+n)^u (-l+m+n)^{2q}} - 2\sum_{\nu=0}^{p}\phi(2p-2\nu)\\
&\quad\times\sum_{\mu=0}^{2\nu}\binom{2q-1+2\nu-\mu}{2\nu-\mu}\frac{(-i\theta)^\mu}{\mu!}\sum_{m,n=1}^{\infty}\frac{(-1)^{m+n}x^m y^n e^{i(m+n)\theta}}{m^s n^t (m+n)^{u+2q+2\nu-\mu}}\\
&\quad-2\sum_{\nu=0}^{q}\phi(2q-2\nu)\sum_{\mu=0}^{2\nu}\binom{2p-1+2\nu-\mu}{2\nu-\mu}\\
&\quad\times\frac{(i\theta)^\mu}{\mu!}\sum_{m,n=1}^{\infty}\frac{x^m y^n}{m^s n^t (m+n)^{u+2p+2\nu-\mu}} = 0\,(\theta \in [-\pi,\pi]).
\end{aligned} \tag{3.17}$$

Now we replace y by $-ye^{-i\theta}$ in (3.17). This is an important procedure to produce a factor in the denominator of each summand, which corresponds to the root. In fact, we consider

$$
\begin{aligned}
(3.18)\qquad & \widetilde{F}_0(\theta; 2p, s, t, u, 2q; x, y) \\
:= & \sum_{l,m,n=1}^{\infty} \frac{(-1)^{l+m} x^m y^n e^{i(l+m)\theta}}{l^{2p} m^s n^t (m+n)^u (l+m+n)^{2q}} \\
& + \sum_{\substack{l,m,n=1 \\ l\neq m+n}}^{\infty} \frac{(-1)^{l+m} x^m y^n e^{i(-l+m)\theta}}{l^{2p} m^s n^t (m+n)^u (-l+m+n)^{2q}} - 2\sum_{\nu=0}^{p} \phi(2p-2\nu) \\
& \times \sum_{\mu=0}^{2\nu} \binom{2q-1+2\nu-\mu}{2\nu-\mu} \frac{(-i\theta)^\mu}{\mu!} \\
& \times \sum_{m,n=1}^{\infty} \frac{(-1)^m x^m y^n e^{im\theta}}{m^s n^t (m+n)^{u+2q+2\nu-\mu}} - 2\sum_{\nu=0}^{q} \phi(2q-2\nu) \\
& \times \sum_{\mu=0}^{2\nu} \binom{2p-1+2\nu-\mu}{2\nu-\mu} \frac{(i\theta)^\mu}{\mu!} \sum_{m,n=1}^{\infty} \frac{(-1)^n x^m y^n e^{-in\theta}}{m^s n^t (m+n)^{u+2p+2\nu-\mu}} \\
= & \ 0 \quad (\theta \in [-\pi, \pi]).
\end{aligned}
$$

We can apply the same argument about F_0 as mentioned above to $\widetilde{F}_0$. In fact, for p, q, s, t, u, x, y as fixed above and for $d \in \mathbb{N}_0$, we put

$$
\begin{aligned}
(3.19)\quad & \widetilde{G}_d(\theta; 2p, s, t, u, 2q; x, y) \\
:= & \frac{1}{i^d}\Bigg[\sum_{l,m,n=1}^{\infty} \frac{(-1)^{l+m} x^m y^n e^{i(l+m)\theta}}{l^{2p} m^s n^t (l+m)^d (m+n)^u (l+m+n)^{2q}} \\
& \quad + \sum_{\substack{l,m,n=1 \\ l\neq m \\ l\neq m+n}}^{\infty} \frac{(-1)^{l+m} x^m y^n e^{i(-l+m)\theta}}{l^{2p} m^s n^t (-l+m)^d (m+n)^u (-l+m+n)^{2q}} \\
& - 2\sum_{\nu=0}^{p} \phi(2p-2\nu) \sum_{\mu=0}^{2\nu} \binom{2q-1+2\nu-\mu}{2\nu-\mu} \sum_{\rho=0}^{\mu} \binom{d-1+\mu-\rho}{\mu-\rho} \\
& \qquad \times \frac{(-i\theta)^\rho}{\rho!} \sum_{m,n=1}^{\infty} \frac{(-1)^m x^m y^n e^{im\theta}}{m^{s+d+\mu-\rho} n^t (m+n)^{u+2q+2\nu-\mu}}
\end{aligned}
$$

$$- 2(-1)^d \sum_{\nu=0}^{q} \phi(2q-2\nu) \sum_{\mu=0}^{2\nu} \binom{2p-1+2\nu-\mu}{2\nu-\mu} \sum_{\rho=0}^{\mu} \binom{d-1+\mu-\rho}{\mu-\rho}$$
$$\times \frac{(i\theta)^\rho}{\rho!} \sum_{m,n=1}^{\infty} \frac{(-1)^n x^m y^n e^{-in\theta}}{m^s n^{t+d+\mu-\rho}(m+n)^{u+2p+2\nu-\mu}} \Bigg] \quad (\theta \in [-\pi, \pi]).$$

Let

$$\widetilde{C}_0 = \widetilde{C}_0(2p, s, t, u, 2q; x, y) = -\sum_{m,n=1}^{\infty} \frac{x^m y^n}{m^{s+2p} n^{t+2q}(m+n)^u}.$$

By $(d/d\theta)\widetilde{G}_d(\theta; 2p, s, t, u, 2q; x, y) = \widetilde{G}_{d-1}(\theta; 2p, s, t, u, 2q; x, y)$ $(d \in \mathbb{N})$ similarly to Lemma 3.1, we can obtain the following lemma.

Lemma 3.2. *With the above notation, for $d \in \mathbb{N}_0$, we have*

$$i^d \widetilde{G}_d(\theta; 2p, s, t, u, 2q; x, y) = \sum_{j=0}^{d} \widetilde{C}_{d-j} \frac{(i\theta)^j}{j!}, \tag{3.20}$$

where $\widetilde{C}_k = \widetilde{C}_k(2p, s, t, u, 2q; x, y)$ $(0 \le k \le d)$ are certain constants.

In the third and the fourth terms on the right-hand side of (3.19), we change the running indices μ and ρ $(0 \le \mu \le 2\nu,\ 0 \le \rho \le \mu)$ to ρ and σ $(0 \le \rho \le 2\mu,\ 0 \le \sigma \le 2\nu - \rho)$, by putting $\sigma = 2\nu - \mu$. Then we have

$$\begin{aligned} (3.21) \qquad & \widetilde{G}_d(\theta; 2p, s, t, u, 2q; x, y) \\ &= \frac{1}{i^d}\Bigg[\sum_{l,m,n=1}^{\infty} \frac{(-1)^{l+m} x^m y^n e^{i(l+m)\theta}}{l^{2p} m^s n^t (l+m)^d (m+n)^u (l+m+n)^{2q}} \\ &\qquad + \sum_{\substack{l,m,n=1 \\ l \ne m \\ l \ne m+n}}^{\infty} \frac{(-1)^{l+m} x^m y^n e^{i(-l+m)\theta}}{l^{2p} m^s n^t (-l+m)^d (m+n)^u (-l+m+n)^{2q}} \end{aligned}$$

$$
\begin{aligned}
&-2\sum_{\nu=0}^{p}\phi(2p-2\nu)\sum_{\rho=0}^{2\nu}\sum_{\sigma=0}^{2\nu-\rho}\binom{2q-1+\sigma}{\sigma}\binom{d-1+2\nu-\sigma-\rho}{2\nu-\sigma-\rho}\\
&\qquad\times\frac{(-i\theta)^{\rho}}{\rho!}\sum_{m,n=1}^{\infty}\frac{(-1)^m x^m y^n e^{im\theta}}{m^{s+d+2\nu-\sigma-\rho}n^t(m+n)^{u+2q+\sigma}}\\
&-2(-1)^d\sum_{\nu=0}^{q}\phi(2q-2\nu)\sum_{\rho=0}^{2\nu}\sum_{\sigma=0}^{2\nu-\rho}\binom{2p-1+\sigma}{\sigma}\\
&\qquad\times\binom{d-1+2\nu-\sigma-\rho}{2\nu-\sigma-\rho}\frac{(i\theta)^{\rho}}{\rho!}\sum_{m,n=1}^{\infty}\\
&\qquad\times\left.\frac{(-1)^n x^m y^n e^{-in\theta}}{m^s n^{t+d+2\nu-\sigma-\rho}(m+n)^{u+2p+\sigma}}\right]\quad(\theta\in[-\pi,\pi]).
\end{aligned}
$$

For simplicity, we denote by $\zeta_2(s_1,s_2,s_3;x,y;A_2)$ the double polylogarithm of type A_2 defined by

$$
\zeta_2(s_1,s_2,s_3;x,y;A_2)=\sum_{m,n=1}^{\infty}\frac{x^m y^n}{m^{s_1}n^{s_3}(m+n)^{s_2}}. \tag{3.22}
$$

By (3.20) and (3.21) with $d=2r+1$ for $r\in\mathbb{N}_0$, we have

(3.23)

$$
\begin{aligned}
&\frac{i^{2r+1}}{2(i\pi)}\left\{\widetilde{G}_{2r+1}(\pi;2p,s,t,u,2q;x,y)-\widetilde{G}_{2r+1}(-\pi;2p,s,t,u,2q;x,y)\right\}\\
&=2\sum_{\nu=0}^{p}\phi(2p-2\nu)\sum_{\mu=0}^{\nu-1}\sum_{\sigma=0}^{2\nu-2\mu-1}\binom{2q-1+\sigma}{\sigma}\binom{2r-1+2\nu-\sigma-2\mu}{2\nu-\sigma-2\mu-1}\\
&\qquad\times\frac{(i\pi)^{2\mu}}{(2\mu+1)!}\zeta_2(s+2r+2\nu-\sigma-2\mu,u+2q+\sigma,t;x,y;A_2)\\
&+2\sum_{\nu=0}^{q}\phi(2q-2\nu)\sum_{\mu=0}^{\nu-1}\sum_{\sigma=0}^{2\nu-2\mu-1}\binom{2p-1+\sigma}{\sigma}\binom{2r-1+2\nu-\sigma-2\mu}{2\nu-\sigma-2\mu-1}\\
&\qquad\times\frac{(i\pi)^{2\mu}}{(2\mu+1)!}\zeta_2(s,u+2p+\sigma,t+2r+2\nu-\sigma-2\mu;x,y;A_2)\\
&=\sum_{\nu=0}^{r}\widetilde{C}_{2r-2\nu}\frac{(i\pi)^{2\nu}}{(2\nu+1)!}.
\end{aligned}
$$

Applying [11, Lemma 2.1] to the second member of (3.23), we have

(3.24)

$$
\begin{aligned}
&-\sum_{\sigma=0}^{2p-1}\binom{2q-1+\sigma}{\sigma}\binom{2p+2r-\sigma-1}{2p-\sigma-1}\\
&\qquad\times\zeta_2(s+2p+2r-\sigma,u+2q+\sigma,t;x,y;A_2)\\
&-\sum_{\sigma=0}^{2q-1}\binom{2p-1+\sigma}{\sigma}\binom{2q+2r-\sigma-1}{2q-\sigma-1}\\
&\qquad\times\zeta_2(s,u+2p+\sigma,t+2q+2r-\sigma;x,y;A_2)=\sum_{\nu=0}^{r}\widetilde{C}_{2r-2\nu}\frac{(i\pi)^{2\nu}}{(2\nu+1)!}.
\end{aligned}
$$

On the other hand, by (3.20) and (3.21) with $d=2r$, we have

(3.25)

$$
\begin{aligned}
&\frac{i^{2r}}{2}\left\{\widetilde{G}_{2r}(\pi;2p,s,t,u,2q;x,y)+\widetilde{G}_{2r}(-\pi;2p,s,t,u,2q;x,y)\right\}\\
&=\sum_{l,m,n=1}^{\infty}\frac{x^my^n}{l^{2p}m^sn^t(l+m)^{2r}(m+n)^u(l+m+n)^{2q}}\\
&\quad+\sum_{\substack{l,m,n=1\\ l\neq m\\ l\neq m+n}}^{\infty}\frac{x^my^n}{l^{2p}m^sn^t(-l+m)^{2r}(m+n)^u(-l+m+n)^{2q}}\\
&\quad-2\sum_{\nu=0}^{p}\phi(2p-2\nu)\sum_{\mu=0}^{\nu}\sum_{\sigma=0}^{2\nu-2\mu}\binom{2q-1+\sigma}{\sigma}\binom{2r-1+2\nu-\sigma-2\mu}{2\nu-\sigma-2\mu}\\
&\qquad\times\frac{(i\pi)^{2\mu}}{(2\mu)!}\zeta_2(s+2r+2\nu-\sigma-2\mu,u+2q+\sigma,t;x,y;A_2)\\
&\quad-2\sum_{\nu=0}^{q}\phi(2q-2\nu)\sum_{\mu=0}^{\nu}\sum_{\sigma=0}^{2\nu-2\mu}\binom{2p-1+\sigma}{\sigma}\binom{2r-1+2\nu-\sigma-2\mu}{2\nu-\sigma-2\mu}\\
&\qquad\times\frac{(i\pi)^{2\mu}}{(2\mu)!}\zeta_2(s,u+2p+\sigma,t+2r+2\nu-\sigma-2\mu;x,y;A_2)\\
&=\sum_{\nu=0}^{r}\widetilde{C}_{2r-2\nu}\frac{(i\pi)^{2\nu}}{(2\nu)!}.
\end{aligned}
$$

Applying [11, Lemma 2.1] to the third and the fourth terms on the second member of (3.25), and applying [23, Lemma 4.4] to (3.24), (3.25), we obtain the following.

Proposition 3.3. *With the above notation,*

$$
\begin{aligned}
(3.26)\quad & \sum_{l,m,n=1}^{\infty} \frac{x^m y^n}{l^{2p} m^s n^t (l+m)^{2r} (m+n)^u (l+m+n)^{2q}} \\
&+ \sum_{\substack{l,m,n=1 \\ l\neq m \\ l\neq m+n}}^{\infty} \frac{x^m y^n}{l^{2p} m^s n^t (-l+m)^{2r} (m+n)^u (-l+m+n)^{2q}} \\
&= 2\sum_{\rho=0}^{p} \zeta(2\rho) \sum_{\sigma=0}^{2p-2\rho} \binom{2q-1+\sigma}{\sigma} \binom{2p+2r-2\rho-\sigma-1}{2p-2\rho-\sigma} \\
&\qquad \times \zeta_2(s+2p+2r-2\rho-\sigma, u+2q+\sigma, t; x, y; A_2) \\
&+ 2\sum_{\rho=0}^{q} \zeta(2\rho) \sum_{\sigma=0}^{2q-2\rho} \binom{2p-1+\sigma}{\sigma} \binom{2q+2r-2\rho-\sigma-1}{2q-2\rho-\sigma} \\
&\qquad \times \zeta_2(s, u+2q+\sigma, t+2q+2r-2\rho-\sigma; x, y; A_2) \\
&+ 2\sum_{\rho=0}^{r} \zeta(2\rho) \sum_{\sigma=0}^{2p-1} \binom{2q-1+\sigma}{\sigma} \binom{2p+2r-2\rho-\sigma-1}{2p-\sigma-1} \\
&\qquad \times \zeta_2(s+2p+2r-2\rho-\sigma, u+2q+\sigma, t; x, y; A_2) \\
&+ 2\sum_{\rho=0}^{r} \zeta(2\rho) \sum_{\sigma=0}^{2q-1} \binom{2p-1+\sigma}{\sigma} \binom{2q+2r-2\rho-\sigma-1}{2q-\sigma-1} \\
&\qquad \times \zeta_2(s, u+2q+\sigma, t+2q+2r-2\rho-\sigma; x, y; A_2).
\end{aligned}
$$

Let $(x, y) = (1, 1)$ and calculate the second term on the left-hand side. Then we obtain the main result in this section.

Theorem 3.4. *For $p, q \in \mathbb{N}$ and $r \in \mathbb{N}_0$,*

$$\begin{aligned}(3.27)\qquad &\zeta_3(2p,2r,2q,s,u,t;A_3)+\zeta_3(2p,s,u,2r,2q,t;A_3)\\ &+\zeta_3(2q,2r,2p,t,u,s;A_3)+\zeta_3(2q,t,u,2r,2p,s;A_3)\\ &=2\sum_{\rho=0}^{p}\zeta(2\rho)\sum_{\sigma=0}^{2p-2\rho}\binom{2q-1+\sigma}{\sigma}\binom{2p+2r-2\rho-\sigma-1}{2p-2\rho-\sigma}\\ &\quad\times\zeta_2(s+2p+2r-2\rho-\sigma,u+2q+\sigma,t;A_2)\\ &+2\sum_{\rho=0}^{q}\zeta(2\rho)\sum_{\sigma=0}^{2q-2\rho}\binom{2p-1+\sigma}{\sigma}\binom{2q+2r-2\rho-\sigma-1}{2q-2\rho-\sigma}\\ &\quad\times\zeta_2(s,u+2q+\sigma,t+2q+2r-2\rho-\sigma;A_2)\\ &+2\sum_{\rho=0}^{r}\zeta(2\rho)\sum_{\sigma=0}^{2p-1}\binom{2q-1+\sigma}{\sigma}\binom{2p+2r-2\rho-\sigma-1}{2p-\sigma-1}\\ &\quad\times\zeta_2(s+2p+2r-2\rho-\sigma,u+2q+\sigma,t;A_2)\\ &+2\sum_{\rho=0}^{r}\zeta(2\rho)\sum_{\sigma=0}^{2q-1}\binom{2p-1+\sigma}{\sigma}\binom{2q+2r-2\rho-\sigma-1}{2q-\sigma-1}\\ &\quad\times\zeta_2(s,u+2q+\sigma,t+2q+2r-2\rho-\sigma;A_2)\end{aligned}$$

holds for all $(s,t,u) \in \mathbb{C}^3$ except for singularities of all functions on the both sides of (3.27).

Example 3.5. Putting $(p,q,r) = (1,1,1)$ and $(s,t,u) = (2,2,2)$ in (3.27), we have

(3.28)

$$\begin{aligned}4\zeta_3(2,2,2,2,2,2;A_3) &= 8\zeta(2)\{\zeta_2(4,4,2;A_2)+\zeta_2(3,5,2;A_2)\}\\ &-6\{2\zeta_2(6,4,2;A_2)+2\zeta_2(5,5,2;A_2)+\zeta_2(4,6,2;A_2)\}.\end{aligned}$$

Using $\zeta_2(k,m,l;A_2) = \zeta_2(k,m-1,l+1;A_2) - \zeta_2(k-1,m,l+1;A_2)$ (see [19]) repeatedly, we see that

$$\zeta_2(4,4,2;A_2)+\zeta_2(3,5,2;A_2) = \frac{1}{2}\{2\zeta_2(4,3,3;A_2)-\zeta_2(3,4,3;A_2)\}.$$

From [20, Theorem 2], the right-hand side is equal to $\pi^{10}/935550$. Similarly we have

$$\begin{aligned} &2\zeta_2(6,4,2;A_2) + 2\zeta_2(5,5,2;A_2) + \zeta_2(4,6,2;A_2) \\ &\quad = 2\zeta_2(6,3,3;A_2) - \zeta_2(3,6,3;A_2) - \frac{1}{2}\zeta_2(4,4,4;A_2). \end{aligned}$$

From [20, Theorem 2] and [18, Theorem 4.1], we see that the right-hand side is equal to $887\pi^{12}/3831077250$. Therefore we obtain Witten's volume formula

$$\zeta_3(2,2,2,2,2,2;A_3) = \frac{23}{2554051500}\pi^{12}. \tag{3.29}$$

This implies that (3.27) can be regarded as a continuous relation including Witten's volume formula.

Remark 3.1. If we start from (3.3) instead of (3.2) in (3.4), then we can construct a general functional relation for

$$\begin{aligned} &\zeta_3(a,c,b,s,u,t;A_3) + (-1)^a\zeta_3(a,s,u,c,b,t;A_3) \\ &\quad + (-1)^{a+b+c}\zeta_3(b,c,a,t,u,s:A_3) + (-1)^{a+c}\zeta_3(b,t,u,c,a,s;A_3) \end{aligned}$$

in terms of $\zeta_2(s_1,s_2,s_3;A_2)$ and $\zeta(s)$, where $a,b \in \mathbb{N}$, $c \in \mathbb{N}_0$ and $s,t,u \in \mathbb{C}$. For example, in the case $(a,b,c) = (1,3,0)$, we can obtain

$$\begin{aligned} &\zeta_3(1,0,3,s,u,t;A_3) - \zeta_3(1,s,u,0,3,t;A_3) \\ &\quad + \zeta_3(3,0,1,t,u,s;A_3) - \zeta_3(3,t,u,0,1,s;A_3) \\ &\quad = 4\zeta_2(s,u+4,t;A_2) + \zeta_2(s+1,u,t+3;A_2) - 2\zeta(2)\zeta_2(s,u+2,t;A_2). \end{aligned}$$

Remark 3.2. As described in this section, we produced relation formulas for $\zeta_3(\,\cdot\,;A_3)$ by uniting those for $\zeta_2(\,\cdot\,;A_2)$ and $\zeta_1(\,\cdot\,;A_1)$. This corresponds to the fact that the fundamental root system of A_3 can be produced by adding a fundamental root to that of A_2 in the Dynkin diagram. This strategy can be applied to various other cases. For instance, in [5], we deduced relation formulas for $\zeta_2(\,\cdot\,;B_2)$ from those of $\zeta_2(\,\cdot\,;A_2)$ and $\zeta_1(\,\cdot\,;A_1)$. Similarly, it is possible to deduce relation formulas for $\zeta_3(\,\cdot\,;B_3)$ and $\zeta_3(\,\cdot\,;C_3)$ from those of $\zeta_3(\,\cdot\,;A_3)$, or alternatively, from $\zeta_2(\,\cdot\,;B_2)$ and $\zeta_1(\,\cdot\,;A_1)$. In [5], the process of the latter way was described in detail. These correspondences between functional relations for zeta-functions and the structure of root systems seems to be fascinating.

References

[1] T. M. Apostol, Introduction to Analytic Number Theory, Springer, 1976.

[2] N. Bourbaki, Groupes et Algèbres de Lie, Chapitres 4, 5 et 6, Hermann, Paris, 1968.

[3] P. E. Gunnells and R. Sczech, Evaluation of Dedekind sums, Eisenstein cocycles, and special values of L-functions, Duke Math. J. **118** (2003), 229-260.

[4] Y. Komori, K. Matsumoto and H. Tsumura, On Witten multiple zeta-functions associated with semisimple Lie algebras II, preprint, submitted for publication.

[5] Y. Komori, K. Matsumoto and H. Tsumura, On Witten multiple zeta-functions associated with semisimple Lie algebras III, preprint, submitted for publication.

[6] Y. Komori, K. Matsumoto and H. Tsumura, On Witten multiple zeta-functions associated with semisimple Lie algebras IV, in preparation.

[7] K. Matsumoto, Asymptotic expansions of double zeta-functions of Barnes, of Shintani, and Eisenstein series, Nagoya Math J. **172** (2003), 59-102.

[8] K. Matsumoto, On the analytic continuation of various multiple zeta-functions, in 'Number Theory for the Millennium II, Proc. Millennial Conference on Number Theory', M. A. Bennett et al (eds.), A K Peters, 2002, pp. 417-440.

[9] K. Matsumoto, The analytic continuation and the asymptotic behaviour of certain multiple zeta-functions I, J. Number Theory **101** (2003), 223-243.

[10] K. Matsumoto, On Mordell-Tornheim and other multiple zeta-functions, Proceedings of the Session in analytic number theory

and Diophantine equations (Bonn, January-June 2002), D. R. Heath-Brown and B. Z. Moroz (eds.), Bonner Mathematische Schriften Nr. 360, Bonn 2003, n.25, 17pp.

[11] K. Matsumoto, T. Nakamura, H. Ochiai and H. Tsumura, On value-relations, functional relations and singularities of Mordell-Tornheim and related triple zeta-functions, preprint, submitted for publication.

[12] K. Matsumoto and H. Tsumura, On Witten multiple zeta-functions associated with semisimple Lie algebras I, Ann. Inst. Fourier (Grenoble) **56** (2006), 1457-1504.

[13] K. Matsumoto and H. Tsumura, Functional relations for various multiple zeta-functions, ('Analytic Number Theory', Kyoto, 2005), RIMS K = oky = uroku No. 1512 (2006), 179-190.

[14] K. Matsumoto and H. Tsumura, A new method of producing functional relations among multiple zeta-functions, Tokyo Metropolitan University Preprint Series 2006 : No.1.

[15] L. J. Mordell, On the evaluation of some multiple series, J. London Math. Soc. **33** (1958), 368-371.

[16] T. Nakamura, A functional relation for the Tornheim double zeta function, Acta Arith., to appear.

[17] H. Samelson, Notes on Lie Algebras, Universitext, Springer-Verlag, 1990.

[18] M. V. Subbarao and R. Sitaramachandrarao, On some infinite series of L. J. Mordell and their analogues, Pacific J. Math. **119** (1985), 245-255.

[19] L. Tornheim, Harmonic double series, Amer. J. Math. **72** (1950), 303-314.

[20] H. Tsumura, On some combinatorial relations for Tornheim's double series, Acta Arith. **105** (2002), 239-252.

[21] H. Tsumura, Evaluation formulas for Tornheim's type of alternating double series, Math. Comp. **73** (2004), 251-258.

[22] H. Tsumura, On Witten's type of zeta values attached to $SO(5)$, Arch. Math. (Basel) **84** (2004), 147-152.

[23] H. Tsumura, On functional relations between the Mordell-Tornheim double zeta functions and the Riemann zeta function, Math. Proc. Cambridge Philos. Soc., to appear.

[24] E. Witten, On quantum gauge theories in two dimensions, Comm. Math. Phys. **141** (1991), 153-209.

[25] D. Zagier, Values of zeta functions and their applications, in Proc. First Congress of Math., Paris, vol. II, Progress in Math., **120**, Birkhäuser, 1994, 497-512.

Y. Komori
Graduate School of Mathematics,
Nagoya University,
Chikusa-ku, Nagoya 464-8602
Japan
Email: komori@math.nagoya-u.ac.jp

K. Matsumoto
Graduate School of Mathematics,
Nagoya University,
Chikusa-ku, Nagoya 464-8602
Japan
Email: kohjimat@math.nagoya-u.ac.jp

H. Tsumura
Department of Mathematics and Information Sciences,
Tokyo Metropolitan University
1-1, Minami-Ohsawa, Hachioji,
Tokyo 192-0397
Japan
Email: tsumura@comp.metro-u.ac.jp

Sums of Kloosterman Sums Revisited

Y. MOTOHASHI

The aim of the present article is to show that the Spectral–Kloosterman and the Kloosterman–Spectral sum formulas are practically equivalent to each other. Previously, only the procedure for deriving the latter from the former was known, which is the inversion due to N.V. Kuznetsov [10][11]. We shall not only prove it in our own way but also establish a reverse procedure. Our motivation is detailed in the third and the final sections.

Technically, our discussion is focused on two Bessel transforms which appear in those sum formulas. These transforms, especially their inversions, seem to be best understood as special instances of the Plancherel theorem on Lie groups as discussed by R.W. Bruggeman [2]. We shall, however, employ an alternative argument which appears to demand less background knowledge, yet it has a certain merit in applications of the sum formula. We shall restrict ourselves to the most basic instance offered by the full modular group, solely for the sake of simplicity. Notations are introduced where they are needed for the first time, and will continue to be effective thereafter.

1 Automorphic forms

To begin with, we shall collect elements of the theory of Γ-automorphic representations of G, with

$$G = \mathrm{PSL}_2(\mathbb{R}), \qquad \Gamma = \mathrm{PSL}_2(\mathbb{Z}). \tag{1.1}$$

We write

$$\mathrm{n}[x] = \begin{bmatrix} 1 & x \\ & 1 \end{bmatrix}, \quad \mathrm{a}[y] = \begin{bmatrix} \sqrt{y} & \\ & 1/\sqrt{y} \end{bmatrix}, \quad \mathrm{k}[\theta] = \begin{bmatrix} \cos\theta & \sin\theta \\ -\sin\theta & \cos\theta \end{bmatrix}. \tag{1.2}$$

Let $N = \{\mathrm{n}[x] : x \in \mathbb{R}\}$, $A = \{\mathrm{a}[y] : y > 0\}$, and $K = \{\mathrm{k}[\theta] : \theta \in \mathbb{R}/\pi\mathbb{Z}\}$, so that $G = NAK$ be the Iwasawa decomposition of G. We read it as $G \ni \mathrm{g} = \mathrm{nak} = \mathrm{n}[x]\mathrm{a}[y]\mathrm{k}[\theta]$. A Haar measure on G is defined by $dg = (\pi y^2)^{-1}dxdyd\theta$. The space $L^2(\Gamma\backslash G)$ is composed of all left Γ-automorphic functions on G, vectors for short, which are square integrable over $\Gamma\backslash G$ against dg. Elements of G act unitarily on vectors from the right, and we have the orthogonal decomposition into invariant subspaces

$$L^2(\Gamma\backslash G) = \mathbb{C}\cdot 1 \oplus {}^c\!L^2(\Gamma\backslash G) \oplus {}^e\!L^2(\Gamma\backslash G), \tag{1.3}$$

where ${}^c\!L^2$ is the cuspidal subspace, and ${}^e\!L^2$ is spanned by integrals of Eisenstein series.

The cuspidal subspace splits into irreducible subspaces:

$${}^c\!L^2(\Gamma\backslash G) = \overline{\oplus V}; \quad \Omega|_V = \left(\nu_V^2 - \tfrac{1}{4}\right)\cdot 1, \tag{1.4}$$

where $\Omega = y^2\left(\partial_x^2 + \partial_y^2\right) - y\partial_x\partial_\theta$ is the Casimir operator. With (1.1), we can restrict our attention to two cases: either $\nu_V \in (0, +\infty)i$ or $\nu_V \in \mathbb{N} - \frac{1}{2}$. According to the right action of K, the space V is decomposed into K-irreducible subspaces

$$V = \overline{\bigoplus_{p=-\infty}^{\infty} V_p}, \quad \dim V_p \le 1. \tag{1.5}$$

If it is not trivial, V_p is spanned by a Γ-automorphic function φ_p such that $\varphi_p(\mathrm{gk}[\theta]) = \exp(2ip\theta)\varphi_p(\mathrm{g})$; it is called a Γ-automorphic form of spectral parameter ν_V and weight $2p$.

Let us assume temporarily that V belongs to the unitary principal series, i.e., $\nu_V \in (0, +\infty)i$ under (1.1). Then $\dim V_p = 1$ for all $p \in \mathbb{Z}$, and there exists a complete orthonormal system $\left\{\varphi_p \in V_p : p \in \mathbb{Z}\right\}$ of V such that

$$\varphi_p(\mathrm{g}) = \sum_{\substack{n=-\infty \\ n\neq 0}}^{\infty} \frac{\varrho_V(n)}{\sqrt{|n|}} \mathcal{A}_{\mathrm{sgn}(n)}\phi_p(\mathrm{a}[|n|]\mathrm{g}; \nu_V), \tag{1.6}$$

where $\phi_p(\mathrm{g}; \nu) = y^{\nu+\frac{1}{2}}\exp(2ip\theta)$, and $\mathcal{A}_u$ is the Jacquet operator:

$$\mathcal{A}_u\phi_p(\mathrm{g}; \nu) = \int_{-\infty}^{\infty} \exp(-2\pi i u v)\phi_p(\mathrm{wn}[v]\mathrm{g}; \nu)dv, \quad \mathrm{w} = \mathrm{k}\left[\tfrac{1}{2}\pi\right]. \tag{1.7}$$

In particular, we have the expansion

$$\varphi_0(\mathrm{g}) = \frac{2\pi^{\nu_V+\frac{1}{2}}}{\Gamma\left(\nu_V+\frac{1}{2}\right)}\sqrt{y}\sum_{\substack{n=-\infty\\ n\neq 0}}^{\infty}\varrho_V(n)K_{\nu_V}(2\pi|n|y)\exp(2\pi i n x), \tag{1.8}$$

with K_ν standing for the K-Bessel function of order ν. This is a cusp-form of unit length on the hyperbolic upper half plane G/K.

Let us consider a V in the discrete series; thus $\nu_V = q-\frac{1}{2}$, with a $q\in\mathbb{N}$. We have, in place of (1.5),

$$\text{either}\quad V=\overline{\bigoplus_{p=q}^{\infty} V_p}\quad\text{or}\quad V=\overline{\bigoplus_{p=-\infty}^{-q} V_p}, \tag{1.9}$$

with $\dim V_p = 1$, corresponding to the holomorphic and the antiholomorphic discrete series, respectively. The involution $\mathrm{g}=\mathrm{nak}\mapsto \mathrm{n}^{-1}\mathrm{ak}^{-1}$ maps one to the other. In the holomorphic case, we have a complete orthonormal system $\{\varphi_p : p\geq q\}$ in V such that

$$\varphi_p(\mathrm{g}) = \pi^{\frac{1}{2}-q}\left(\frac{\Gamma(p+q)}{\Gamma(p-q+1)}\right)^{\frac{1}{2}}\sum_{n=1}^{\infty}\frac{\varrho_V(n)}{\sqrt{n}}\mathcal{A}_1\phi_p(\mathrm{a}[n]\mathrm{g};\nu_V). \tag{1.10}$$

In particular, we have

$$\varphi_q(\mathrm{g}) = (-1)^q\frac{2^{2q}\pi^{q+\frac{1}{2}}}{\sqrt{\Gamma(2q)}}\exp(2iq\theta)y^q\sum_{n=1}^{\infty}\varrho_V(n)n^{q-\frac{1}{2}}\exp(2\pi i n(x+iy)). \tag{1.11}$$

This infinite sum is a holomorphic cusp-form of weight $2q$ on G/K, the length of which is equal to $2^{-2q}\pi^{-q-\frac{1}{2}}\sqrt{\Gamma(2q)}$.

We shall not deal with irreducible subspaces of complementary series type, for there are no such subspaces under (1.1), as indicated above. We note relevantly that our argument below can be extended to general cofinite groups with cusps, without essential changes if the corresponding Kloosterman sums satisfy a Weil type bound; see a remark following (2.3) below.

Let V be generic as in (1.4). It should be observed that the coefficients $\varrho_V(n)$ in either (1.6) or (1.10) are normalized in such a way that they do not depend on the weight. We recall the basic bound: For any $m\in\mathbb{N}$

$$\sum_{\substack{V\\ |\nu_V|\leq N}}|\varrho_V(m)|^2 \ll N^2 + m^{\frac{1}{2}+\varepsilon} \tag{1.12}$$

as N tends to $+\infty$, where ε is an arbitrary small positive constant, solely on which the implicit constant may depend. This is an extended version of [12, (2.3.2)] under our present normalization; those V in discrete series are to be treated with Petersson's sum formula (see (4.5) below).

2 Sum formulas

We are going to display the two versions of the sum formula involving Kloosterman sums. We stress again that the aim of the present work is to discuss the relation of these two but not to develop any proof of the sum formula itself; see the remarks (1)–(3) in the final section. Our notation depends on (1.6) and (1.10) and is different from those employed in [1][2][10][11][12], for instance. We shall skip over convergence issues, as they will be touched on in later sections.

We define a Kloosterman sum attached to Γ by

$$S(m,n;\ell) = \sum_{\substack{w \bmod \ell \\ (w,\ell)=1}} \exp\left(\frac{2\pi i}{\ell}(mw + n\underline{w})\right), \quad w\underline{w} \equiv 1 \bmod \ell, \tag{2.1}$$

with arbitrary integers $\ell > 0$, and m, n. We put, for $x > 0$,

$$\begin{aligned} \mathrm{B}h(x) &= \frac{1}{2\pi i}\int_{(0)} \frac{r}{\cos(\pi r)} J_{2r}(x)h(r)dr \\ &= \frac{i}{4\pi}\int_{(0)} \frac{J_{-2r}(x) - J_{2r}(x)}{\sin(\pi r)} r\tan(\pi r)h(r)dr, \end{aligned} \tag{2.2}$$

where J_ν stands for the J-Bessel function of order ν; and (η) denotes the vertical line passing through the point η. We require that

$$h \text{ be even, regular, } |h(r)| \ll (|r|+1)^{-2-\varepsilon}, \text{ provided } |\mathrm{Re}\, r| \le \frac{1}{4} + \varepsilon, \tag{2.3}$$

where the implicit constant may depend on ε. Note that this width of the vertical strip is relevant to the Weil bound for $S(m,n;\ell)$; the sum over ℓ in (2.4) below converges absolutely, since $\mathrm{B}h(x) \ll x^{\frac{1}{2}+\varepsilon}$ for x close to 0, as is implied by an appropriate shift of the contour in the first integral of (2.2), provided (2.3) holds. Thus the width has to be altered according to the choice of underlying discrete subgroups; in general, the condition

$|\mathrm{Re}\, r| \le \frac{1}{2} + \varepsilon$ is appropriate, on which as well our argument works fine with minor changes.

With this, the first sum formula that we are concerned with asserts that it holds that for any $m, n \in \mathbb{N}$

$$\begin{aligned}&\sum_V{}^{\mathrm{u}} \varrho_V(m)\overline{\varrho_V(n)}h(\nu_V) + \frac{1}{4\pi i}\int_{(0)} \frac{\sigma_{2r}(m)\sigma_{2r}(n)}{(mn)^r\zeta(1+2r)\zeta(1-2r)}h(r)dr = \\ &\delta_{m,n}\frac{i}{4\pi^2}\int_{(0)} r\tan(\pi r)h(r)dr + \sum_{\ell=1}^{\infty}\frac{1}{\ell}S(m,n;\ell)\mathrm{B}h\left(\frac{4\pi}{\ell}\sqrt{mn}\right).\end{aligned} \tag{2.4}$$

Here $\sum^{\mathrm{u}}$ indicates that V belong to the unitary principal series; ζ the Riemann zeta-function, $\sigma_\alpha(n) = \sum_{d|n} d^\alpha$, and $\delta_{m,n}$ the Kronecker delta. This is a version of the Spectral–Kloosterman sum formula established independently by Bruggeman [1] and by Kuznetsov [10][11]. A relatively accessible proof is given in [12]; as to the condition (2.3) see [*ibid*, pp. 80–81].

To state (2.4) alternatively, we suppose that the domain of h contains $\mathbb{N} - \frac{1}{2}$ also, and augment (2.3) by

$$h\left(q - \tfrac{1}{2}\right) \ll q^{-2-\varepsilon}, \quad q \in \mathbb{N}. \tag{2.5}$$

We put

$$\mathrm{L}h(x) = \mathrm{B}h(x) + \frac{1}{2\pi}\sum_{q=1}^{\infty}(-1)^q(2q-1)h\left(q-\tfrac{1}{2}\right)J_{2q-1}(x), \tag{2.6}$$

$$\Delta(h) = i\int_{(0)} r\tan(\pi r)h(r)dr + \sum_{q=1}^{\infty}(2q-1)h\left(q-\tfrac{1}{2}\right). \tag{2.7}$$

Then we have, in place of (2.4),

$$\begin{aligned}&\sum_V \varrho_V(m)\overline{\varrho_V(n)}h(\nu_V) + \frac{1}{4\pi i}\int_{(0)} \frac{\sigma_{2r}(m)\sigma_{2r}(n)}{(mn)^r\zeta(1+2r)\zeta(1-2r)}h(r)dr \\ &= \frac{\delta_{m,n}}{(2\pi)^2}\Delta(h) + \sum_{\ell=1}^{\infty}\frac{1}{\ell}S(m,n;\ell)\mathrm{L}h\left(\frac{4\pi}{\ell}\sqrt{mn}\right),\end{aligned} \tag{2.8}$$

with V running over all irreducible cuspidal subspaces enlisted in (1.4). The equivalence of (2.4) and (2.8), under (2.3) $\cap$ (2.5), will be shown in Section 4.

The maps L and Δ came later from the effort to understand Kuznetsov's inversion of (2.4), i.e., the second version of the sum formula (2.12)/(2.13) below in the framework of the spectral decomposition of $L^2(\Gamma\backslash G)$, as developed in [2]; their origin can be traced back to [1] and [10][11]. They might appear irrelevant to (2.4) itself, for one may employ in fact any function on $\mathbb{N}-\frac{1}{2}$ as far as the decay condition is satisfied. For more discussion see the remark (1) in the final section.

Next, let us introduce another Bessel transform

$$\mathrm{K}f(r) = 2\pi \int_0^\infty \frac{J_{-2r}(x) - J_{2r}(x)}{\sin(\pi r)} f(x)\frac{dx}{x}, \tag{2.9}$$

which is to be compared with the second line of (2.2); one may call K the adjoint integral transformation of B. Here we require that there exists a sufficiently large constant $A > 0$ such that for $a \le A$

$$f^{(a)}(x) \ll (x + 1/x)^{-A}, \quad x > 0, \tag{2.10}$$

where the implied constant may depend on A. Also, we put

$$\mathrm{P}f(x) = f(x) - \frac{1}{2\pi}\sum_{q=1}^\infty (-1)^q(2q-1)\mathrm{K}f\left(q-\tfrac{1}{2}\right) J_{2q-1}(x). \tag{2.11}$$

Then the second sum formula we are concerned with is

$$\begin{aligned}&\sum_{\ell=1}^\infty \frac{1}{\ell} S(m,n;\ell)\mathrm{P}f\left(\frac{4\pi}{\ell}\sqrt{mn}\right) + \delta_{m,n}\frac{i}{4\pi^2}\int_{(0)} r\tan(\pi r)\mathrm{K}f(r)dr\\ &=\sum_V{}^{*} \varrho_V(m)\overline{\varrho_V(n)}\mathrm{K}f(\nu_V)\\ &\qquad + \frac{1}{4\pi i}\int_{(0)} \frac{\sigma_{2r}(m)\sigma_{2r}(n)}{(mn)^r\zeta(1+2r)\zeta(1-2r)}\mathrm{K}f(r)dr,\end{aligned} \tag{2.12}$$

where $m, n \in \mathbb{N}$ are arbitrary. Corresponding to (2.8), we have the alternative expression:

$$\begin{aligned}\sum_{\ell=1}^\infty \frac{1}{\ell} S(m,n;\ell) f\left(\frac{4\pi}{\ell}\sqrt{mn}\right) &= \sum_V \varrho_V(m)\overline{\varrho_V(n)}\mathrm{K}f(\nu_V)\\ &\quad + \frac{1}{4\pi i}\int_{(0)} \frac{\sigma_{2r}(m)\sigma_{2r}(n)}{(mn)^r\zeta(1+2r)\zeta(1-2r)}\mathrm{K}f(r)dr.\end{aligned} \tag{2.13}$$

The equivalence of (2.12) and (2.13) is analogous to that of (2.4) and (2.8), apart from the fact that we need to incorporate

$$\Delta \mathrm{K} = 0, \tag{2.14}$$

as we shall show in Section 4 on the condition (2.10). The spectral decomposition (2.13) is a version of the Kloosterman–Spectral sum formula of Kuznetsov [11, Theorem 2 supplemented with (6.33)].

The condition (2.10) is introduced solely for the sake of simplicity; it is immaterial to discuss how large the constant A should be. In fact, (2.10) could be either relaxed or strengthened considerably. For instance, the mild condition [12, (2.4.6)] should work well and also one may start with an arbitrary f which is smooth and compactly supported in $(0, +\infty)$, only we need to reduce respective situations to (2.10) appealing to a usual approximation procedure. In the same token, the condition on h could be highly strengthened; (2.3) appears presently to be the least restrictive.

3 The inversion problem

Historically, (2.4) was proved first and (2.12) second, by Kuznetsov [10]. It is stated in [11, three lines above (6.32)] that they are, nevertheless, *essentially* equivalent. To endorse this fundamental assertion, it is proved in [*ibid*, (6.32) and Appendix] that

$$\mathrm{P}f(x) = -i \int_{(0)} \frac{r}{\cos(\pi r)} J_{2r}(x) \int_0^\infty \frac{J_{-2r}(y) - J_{2r}(y)}{\sin(\pi r)} f(y) \frac{dy}{y} dr; \tag{3.1}$$

namely,

$$\mathrm{BK} = \mathrm{P}, \tag{3.2}$$

on (2.10). In fact, a condition less restrictive than (2.10) is assumed there, but such an aspect is irrelevant to our present purpose as remarked already. We shall show at the end of the next section that the image $\mathrm{K}f$ is in the space (2.3) $\cap$ (2.5), and (2.8) should hold with $h = \mathrm{K}f$. On noting (2.14), this procedure yields (2.13), for $\mathrm{LK} - \mathrm{BK} = 1 - \mathrm{P}$ by definition, and (3.2) means that on (2.10)

$$\mathrm{LK} = 1. \tag{3.3}$$

Provided (2.10), say, holds, the Kloosterman-Spectral sum formula follows from the Spectral-Kloosterman sum formula.

However, it does not appear that the endorsement from the opposite direction, i.e.,

$$\mathrm{KL} = 1 \tag{3.4}$$

on (2.3) ∩ (2.5) has ever been asserted and proved either in [11] or anywhere else. This would practically complete the proof of the equivalence in question.

We shall establish

Theorem. *The identities* (3.3), *and* (3.4) *hold. The sum formulas* (2.8) *and* (2.13) *are equivalent to each other; that is, they can be derived from each other.*

This assertion depends on the conditions (2.3) ∩ (2.5) and (2.10) implicitly and in a somewhat peculiar way. In fact, the map L does not generally take h in the space (2.3) ∩ (2.5) to the space (2.10), which is in a sharp contrast against the property of K under (2.10) mentioned above. Nonetheless, to the image Lh the map K is applicable, and (3.4) holds. This causes a complication in our derivation of (2.8) from (2.13). Thus we first modify both h and L or rather B so that we may apply (2.13) to the result; then a limiting procedure is employed to reach (2.8) with the aid of (3.4). In the course of discussion, we shall actually prove, solely on (2.3),

$$\mathrm{KB} = 1, \tag{3.5}$$

which is slightly deeper than (3.4), for it implies that Kuznetsov's equivalence assertion on (2.4) and (2.12) is indeed correct. Details are developed in Section 5.

It should be stressed that we know already that the relations (3.3) and (3.4) are valid in L^2-sense. To explain this, let us equip the set $Y = \mathbb{R}i \cup \left\{\mathbb{N} - \frac{1}{2}\right\}$ with the measure $d\tau$ defined by

$$\int_Y h d\tau = \frac{1}{8\pi^2}\Delta(h). \tag{3.6}$$

Then in [2, Propositions 14.2.6 and 14.2.8] Bruggeman showed that K and

L extend to the unitary isomorphisms

$$\begin{aligned} \mathrm{K_u} &: L^2((0,+\infty), d^\times) \longrightarrow L^2(Y, d\tau), \\ \mathrm{L_u} &: L^2(Y, d\tau) \longrightarrow L^2((0,+\infty), d^\times), \end{aligned} \tag{3.7}$$

with $d^\times x = dx/x$; moreover,

$$\mathrm{K_u L_u} = 1, \quad \mathrm{L_u K_u} = 1. \tag{3.8}$$

It remains thus to prove (3.8) on spaces of test functions for which the resulting integrals converge absolutely. This is not merely a theoretical exercise. When we apply the sum formula, we need test functions that are well defined on the relevant sets. Relations in L^2-sense such as (3.8) are useless for that purpose. We may then try to work in practical spaces like those introduced above and proceed weakly in $L^2(Y, d\tau)$ and $L^2\,((0,+\infty), d^\times)$, being aided with (3.8). However, this leads us inevitably to an estimation procedure that is shared to a certain extent by our lowbrow argument to prove the theorem. As a byproduct of our discussion, we shall be able to see how Bh deviates from the reasonable space (2.10), which appears to have applications to the circle of problems on Hecke L-functions that we dealt with recently in [14] and [15].

4 Proof (1)

We shall begin with the equivalence of (2.4) and (2.8). It suffices to show that

$$\begin{aligned} &\sum_{\substack{V \\ \text{discrete series}}} \varrho_V(m)\overline{\varrho_V(n)}h(\nu_V) \\ &= \frac{1}{2\pi}\sum_{q=1}^{\infty}(-1)^q(2q-1)h\left(q-\tfrac{1}{2}\right)\left(\sum_{\ell=1}^{\infty}\frac{1}{\ell}S(m,n;\ell)J_{2q-1}\left(\frac{4\pi}{\ell}\sqrt{mn}\right)\right) \\ &+ \frac{\delta_{m,n}}{(2\pi)^2}\sum_{q=1}^{\infty}(2q-1)h\left(q-\tfrac{1}{2}\right). \end{aligned} \tag{4.1}$$

The left side converges absolutely because of (1.12) and the decay condition in (2.3). The double sum on the right converges also absolutely, as we have the Weil bound for $S(m,n;\ell)$ and

$$|J_q(x)| \le \min\left(1, \frac{1}{q!}\left(\frac{x}{2}\right)^q\right), \tag{4.2}$$

uniformly in $x > 0$ and $q \in \mathbb{N}$, which follows from a combination of

$$J_q(x) = \frac{1}{\pi}\int_0^\pi \cos(q\theta - x\sin\theta)d\theta, \tag{4.3}$$

$$J_\nu(x) = \frac{2}{\sqrt{\pi}\Gamma(\nu+\frac{1}{2})}\left(\frac{x}{2}\right)^\nu \int_0^1 (1-t^2)^{\nu-\frac{1}{2}}\cos(xt)dt, \mathrm{Re}\,\nu > -\frac{1}{2}. \tag{4.4}$$

The convergence of the last sum in (4.1) is obvious, because of (2.5). In passing, we note that the convergence of (2.11) is confirmed analogously.

With this, we invoke Petersson's sum formula under the normalization (1.10); thus the last sum over ℓ is equal to

$$\frac{1}{2\pi}(-1)^{q-1}\delta_{m,n} + (-1)^q\frac{2\pi}{(2q-1)}\sum_{\substack{V \\ \nu_V = q-\frac{1}{2}}} \varrho_V(m)\overline{\varrho_V(n)}, \tag{4.5}$$

where V are in the holomorphic discrete series (see [12, (2.2.9)]). Hence we get (4.1); as remarked above, we see that any value of $h\left(q-\frac{1}{2}\right)$ works as far as (2.5) holds.

We shall prove (2.14), which states that on (2.10)

$$\int_{(0)} r\tan(\pi r)\mathrm{K}f(r)dr = i\sum_{q=1}^\infty (2q-1)\mathrm{K}f\left(q-\tfrac{1}{2}\right). \tag{4.6}$$

We return to the definition (2.9). Shifting the contour to $\left(\frac{5}{4}\right)$, we see that the left side is equal to

$$i\mathrm{K}f\left(\tfrac{1}{2}\right) - 4\pi\int_{(\frac{5}{4})}\frac{r}{\cos(\pi r)}\int_0^\infty J_{2r}(x)f(x)\frac{dx}{x}dr. \tag{4.7}$$

This double integral converges absolutely; in fact (4.4) gives the uniform bound $J_{2r}(x) \ll (x/|r|)^{5/2}e^{\pi|r|}$. Exchanging the order of integration, and computing the new inner integral with residue calculus, which is possible with x fixed as (4.4) implies, we find that (4.7) is equal to

$$\begin{aligned} 4\pi i \int_0^\infty f(x)\left(\sum_{q=1}^\infty (-1)^q(2q-1)J_{2q-1}(x)\right)\frac{dx}{x} \\ = -4\pi i \int_0^\infty J_0(x)f(x)dx, \end{aligned} \tag{4.8}$$

where we have used the relation

$$J_{\nu-1}(x) + J_{\nu+1}(x) = (\nu/x)J_\nu(x). \tag{4.9}$$

Similarly we have that the right side of (4.6) is equal to

$$-4\pi i \int_0^\infty J_0(x)f(x)dx + 4\pi i \lim_{Q\to\infty}(-1)^Q \int_0^\infty J_{2Q}(x)f(x)dx. \tag{4.10}$$

The limit can be taken inside the integral because of (4.2). Hence, (4.6) follows from (4.8) and (4.10); and the proof of the equivalence of (2.12) and (2.13) ends.

Next, we shall prove Kuznetsov's inversion formula (3.1) in a manner different from his that is developed in [11]. To this end, let us put

$$D_s^\pm = -y^2\left(\frac{d}{dy}\right)^2 + s(s-1) \mp y^2, \quad y \in (0, +\infty). \tag{4.11}$$

We claim that the resolvent kernels of $y^{-2}D_s^\pm$ give rise to

$$\begin{aligned} &(+)\text{: Kuznetsov inversion formula,} \\ &(-)\text{: Kontorovitch–Lebedev inversion formula.} \end{aligned} \tag{4.12}$$

In fact, the minus case is treated in [12, Lemma 2.6], and is pertinent to the analogous sum formula involving Kloosterman sums of opposite sign, i.e., $S(-m, n; \ell)$ with $m, n \in \mathbb{N}$. We are going to show that the argument there works for the plus case as well. Thus, we observe that the differential equation $D_s^+ f(y) = 0$ has two linearly independent solutions:

$$\begin{aligned} j_{s-\frac{1}{2}}(y) &= \sqrt{y}J_{s-\frac{1}{2}}(y), \\ k_{s-\frac{1}{2}}(y) &= \frac{\pi\sqrt{y}}{2\cos\pi s}\left(J_{s-\frac{1}{2}}(y) - J_{\frac{1}{2}-s}(y)\right). \end{aligned} \tag{4.13}$$

Their Wronskian is equal to -1. Hence a resolvent kernel R_s^+ of $y^{-2}D_s^+$ is given by

$$R_s^+(v,y) = \begin{cases} j_{s-\frac{1}{2}}(v)k_{s-\frac{1}{2}}(y), & v \le y, \\ j_{s-\frac{1}{2}}(y)k_{s-\frac{1}{2}}(v), & v \ge y. \end{cases} \tag{4.14}$$

To eliminate the separation of cases, we put

$$\begin{aligned} C_s(v,y) &= \frac{1}{i}\int_{(0)} \frac{r}{r^2 - \left(s-\frac{1}{2}\right)^2} \sin(\pi r)\, k_r(v)k_r(y)dr \\ &= \pi i\sqrt{v}\int_{(0)} \frac{r}{r^2 - \left(s-\frac{1}{2}\right)^2} J_r(v)k_r(y)dr. \end{aligned} \tag{4.15}$$

The power series expansion of $J_{\pm r}$ gives that the integrand is $\ll (|r|^2+1)^{-1}$, and the integral is absolutely convergent. Move the line of integration to $\operatorname{Re} r = 2Q$, with a large $Q \in \mathbb{N}$. We have, for $2Q > \operatorname{Re} s > \frac{1}{2}$, $v \le y$,

$$\begin{aligned} C_s(v,y) = \pi^2 R_s^+(v,y) &+ 2\pi^2\sqrt{vy}\sum_{q=1}^{Q} \frac{2q-1}{(2q-1)^2 - \left(s-\frac{1}{2}\right)^2} \\ &\times J_{2q-1}(v)J_{2q-1}(y) + \pi i\sqrt{v}\int_{(2Q)} \cdots dr. \end{aligned} \tag{4.16}$$

Again by the series expansion of $J_{\pm r}$, this integral is $\ll (v/y)^{2Q}/Q$, which vanishes as $Q \to \infty$; and we have

$$\begin{aligned} R_s^+(v,y) = &\frac{1}{\pi^2 i}\int_{(0)} \frac{r}{r^2 - \left(s-\frac{1}{2}\right)^2} \sin(\pi r)k_r(v)k_r(y)dr \\ &- 2\sqrt{vy}\sum_{q=1}^{\infty} \frac{2q-1}{(2q-1)^2 - \left(s-\frac{1}{2}\right)^2} J_{2q-1}(v)J_{2q-1}(y). \end{aligned} \tag{4.17}$$

The same holds for $v \ge y$ as well, because of symmetry.

Now, we have, for a smooth function f which is compactly supported in $(0,+\infty)$,

$$v^2 f(v) = D_s^+ R_s^+ f(v). \tag{4.18}$$

Insert the last expression for the resolvent kernel, change the order of integration, and take the operator D_s^+ inside the outer integral. In the resulting identity replace $f(v)$, $f(y)$ by $f(v)v^{-3/2}$, $f(y)y^{-3/2}$, respectively. Then we obtain

$$\begin{aligned} f(v) = &- i \int_{(0)} \frac{r}{\cos(\pi r)} J_{2r}(v) \int_0^\infty \frac{J_{-2r}(y) - J_{2r}(y)}{\sin(\pi r)} f(y) \frac{dy}{y} dr \\ &+ 2 \sum_{q=1}^\infty (2q-1) J_{2q-1}(v) \int_0^\infty f(y) J_{2q-1}(y) \frac{dy}{y}, \end{aligned} \tag{4.19}$$

which ends our proof of (3.1) essentially.

What remains is to check if f can be as in (2.10); and it depends on the nature of convergence on the right side or that of the image Kf. For this sake, let f^* be the Mellin transform of f satisfying (2.10). We have, for $r \in \mathbb{R}i$,

$$\mathrm{K}f(r) = -i \int_{(-\frac{1}{4})} f^*(s) \int_0^\infty \frac{J_{2r}(x) - J_{-2r}(x)}{\sin(\pi r)} x^{-s-1} dx ds; \tag{4.20}$$

the necessary absolute convergence is endorsed by a well-known asymptotic expansion for J-Bessel functions [16, eq. (1) on p. 199]. Thus we have, by [*ibid*, eq. (1) on p. 391],

$$\mathrm{K}f(r) = \frac{2}{\pi i} \int_{(\frac{1}{8})} 2^{-2s} \cos(\pi s) \Gamma(s+r) \Gamma(s-r) f^*(-2s) ds. \tag{4.21}$$

Analogously we have, for any $q \in \mathbb{N}$,

$$\mathrm{K}f\left(q - \tfrac{1}{2}\right) = \int_{(\frac{1}{8})} 2^{-2s} \frac{\Gamma\left(q + s - \frac{1}{2}\right)}{\Gamma\left(q - s + \frac{1}{2}\right)} f^*(-2s) ds. \tag{4.22}$$

Since (2.10) implies that $f^*(s)$ is regular for $|\mathrm{Re}\, s| < \frac{1}{4}A$, and $\ll (|s|+1)^{-\frac{1}{4}A}$ there, appropriate shifts of the contours in the last two integrals yield that Kf satisfies (2.3) $\cap$ (2.5) amply, and thus (4.19) holds well via a usual procedure of approximation. This finishes our proof of (3.1) under (2.10). In particular, the sum formula (2.8) with $h = \mathrm{K}f$ gives rise to (2.13), since (2.14) has been already proved. Hence we have proved the Kuznetsov inversion of the Spectral–Kloosterman sum formula, on (2.10).

5 Proof (2)

We turn to the opposite direction of operations, that is, KL or rather KB. We shall try to extend [12, Lemma 2.10] that concerns the reverse of the minus case in (4.12).

Let h satisfy (2.3) but not necessarily (2.5). By definition, we have, for $\nu \in \mathbb{R}i$,

$$\mathrm{KB}h(\nu) = -i \int_0^\infty \frac{J_{-2\nu}(x) - J_{2\nu}(x)}{\sin(\pi\nu)} \int_{(0)} \frac{r}{\cos(\pi r)} J_{2r}(x) h(r) dr \frac{dx}{x}. \tag{5.1}$$

The image Bh might not satisfy (2.10), but K is applicable to it. In fact, shifting the contour of the inner-integral to $\left(\frac{1}{8}\right)$, we see that the resulting double integral is absolutely convergent; note that we need (4.4) for $J_{2r}(x)$ and the asymptotic expansion for $J_{\pm 2\nu}(x)$ as $x \to \infty$. With this observation, we consider then

$$\begin{aligned} &M(s,\nu) \\ &= -i \int_0^\infty \frac{J_{-2\nu}(x) - J_{2\nu}(x)}{\sin(\pi\nu)} \left(\int_{\left(\frac{1}{8}\right)} \frac{r}{\cos(\pi r)} J_{2r}(x) h(r) dr \right) x^{2s-1} dx. \end{aligned} \tag{5.2}$$

When $|\mathrm{Re} s| < \frac{1}{8}$, this is absolutely and uniformly convergent; thus $M(s,\nu)$ is regular there, and

$$M(0,\nu) = \mathrm{KB}h(\nu), \quad \nu \in \mathbb{R}i. \tag{5.3}$$

Then, we assume that $0 < \mathrm{Re}\, s < \frac{1}{8}$. Exchanging the order of integration, we have

$$\begin{aligned} &M(s,\nu) \\ &= \frac{1}{i\sin(\pi\nu)} \int_{\left(\frac{1}{8}\right)} \frac{r}{\cos(\pi r)} h(r) \int_0^\infty J_{2r}(x)(J_{-2\nu}(x) - J_{2\nu}(x)) x^{2s-1} dx dr. \end{aligned} \tag{5.4}$$

The inner integral can be computed explicitly using the Mellin–Parseval reasoning or rather invoking [16, eq. (2) on p. 403]; that is,

$$\begin{aligned} &\int_0^\infty J_{2r}(x) J_{2\nu}(x) x^{2s-1} dx \\ &= \frac{2^{2s-1}\Gamma(1-2s)\Gamma(r+s+\nu)}{\Gamma(1-s-r+\nu)\Gamma(r-s+\nu+1)\Gamma(r-s-\nu+1)}. \end{aligned} \tag{5.5}$$

Inserting this into (5.4), we get

$$\begin{aligned} &M(s,\nu)\\ &=\frac{2^{2s}}{\pi i}\Gamma(1-2s)\int_{(\frac{1}{8})}\frac{rh(r)\cos(\pi(r+s))\Gamma(r+s+\nu)\Gamma(r+s-\nu)}{\cos(\pi r)\Gamma(r-s+\nu+1)\Gamma(r-s-\nu+1)}dr. \end{aligned} \tag{5.6}$$

We move the contour to $\left(-\frac{1}{4}\right)$. Computing the residues at $r=-s\pm\nu$ and letting s tend to 0 in the result, we have

$$M(0,\nu)=2h(\nu)+\frac{1}{\pi i}\int_{(-\frac{1}{4})}\frac{rh(r)}{r^2-\nu^2}dr. \tag{5.7}$$

Moving further the contour to $\left(\frac{1}{4}\right)$, we see that this integral is equal to $-i\pi h(\nu)$. Hence we find that

$$M(0,\nu)=h(\nu),\quad \nu\in\mathbb{R}i. \tag{5.8}$$

This and (5.3) give (3.5) under (2.3).

In passing, we observe that analytic continuation of (5.5) implies

$$\int_0^\infty J_{2q-1}(x)J_{2\nu}(x)\frac{dx}{x}=\frac{(-1)^{q-1}}{2\pi}\frac{\cos(\pi\nu)}{\left(q-\frac{1}{2}\right)^2-\nu^2}. \tag{5.9}$$

Since this is even in ν, we have readily

$$\mathrm{KB}h\left(q-\tfrac{1}{2}\right)=0,\quad q\in\mathbb{N}. \tag{5.10}$$

Also, the equality $\mathrm{KB}h(\nu)=h(\nu)$ itself extends off the imaginary axis, to a certain extent. This is not relevant to our present purpose but to the situation where one has to take into account the possible existence of the contribution of complementary series representations; then the line of integration in the inner integral of (5.2) has to be adapted, provided the vertical strip in (2.3) is widened appropriately.

It remains to derive (2.4) from (2.12) on the assumption (2.3). To this end, let $\beta>0$ be arbitrary and $h_\beta(r)=h(r)\exp(\beta r^2)$; note that $h_\beta(r)\ll\exp(-\beta|r|^2)$. Also, let $Q\in\mathbb{N}$ be arbitrary and put

$$\tilde{\mathrm{B}}h_\beta(x)=\mathrm{B}h_\beta(x)-\sum_{q=1}^{Q}\gamma_{q,\beta}J_{2q-1}(x), \tag{5.11}$$

where $\gamma_{q,\beta}$ are to satisfy

$$\begin{aligned}&\sum_{q=1}^{Q}(-1)^{q-1}\left(q-\tfrac{1}{2}\right)^{2k}\gamma_{q,\beta}\\&=\frac{1}{2\pi i}\int_{(0)}h_\beta(r)r^{2k+1}\tan(\pi r)dr,\quad 0\le k\le Q-1,\end{aligned}\tag{5.12}$$

which is obviously soluble. Later in the present section, it is proved that as $x\to+\infty$

$$\left(\frac{d}{dx}\right)^a\tilde{\mathrm{B}}h_\beta(x)\ll x^{-\frac{1}{4}Q},\quad a\le\frac{1}{4}Q,\tag{5.13}$$

provided Q is large. We then put $f(x)=\omega_\alpha(x)\tilde{\mathrm{B}}h_\beta(x)$, where $\omega_\alpha(x)=\exp(-\alpha/x)$, $\alpha>0$. On noting that the left side of (5.13) is $\ll x^{\frac{1}{2}-a}$ as $x\to 0^+$ and that Q is arbitrary, this f satisfies (2.10) and we can apply (2.12) as it is supposed now to hold. We shall consider the situation as $\alpha\to 0^+$ and $\beta\to 0^+$.

The Kloosterman-sum side, i.e.,

$$\sum_{\ell=1}^{\infty}\frac{1}{\ell}S(m,n;\ell)\mathrm{P}\omega_\alpha\tilde{\mathrm{B}}h_\beta\left(\frac{4\pi}{\ell}\sqrt{mn}\right)\tag{5.14}$$

is easy to handle. In fact, we have seen that $\tilde{\mathrm{B}}h_\beta(x)$ decays appropriately at both ends of $(0,+\infty)$, whence for $q\in\mathbb{N}$

$$\left|\mathrm{K}\omega_\alpha\tilde{\mathrm{B}}h_\beta\left(q-\tfrac{1}{2}\right)\right|\le 4\pi\int_0^\infty\left|\tilde{\mathrm{B}}h_\beta(x)\right|\frac{dx}{x}\tag{5.15}$$

is finite, which means in turn that the sum (5.14) converges uniformly for $\alpha\ge 0$. Since (5.10) gives that

$$\mathrm{P}\tilde{\mathrm{B}}=\mathrm{PB}=\mathrm{B},\tag{5.16}$$

the value of (5.14) at $\alpha=0$ is equal to the Kloosterman-sum part of $(2.4)_\beta$, the result of replacing h by h_β in (2.4).

On the other hand, we have, invoking the Mellin transform of ω_α,

$$\mathrm{K}\omega_\alpha\tilde{\mathrm{B}}h_\beta(\nu)=\frac{1}{\pi i}\int_{(\varepsilon)}\alpha^{-2s}\Gamma(2s)\tilde{M}_\beta(s,\nu)ds,\quad \nu\in\mathbb{R}i,\tag{5.17}$$

where

$$\begin{aligned}\tilde{M}_\beta(s,\nu) = M_\beta(s,\nu) - \frac{2^{2s-1}}{\Gamma(2s)\cos(\pi s)} \sum_{q=1}^{Q} (-1)^q \gamma_{q,\beta} \\ \times \frac{\Gamma\left(q+s+\nu-\frac{1}{2}\right)\Gamma\left(q+s-\nu-\frac{1}{2}\right)}{\Gamma\left(q-s+\nu+\frac{1}{2}\right)\Gamma\left(q-s-\nu+\frac{1}{2}\right)}.\end{aligned} \tag{5.18}$$

It is expedient to observe here that $M_\beta(s,\nu)$ is of polynomial order in $|s|+|\nu|$, as far as $|\mathrm{Re}\, s| < \frac{1}{8}$ and $\nu \in \mathbb{R}i$. In fact, returning to $(5.6)_\beta$ with the same convention as above, we apply Stirling's formula and the trivial bound for trigonometrical functions. We may assume without loss of generality that $r, s, \nu \in \mathbb{R}i$; then the argument of the exponentiated factor is

$$\begin{aligned}&-\frac{\pi}{2}\left(|r+s+\nu|+|r+s-\nu|-|r-s+\nu|-|r-s-\nu|\right) \\ &\qquad\qquad + \pi\left(|r+s|-|r|-|s|\right) \\ &= -\pi\left\{\max\left(|r+s|,|\nu|\right)-\max\left(|r-s|,|\nu|\right)\right\} \\ &\qquad\qquad + \pi\left(|r+s|-|r|-|s|\right) \le 0,\end{aligned} \tag{5.19}$$

and the assertion follows. Thus we see immediately that the integrand of (5.17) decays exponentially in $|s|$, while being of polynomial order in $|\nu|$. With this, shifting the contour to $(-\varepsilon)$, we get the residue $M_\beta(0,\nu) = h_\beta(\nu)$, which gives rise to the terms of $(2.4)_\beta$ other than the Kloosterman part. For a large ν, the new integral is truncated to the segment $[-\varepsilon - |\nu|^\varepsilon i, -\varepsilon + |\nu|^\varepsilon i]$, and the integral $(5.6)_\beta$ with s in this range to the segment $\left[\frac{1}{8} - |\nu|^\varepsilon i, \frac{1}{8} + |\nu|^\varepsilon i\right]$. The error thus caused is clearly $\ll \alpha^{2\varepsilon}\exp(-|\nu|^\varepsilon)$, and Stirling's formula implies that the remaining part is $\ll \alpha^{2\varepsilon}|\nu|^{-2-4\varepsilon}$, where implicit constants may depend on β. Since we have (1.12), the relevant contribution vanishes as $\alpha \to 0^+$. Hence we have derived $(2.4)_\beta$ from (2.12) under the condition (2.3). Invoking $\mathrm{B}h(x) \ll x^{\frac{1}{2}+\varepsilon}$ with x close to 0 and taking limit as $\beta \to 0^+$, we end the proof of our theorem, provided (5.13) is confirmed.

Now, to prove (5.13) we adopt an argument employed in [14, Section 3]. In what follows, we let g stand for h_β; thus g is regular in $|\mathrm{Re}\, r| \le \frac{1}{4} + \varepsilon$ and decays wildly there. We observe first that

$$\mathrm{B}g(x) = -\frac{1}{(2\pi)^2}\int_{(-\frac{1}{8})} \hat{g}(s)\left(\frac{x}{2}\right)^{-2s} ds, \tag{5.20}$$

where $\hat{g}(s)$ is understood to be an analytic continuation of

$$\hat{g}(s) = \int_{(0)} \frac{\Gamma(s+r)}{\Gamma(1-s+r)} \frac{rg(r)}{\cos(\pi r)} dr, \quad \mathrm{Re}\, s > 0. \tag{5.21}$$

In fact, we may shift the latter contour to $\left(\frac{1}{4}\right)$, getting a continuation to $\mathrm{Re}\, s > -\frac{1}{4}$. We insert the result into the right side of (5.20); the double integral is absolutely convergent, and exchanging the order of integration we obtain the identity (5.20) via [16, eq. (7) on p. 192].

Assuming that $\mathrm{Re}\, s > 0$ and $\mathrm{Im}\, s > 0$, we move the last contour to $\mathcal{L}$ which is the result of connecting, with straight lines, the points $\frac{1}{4} - \infty i$, $\frac{1}{4} - ci$, $-\frac{1}{4} + ci$, $-\frac{1}{4} + \infty i$ in this order, where $c > 0$ is arbitrary. An analytic continuation of $\hat{g}(s)$ to the right of $\mathcal{L}$ is obtained. On noting that g is even and $\mathcal{L}$ symmetric, we have

$$\hat{g}(s) = -\frac{1}{\pi} \cos(\pi s) \int_{\mathcal{L}} \Gamma(s+r)\Gamma(s-r) g(r) r \tan(\pi r) dr. \tag{5.22}$$

Let us suppose that $s = \sigma + ti$ is such that t tends to $+\infty$ while $\sigma > -\frac{1}{4}$ remains bounded. Then an appropriate deform of $\mathcal{L}$ and a truncation give

$$\hat{g}(s) = -\frac{1}{\pi} \cos(\pi s) \int_{-t^{\varepsilon} i}^{t^{\varepsilon} i} rg(r) \tan(\pi r) \Gamma(s+r)\Gamma(s-r) dr + O\left(e^{-t^{\varepsilon}}\right). \tag{5.23}$$

Here we have, by Stirling's formula,

$$\begin{aligned} \Gamma(s+r)\Gamma(s-r) = &-2\pi i t^{2\sigma-1+2it} \exp(\pi i \sigma - 2it - \pi t) \\ &\times \left\{ 1 + \sum_{k=1}^{Q-1} t^{-k} p_k(r^2) + O\left(\left((|r|+1)^2 t^{-1} \right)^Q \right) \right\}, \end{aligned} \tag{5.24}$$

with any fixed $Q \in \mathbb{N}$, where the polynomial $p_k(\xi)$ is of degree k in ξ; p_k depends on σ, and the implicit constant on σ and Q. Thus we have

$$\begin{aligned} \hat{g}(s) = & i t^{2\sigma-1+2it} e^{-2it} \\ &\times \int_{(0)} rg(r) \tan(\pi r) \left\{ 1 + \sum_{k=1}^{Q-1} t^{-k} p_k\left(r^2\right) \right\} dr + O(t^{2\sigma-Q}). \end{aligned} \tag{5.25}$$

By (5.12), we may write this as

$$\hat{g}(s) = 2\pi t^{2\sigma-1+2it} e^{-2it} \sum_{q=1}^{Q} (-1)^q \gamma_{q,\beta} \left\{ 1 + \sum_{k=1}^{Q-1} t^{-k} p_k \left(\left(q - \tfrac{1}{2} \right)^2 \right) \right\} + O(t^{2\sigma-Q}), \tag{5.26}$$

and, by (5.24),

$$\hat{g}(s) = 2\pi i \sum_{q=1}^{Q} \gamma_{q,\beta} \frac{\Gamma\left(q+s-\frac{1}{2}\right)}{\Gamma\left(q-s+\frac{1}{2}\right)} + O\left((|s|+1)^{2\mathrm{Re}\, s - Q} \right), \tag{5.27}$$

which ends the approximation of $\hat{g}(s)$ for the case $\mathrm{Im}\, s > 0$; the case $\mathrm{Im}\, s < 0$ is analogous, and the result is the same as (5.27). This holds with s large in any fixed vertical strip contained in the half plane $\mathrm{Re}\, r > -\frac{1}{4}$.

On the other hand, we have, by (5.11), (5.20), and [16, eq. (7) on p. 192],

$$\tilde{\mathrm{B}} h_\beta(x) = -\frac{1}{(2\pi)^2} \int_{(-\frac{1}{8})} \left\{ \hat{g}(s) - 2\pi i \sum_{q=1}^{Q} \gamma_{q,\beta} \frac{\Gamma\left(q+s-\frac{1}{2}\right)}{\Gamma\left(q-s+\frac{1}{2}\right)} \right\} \left(\frac{x}{2} \right)^{-2s} ds. \tag{5.28}$$

We may shift the contour to $\mathrm{Re}\, s = \frac{1}{4}Q$, because of (5.27); and, differentiating a times inside the integral, we obtain (5.13) and end the proof of the theorem.

6 Concluding remarks

(1) Kuznetsov's proof of (2.12)–(2.13) is in fact an ingenious transformation of a prototype of (2.4) which is a spectral decomposition in $L^2(\Gamma\backslash G/K)$ of inner products of real analytic Poincaré series of weight 0. Thus the appearance in (2.13) of the contribution of holomorphic cusp forms or the discrete series representations was a mysterious phenomenon, in the sense that his method did not provide any structural background for the fact, since $L^2(\Gamma\backslash G/K)$ lacks holomorphic elements. However, as is developed in [2], if we employ instead Poincaré series of a non-zero even integral weight

coupled with the relevant L^2-space, then holomorphic cusp forms play in fact a rôle. One may vary weights and try to view all, and then comes naturally to the space $L^2(\Gamma\backslash G)$ as a proper working place. The maps L and Δ convey such a procedure of extending (2.4) to reach (2.13) via (2.8).

(2) There exists, however, an alternative, direct approach to (2.13) due to J.W. Cogdell and I. Piatetski-Shapiro [7]. Their argument starts with a construction of a particular Γ-automorphic Poincaré series on G whose Fourier expansion is closely related to the Kloosterman part of (2.13). Thus the discrete series representations come into play in a way as natural and indispensable as unitary principal series representations. Technically, the main subject of [7] was to compute projections of the Poincaré series to irreducible subspaces of $L^2(\Gamma\backslash G)$ in such a way that the resulting spectral decomposition is explicit with respect to the seed function which is related to f, that is, the task was to fix the kernel of the relevant integral transform of f. To this end, a use was made of the theory of the Kirillov model; and in particular the Bessel transform K got a clear-cut structural explanation as being an analytic expression of the action in the model of the Weyl element w (see (1.7)) and thus essentially the same as the Bessel function of representations in the sense of [8] and as the Mellin inversion of the Γ-factor of the local functional equation introduced in [9] (see also [13]). In a similar context, it is worth stressing that the theory has yielded a new approach [6] to the fourth moment of the Riemann zeta-function, which is far more transparent than the argument developed by the author in [12], and suggests further developments (see [15]).

(3) However, there has remained a basic problem, that is, the derivation of (2.4) from (2.12) or (2.13); such is highly desirable, as (2.4) is equally fundamental in applications. Section 5 solves this and closes the discussion of [7] as far as the two sum formulas of Kloosterman sums are concerned; to achieve this was a motivation of the present work. In other words, both Bruggeman–Kuznetsov's and Cogdell–Piatetski-Shapiro's approaches come to the same conclusion, if we restrict ourselves to those arithmetic configurations represented by (1.1).

(4) The formula (4.19) is a completion of the Webb–Kapteyn theory of Neumann expansions with J-Bessel functions of odd integral order. See [16, Section 16.4], where a sceptical opinion about the value of relevant works is expressed. This episode makes the inversion formula (4.19) just

more significant, for it is one of the most basic implements which have brought about recent profound changes in Analytic Number Theory.

(5) Most probably our discussion extends to the situation with $G = \mathrm{PSL}_2(\mathbb{C})$ and Γ being Bianchi groups. The extension of the sum formulas (2.8) and (2.13) in the case of the Picard group $\Gamma = \mathrm{PSL}_2(\mathbb{Z}[i])$ is, in fact, fully discussed in [5] but the analogue of the latter formula is deduced from that of the former; thus the procedure is similar to Bruggeman–Kuznetsov's. It is worth extending the argument of [6]. A preliminary work for this purpose has already been done in [4] and [15]. Further extensions to the situation like that dealt with in [3] should be of considerable interest. Lastly, we mention that an inversion of the explicit formula for the fourth moment of the Riemann zeta-function might be discussed along the lines developed in Section 5.

Acknowledgements. The present work incorporates expert comments provided generously by R.W. Bruggeman, which the author appreciates greatly. He is indebted also to the organizers of the conference to which this volume is attributed for the financial aid that made his participation possible.

References

[1] R.W. Bruggeman. Fourier coefficients of cusp forms. Invent. math., **45** (1978), 1–18.

[2] R.W. Bruggeman. Fourier Coefficients of Automorphic Forms. Springer LNM, vol. **865**, Springer-Verlag, Berlin etc 1981.

[3] R.W. Bruggeman and R.J. Miatello. Sum formula for SL_2 over a totally real number field. To appear in Memoirs AMS.

[4] R.W. Bruggeman and Y. Motohashi. A note on the mean value of the zeta and L-functions. XIII. Proc. Japan Acad., **78A** (2002), 87–91.

[5] R.W. Bruggeman and Y. Motohashi. Sum formula for Kloosterman sums and the fourth moment of the Dedekind zeta-function over the Gaussian number field. Functiones et Approximatio, **31** (2003), 7–76.

[6] R.W. Bruggeman and Y. Motohashi. A new approach to the spectral

theory of the fourth moment of the Riemann zeta-function. J. reine angew. Math., **579** (2005), 75–114.

[7] J.W. Cogdell and I. Piatetski-Shapiro. The Arithmetic and Spectral Analysis of Poincaré series. Perspectives in Math., vol. **13**, Academic Press, San Diego 1990.

[8] I.M. Gel'fand, M.I. Graev, and I.I. Pyatetski-Shapiro. Representation Theory and Automorphic Functions. W.B. Saunders Company, Philadelphia 1969.

[9] H. Jacquet and R.P. Langlands. Automorphic Forms on GL(2). Lecture Notes in Math., vol. **114**, Springer-Verlag, Berlin etc. 1970.

[10] N.V. Kuznetsov. Petersson hypothesis for forms of weight zero and Linnik hypothesis. Khabarovsk Complex Res. Inst. Acad. Sci. USSR. Preprint 1977. (Russian)

[11] N.V. Kuznetsov. Petersson hypothesis for parabolic forms of weight zero and Linnik hypothesis. Sums of Kloosterman sums. Math. USSR-Sb., **39** (1981), 299–342.

[12] Y. Motohashi. Spectral Theory of the Riemann Zeta-Function. Cambridge Tracts in Math., vol. **127**, Cambridge Univ. Press, Cambridge 1997.

[13] Y. Motohashi. A note on the mean value of the zeta and L-functions. XII. Proc. Japan Acad., **78A** (2002), 36–41.

[14] Y. Motohashi. A functional equation for the spectral fourth moment of the modular Hecke L-functions. Proc. MPIM-Bonn Special Activity on Analytic Number Theory, Bonn 2002, Bonner Math. Schrift., vol. **130** (2003), 19 pages.

[15] Y. Motohashi. Mean values of zeta-functions via the representation theory. To appear in Proc. Symp. Pure Math., AMS, Workshop on Multiple Zetas, Bretton Woods 2005.

[16] G.N. Watson. A Treatise on the Theory of Bessel Functions. Cambridge Univ. Press, Cambridge 1996.

Yoichi Motohashi
Department of Mathematics,
Nihon University,
Surugadai, Tokyo 101-8308,
Japan
Email: ymoto@math.cst.nihon-u.ac.jp

The Lindelöf Class of L-Functions

K. MURTY

We define a class of functions that contains the Selberg class and which has a natural ring structure.

0 Introduction

About 15 years ago, Selberg [3] formulated a class of L-functions having properties enjoyed by L-functions that arise in arithmetic and geometry. He formulated two key conjectures about this class. A few years later, Ram Murty [2] showed that these conjectures of Selberg have far reaching implications. In particular, he showed that the holomorphy of Artin L-functions belonging to solvable Galois extensions of $\mathbb{Q}$ would follow from these conjectures.

The Selberg class, while capturing many of the features of familiar L-functions, has several shortcomings. One of these is the lack of a ring structure. Ram Murty asked if one could formulate a larger class of L-functions that would have a ring structure. This note is a possible answer to this question. We define the Lindelöf class which contains the Selberg class. This class has a natural ring structure, though we lose the key properties of Euler product and functional equation enjoyed by elements of the Selberg class.

1 The Selberg class

Selberg [3] defined the class $\mathcal{S}$ of L-functions $F(s)$ satisfying the following properties:

(S1) Dirichlet series: can be expressed as a Dirichlet series

$$F(s) = \sum_{n=1}^{\infty} \frac{a_n(F)}{n^s}$$

in the region $Re(s) > 1$ satisfying the following properties:

(S2) Analytic continuation: for some non-negative integer m, $(s-1)^m F(s)$ extends to an entire function;

(S3) Functional equation: there exist non-negative real numbers Q and α_i, complex numbers μ_i with $Re(\mu_i) \geq 0$, and w satisfying $|w| = 1$ such that

$$\Phi(s) := Q^s \prod_i \Gamma(\alpha_i s + \mu_i) F(s) = w\overline{\Phi}(1 - \overline{s});$$

(S4) Euler product: there is an Euler product of the form

$$F(s) = \prod_p F_p(s)$$

where

$$\log F_p(s) = \sum_{k=1}^{\infty} \frac{b_{p^k}}{p^{ks}}$$

with $b_{p^k} = \mathbf{O}(p^{k\theta}), \theta < 1/2$;

(S5) Ramanujan hypothesis: for any $\epsilon > 0$, $a_n(F) = \mathbf{O}(n^{\epsilon})$.

The implied constants in the estimates of (S4) and (S5) depend on F. The parameter Q that appears above in (b) is usually referred to as the "conductor" of F and the degree d_F of F is defined as

$$d_F = 2 \sum_i \alpha_i.$$

It is an interesting conjecture in the Selberg theory that this degree is always an integer.

2 The Lindelöf class

The Selberg class $\mathcal{S}$ is closed under product but not under sums. In this section, our aim is to define a larger class that is closed under sums and products. This larger class will be defined in terms of growth conditions. Towards this end, we begin by defining the μ function.

2.1 The function $\mu_F(\sigma)$

For $F \in \mathcal{S}$, let us define the function μ_F as follows:

$$\mu_F(\sigma) \;=\; \inf\{\lambda \geq 0 \;:\; |F(s)| \;\ll\; (|s|+2)^{\lambda} \text{ for all } s \text{ with } Re(s) = \sigma\}.$$

Note that in this definition, the implied constant is independent of s. Since the Dirichlet series defining $F(s)$ converges absolutely for $Re(s) > 1$ (by (S5)), we have

$$\mu_F(\sigma) \;=\; 0 \;\text{ for } \sigma > 1. \tag{2.1}$$

By the functional equation (S3), we also get that

$$\mu_F(\sigma) \;\leq\; \frac{1}{2} d_F(1-2\sigma) \text{ for } \sigma < 0. \tag{2.2}$$

Indeed, by Stirling's formula

$$\left|\frac{\Gamma(\alpha_i(1-s)+\beta_i)}{\Gamma(\alpha_i s+\beta_i)}\right| \;\ll\; (\alpha_i(|t|+2))^{\alpha_i(1-2\sigma)}.$$

From the Phragmén Lindelöf theorem, for $c \leq \sigma \leq d$, we have

$$\mu_F(\sigma) \;\leq \mu_F(c)\frac{d-\sigma}{d-c} \;+\; \mu_F(d)\frac{\sigma-c}{d-c}. \tag{2.3}$$

Thus, for $0 \leq \sigma \leq 1$, we have

$$\mu_F(\sigma) \;\leq\; \frac{1}{2} d_F(1-\sigma). \tag{2.4}$$

2.2 Nonvanishing in a half plane

From (S1), it follows that F does not vanish in a half plane. Indeed, by (S5), the Dirichlet series defining F converges absolutely for $\sigma > 1$. Then, for $\sigma > 1$, we have

$$|F(s)| \geq 1 - \sum_{n\geq 2} \frac{|a_F(n)|}{n^\sigma}$$

and so for any $\sigma > \alpha > 1$, this is

$$|F(s)| \geq 1 - \frac{1}{2^{\sigma-\alpha}} \sum_{n\geq 2} \frac{|a_F(n)|}{n^\alpha} > 1 - \epsilon$$

if σ is sufficiently large, say $\sigma > \sigma_0$. Hence, $F(s) \neq 0$ in this half plane. It is clear that it is not essential to assume that the Dirichlet series converges absolutely for $\sigma > 1$. From (S1), it converges for $\sigma > 1$ and therefore converges absolutely for $\sigma > 2$. Also, it is not essential that $a_F(1) = 1$. We only need some non-zero Dirichlet coefficient and this exists if $F \neq 0$.

2.3 The number of zeros

Next, let us denote by $N_F(T)$ the number of zeros of $F(s)$ with $0 < \sigma < 1$ and $|t| \leq T$. From (S2), it follows that

$$N_F(T+1) - N_F(T) \ll \log T.$$

Indeed, let σ_0 be such that $F(s) \neq 0$ for $Re(s) > \sigma_0$. Set $n_F(r)$ to be the number of zeros of $F(s)$ in the circle of radius r with center $\sigma_0 + 1 + iT$. Take $R > e\sqrt{(\sigma_0+1)^2+1}$. Then, by Jensen's theorem

$$\int_0^R \frac{n_F(r)}{r} dr = \frac{1}{2\pi} \int_0^{2\pi} \log|F(\sigma_0+1+iT+Re^{i\theta})| d\theta - \log F(\sigma_0+1+iT).$$

In $\sigma_0 + 1 - R \leq \sigma \leq \sigma_0 + 1 + R$, we have for some $A = A(F)$, by (2.2) and (2.3),

$$|(s-1)^m F(s)| \ll_F (|s|+2)^A. \tag{2.5}$$

Hence, on the circle above

$$\log|F(\sigma_0 + 1 + iT + Re^{i\theta})| \leq A \log T + \mathbf{O}(1).$$

Thus,

$$\int_0^R \frac{n_F(r)}{r} dr \ll \log T.$$

In particular
(2.6)

$$N_F(T+1) - N_F(T) \le n_F(\sqrt{(\sigma_0+1)^2+1}) \le \int_0^R \frac{n_F(r)}{r} dr \ll \log T.$$

Hence, for some $c > 0$,

$$N_F(T) \le \frac{c}{2\pi} T \log T.$$

Define

$$\delta_F = \limsup_{T\to\infty} \frac{2\pi N_F(T)}{T \log T}.$$

For $F \in \sigma$, note that $\delta_F = d_F$.

2.4 The class $\mathcal{M}$

Let us define the class $\mathcal{M}$ to be the set of functions $F(s)$ satisfying the following conditions:

(M1) Dirichlet series: For $\sigma > 1$, $F(s)$ is given by the absolutely convergent Dirichlet series $\sum a_F(n)n^{-s}$.

(M2) Analytic continuation: F can be continued as a function of order 1, analytic apart possibly for a pole at $s = 1$.

(M3) Growth condition: The quantity $\mu_F(\sigma)/(1-2\sigma)$ is bounded for $\sigma < 0$.

(M4) Ramanujan hypothesis: $|a_F(n)| \ll n^\epsilon$ for any $\epsilon > 0$.

From the arguments of the previous paragraphs, we see that for an element $F \in \mathcal{M}$, the function μ_F and the constant δ_F are defined.

For an F in $\mathcal{M}$, let us define

$$c_F = \limsup_{\sigma<0} \frac{2\mu_F(\sigma)}{1-2\sigma}.$$

Note that $c_F \geq 0$. Moreover, for $0 \leq \sigma \leq 1$, we have

$$\mu_F(\sigma) \leq \frac{1}{2} c_F (1 - \sigma).$$

Example 1. Consider

$$F(s) = 1 - 2^{-s}.$$

For $\sigma < 0$, we have

$$\mu_F(\sigma) \leq \frac{|\sigma|}{\log \max(|\sigma|, 2)}$$

and in particular, $F \in \mathcal{M}$ and $c_F = 0$.

Example 2. By similar reasoning, any Dirichlet polynomial F is in $\mathcal{M}$ and has $c_F = 0$.

Example 3. If $F(s) = \zeta(s)$, the Riemann zeta function, then $c_F = 1$. Moreover, for any constant a, the function $F(s) = a + \zeta(s)$ is also in $\mathcal{M}$ and has $c_F = 1$.

The following is evident from the definition.

Proposition 1.

$$c_{FG} = c_F + c_G$$

and

$$c_{F+G} \leq \max(c_F, c_G).$$

From these, it is easy to see that $\mathcal{M}$ is a ring.

We also have a relation between the invariants c_F and δ_F.

Proposition 2. *For $F \in \mathcal{M}$, if $c_F = 0$ then $\delta_F = 0$.*

Proof. In (2.5), we can take $A = \epsilon$. Inserting this into (2.6), we see that

$$N_F(T+1) - N_F(T) = \mathbf{o}(\log T).$$

Hence, $N_F(T) = \mathbf{o}(T \log T)$ and so $\delta_F = 0$.

Now we are ready to prove the following result.

Proposition 3. *Suppose $F \in \mathcal{M}$. Then $c_F \geq 1$ or $c_F = 0$.*

The proof is modeled on earlier work of Bochner, Selberg and Conrey-Ghosh [1].

Proof. For $\sigma > 1$, we have

$$\sum_{n=1}^{\infty} a_F(n)e^{-nx} = \frac{1}{2\pi i}\int_{(2)} F(s)\Gamma(s)x^{-s}ds.$$

By (M3) and convexity, F has polynomial growth in $|t|$ in vertical strips. Thus, moving the line of integration to the left and taking into account the possible pole at $s = 1$ of $F(s)$, as well as the poles of $\Gamma(s)$ at $s = 0, -1, -2, \cdots$, we find

$$\sum_{n=1}^{\infty} a_F(n)e^{-nx} = \frac{P(\log x)}{x} + \sum_{n=0}^{\infty} \frac{F(-n)}{n!}(-x)^n \tag{2.7}$$

where P is a polynomial. By (M3),

$$|F(-n)| \ll n^{\frac{1}{2}c_F(1+2n)+\epsilon}.$$

As

$$n! \sim \frac{1}{\sqrt{2\pi n}}\left(\frac{n}{e}\right)^n,$$

the n-th term in the sum on the right of (2.7) is

$$\ll n^{\frac{1}{2}c_F(1+2n)+\epsilon}\sqrt{n}\left(\frac{e|x|}{n}\right)^n = n^{\frac{c_F+1}{2}+\epsilon}\left(\frac{e|x|n^{c_F}}{n}\right)^n.$$

If $c_F < 1$, then the series converges absolutely for all values of x. Hence, the left hand side of (2.7) is analytic in $\mathbb{C}\backslash\{x \le 0 : x \in \mathbb{R}\}$. But it is periodic (with period $2\pi i$) and so it converges for all x. Then, the function

$$f(z) = \sum a_F(n)e^{-nz}$$

is entire and for any real x and any $k \ge 0$, we have

$$a_F(n)e^{-nx} = \frac{1}{2\pi}\int_0^{2\pi} f(x+iy)e^{iny}dy \ll n^{-k}$$

by applying integration by parts repeatedly. Taking $x = 1/n$, we deduce that

$$a_F(n)n^k \ll 1 \text{ for all } k \geq 1.$$

Thus, the series

$$\sum_{n=1}^{\infty} \frac{a_F(n)}{n^s}$$

converges absolutely for all values of s. Hence, $\mu_F(\sigma) = 0$ for all σ.

We note that the above argument also includes a proof of the following result.

Proposition 4. *We have $c_F = 0$ if and only if for all $k \geq 0$, $a_F(n)n^k \ll 1$.*

2.5 The Lindelöf class

Denote by $\mathcal{M}_0$ the subset of $\mathcal{M}$ with $c_F = 0$. It is, in fact, a subring and this can be seen from Proposition 4. It is clear that if $F, G \in \mathcal{M}_0$ then $F - G \in \mathcal{M}_0$. Moreover, if $F, G \in \mathcal{M}_0$ then for any $k \geq 0$, we have

$$a_{FG}(n)n^k = \sum_{d|n} a_F(d)d^k a_G(\frac{n}{d})(\frac{n}{d})^k \ll \sum_{d|n} 1 \ll n^\epsilon$$

for any $\epsilon > 0$. It follows that for any $k \geq 0$,

$$a_{FG}(n)n^k \ll 1$$

and so $FG \in \mathcal{M}_0$.

Now let $\mathcal{M}_{00}$ denote the non-zero elements in $\mathcal{M}_0$. It is a multiplicative subset. The Lindelöf class is defined as the localization

$$\mathcal{L} = \mathcal{M}_{00}^{-1}\mathcal{M}.$$

In particular, elements of $\mathcal{L}$ are analytic for $Re(s) > 1$. However, for $Re(s) \leq 1$, they are meromorphic and may have infinitely many poles. We have inclusions

$$\mathcal{S} \hookrightarrow \mathcal{M} \hookrightarrow \mathcal{L}.$$

Note that $\mathcal{S} \cap \mathcal{M}_0 = \{1\}$. Also, note that by construction, the elements of $\mathcal{M}_{00}$ are units in $\mathcal{L}$.

For an element $G \in \mathcal{L}$, we define c_G as follows. We can write $G = F/u$ where $F \in \mathcal{M}$ and $u \in \mathcal{M}_{00}$. We set

$$c_G = c_F.$$

If we can write $G = H/v$ for some other $H \in \mathcal{M}$ and $v \in \mathcal{M}_{00}$, and $c_H > c_F$, then $c_{Fv-Hu} = c_H$. On the other hand, by the definition of localization, $c_{Fv-Hu} = 0$ and this is a contradiction. Note, in particular, that the only elements G of $\mathcal{L}$ with $c_G = 0$ are units. Conversely, if G is a unit of $\mathcal{L}$ then $c_G = 0$.

2.6 Lindelöf primitive functions

An element $F \in \mathcal{L}$ that is not a unit is *Lindelöf primitive* if it cannot be factored as F_1F_2 with $F_1, F_2 \in \mathcal{L}$ with $c_{F_1}, c_{F_2} < c_F$ and F_1, F_2 non-units.

Proposition 5. *Every element in $\mathcal{L}$ can be factored as a product of a unit and Lindelöf primitive elements.*

Proof. Consider an $F \in \mathcal{L}$ that is not a unit. Then $c_F \geq 1$. If $c_F = 1$, then F is Lindelöf primitive so there is nothing to prove. Assume that we have proved the result for all $G \in \mathcal{L}$ with $c_G < c_F$. If F is not Lindelöf primitive, then it can be written as F_1F_2 with $c_{F_1}, c_{F_2} < c_F$ and F_1, F_2 nonunits. By the induction hypothesis, F_1 and F_2 can be written as a product of primitive elements. The result follows.

The natural question to ask next is whether this factorization is unique, and if so what consequences it would have.

References

[1] B. Conrey and A. Ghosh, On the Selberg class of Dirichlet series: small degrees, *Duke Math. J.*, **72**(1993), 673-693.

[2] M. Ram Murty, Selberg's conjectures and Artin L-functions, *Bull. Amer. Math. Soc.*, **31**(1994), 1-15.

[3] A. Selberg, Old and new conjectures and results about a class of Dirichlet series, *Proc. Amalfi Conf. on Analytic Number Theory*, ed. E. Bombieri et.al., pp. 367-385. (See also Collected Papers, Vol. II, pp. 47-63, Springer-Verlag, 1991.)

V. Kumar Murty
Department of Mathematics,
University of Toronto,
40 St. George Street, Toronto,
Canada
Email: murty@math.toronto.edu

A Proof of the Riemann Hypothesis for the Weng Zeta Function of Rank 3 for the Rationals

M. SUZUKI

0 Introduction

Let F be an algebraic number field. Recently, L. Weng introduced zeta functions of rank n associated to F as a generalization of the Iwasawa-Tate zeta integral from the Arakelov geometric point of view. In the article, we call them the Weng zeta function of rank n. In the case of rank one, the Weng zeta function coincide with the Dedekind zeta function of F. The background and precise definition of Weng zeta functions were published in [4, sec.B.4]. One remarkable fact for Weng zeta functions is that we can prove the Riemann hypothesis in the case of rank 2. It had been done by the author and J.C. Lagarias in [2] for the rational number field and was extended to the case of general number field F by Weng in [4, sec.C.4]. The proof of the Riemann hypothesis for a zeta function of rank 2 depends on the explicit expression for it.

In this spring, Weng obtained an explicit expression for the zeta function of rank 3 in the case of $F = \mathbb{Q}$. It is given in [5] this volume. By using his explicit formula, the author proved the Riemann hypothesis for the zeta function of rank 3 over the rational number field. In the article we give the proof of the Riemann hypohesis for it and the idea of the proof in self-contained fashion, as far as possible.

The zeta function $\widehat{\zeta}_{F,3}(s)$ of rank 3 is obtained by an integral of the completed Epstein zeta function of rank 3 over a moduli space of semi-stable lattices of rank 3 over F. In detail, see [5, sec.5.3, chap.9] (in [5] our $\widehat{\zeta}_{F,3}(s)$ is denoted by $\xi_{F,3}(s)$). In the case $F = \mathbb{Q}$, Theorem 4 in [5] assert that the explicit expression for $\widehat{\zeta}_{\mathbb{Q},3}(s)$ is given by

$$\begin{aligned}\widehat{\zeta}_{\mathbb{Q},3}(s) = &\Big(\frac{\widehat{\zeta}(2)}{3(s-1)} - \frac{1}{2(3s-2)}\Big)\widehat{\zeta}(3s) \\ &+ \frac{1}{9}\Big(\frac{1}{s-1} - \frac{1}{s}\Big)\widehat{\zeta}(3s-1) \\ &- \Big(\frac{\widehat{\zeta}(2)}{3s} - \frac{1}{2(3s-1)}\Big)\widehat{\zeta}(3s-2),\end{aligned} \tag{0.1}$$

where $\widehat{\zeta}(s) = \pi^{-s/2}\Gamma(s/2)\zeta(s)$ and $\zeta(s)$ is the Riemann zeta-function. From the general theory of zeta functions of rank n, the function $\widehat{\zeta}_{F,3}(s)$ satisfies the functional equation $\widehat{\zeta}_{F,3}(s) = \widehat{\zeta}_{F,3}(1-s)$. Hence the Riemann hypothesis for $\widehat{\zeta}_{F,3}(s)$ is that all zeros of $\widehat{\zeta}_{F,3}(s)$ lie on the line $\mathrm{Re}\,(s) = 1/2$. Using (0.1) we can prove the Riemann hypothesis for $\widehat{\zeta}_{\mathbb{Q},3}(s)$.

Theorem 1. *The function $\widehat{\zeta}_{\mathbb{Q},3}(s)$ satisfies the Riemann hypothesis. That is, all zeros of $\widehat{\zeta}_{\mathbb{Q},3}(s)$ lie on the line* $\mathrm{Re}\,(s) = 1/2$ *which is the central line of the functional equation $\widehat{\zeta}_{\mathbb{Q},3}(s) = \widehat{\zeta}_{\mathbb{Q},3}(1-s)$.*

The article is organized as follows. In section 2, we explain the idea of the proof. In section 3, we state and prove Theorem 2 which is a easily version of Theorem 1. We describe the frame of the proof of Theorem 1 by proving Theorem 2 by using Lemma 1 and Lemma 2. In section 4, we give the proof of Theorem 1. It is proved by using Lemma 3 and Lemma 4 whose are modifications of Lemma 1 and Lemma 2. In section 5 we give the proof of Lemma 1 which is also used in the proof of Lemma 3. In section 6, we give the proof of Lemma 5 which is used in the proof of Lemma 4. In section 7, we comment on a possibility of applications of Theorem 1 and Theorem 2 to the original Riemann zeta function.

Acknowledgment. The author thanks professor Lin Weng for his helpful information on zeta functions of rank 3 and encouragement for the proof of the Riemann hypothesis.

1 The idea of the proof

For simplicity, we denote by $Z_3(s)$ the function $\widehat{\zeta}_{\mathbb{Q},3}(s)$:

$$
\begin{aligned}
Z_3(s) = {} & \widehat{\zeta}(2) \cdot \frac{1}{3s-3} \cdot \widehat{\zeta}(3s) - \widehat{\zeta}(2) \cdot \frac{1}{3s} \cdot \widehat{\zeta}(3s-2) \\
& + \frac{1}{3} \cdot \frac{1}{3s-3} \cdot \widehat{\zeta}(3s-1) - \frac{1}{3} \cdot \frac{1}{3s} \cdot \widehat{\zeta}(3s-1) \\
& + \frac{1}{2} \cdot \frac{1}{3s-1} \cdot \widehat{\zeta}(3s-2) - \frac{1}{2} \cdot \frac{1}{3s-2} \cdot \widehat{\zeta}(3s).
\end{aligned} \tag{1.1}
$$

By using the well-known functional equation $\widehat{\zeta}(s) = \widehat{\zeta}(1-s)$, we can directly check that $Z_3(s)$ has the functional equation $Z_3(s) = Z_3(1-s)$. Since $\widehat{\zeta}(s)$ is holomorphic except for two simple poles at $s = 0, 1$ with residues -1 and 1 respectively, the possible poles of $Z_3(s)$ are $s = 0,\ 1/3,\ 2/3,\ 1$. We can easily check that the possible poles $s = 1/3,\ 2/3$ are removable and $s = 0, 1$ are simple poles of $Z_3(s)$. To investigate the zeros of $Z_3(s)$, we consider the entire function

$$
\begin{aligned}
\xi_3(s) &= 3s(3s-1)(3s-2)(3s-3)Z_3(s) \\
&= \frac{\xi(2)}{2} \cdot (3s-2) \cdot \xi(3s) + \frac{\xi(2)}{2} \cdot (1-3s) \cdot \xi(3s-2) \\
&\quad + \xi(3s-1) + \frac{3}{2} \cdot s \cdot \xi(3s-2) + \frac{3}{2} \cdot (1-s) \cdot \xi(3s),
\end{aligned} \tag{1.2}
$$

where $\xi(s)$ is Riemann's xi-function $\xi(s) = s(s-1)\widehat{\zeta}(s)$.

To describe the idea of the proof of Theorem 1, we recall the proof of the Riemann hypothesis for the zeta function $\widehat{\zeta}_{\mathbb{Q},2}(s)$ of rank 2. The key of the proof was in the explicit expression

$$
\xi_2(s) = 2s(2s-1)(2s-2)\widehat{\zeta}_{\mathbb{Q},2}(s) = \xi(2s) - \xi(2s-1).
$$

The right-hand side has the properties

(A) $\xi(2s)$ and $\xi(2s-1)$ are related by the functional equation $\xi(2(1-s)) = \xi(2s-1)$,

(B) all zeros of $\xi(2s)$ lie in the strip $0 < \mathrm{Re}\,(s) < 1/2$.

By using these properties we derived $|\xi(2s-1)/\xi(2s)| < 1$ for $\mathrm{Re}(s) > 1/2$ and $|\xi(2s)/\xi(2s-1)| < 1$ for $\mathrm{Re}(s) < 1/2$. These inequality and (B) gave the Riemann hypothesis for $\widehat{\zeta}_{\mathbb{Q},2}(s)$. Here we note one more important property,

(C) $\xi(2s)$ has the functional equation $\xi(2(\frac{1}{2}-s)) = \xi(2s)$.

The role of (C) was hidden in the back of (B) in the proof of the Riemann hypothesis for $\widehat{\zeta}_{\mathbb{Q},2}(s)$. However the property (C) is the key of the generalization to the proof of rank 3 case.

The idea of our proof of Theorem 1 is to recast the right-hand side of (1.2) into the form $\xi_3(s) = Y(s) + Y(1-s)$ so that all zeros of $Y(s)$ are in some vertical strip on the left of the line $\mathrm{Re}(s) = 1/2$. To obtain such expression, we notice properties (A) and (C) in rank 2 case. If we ignore polynomial factors, the right-hand side of (1.2) consists of $\xi(3s)$, $\xi(3s-1)$ and $\xi(3s-2)$. Using functional equations

$$\xi(3(1-s)) = \xi(3s-2) \tag{1.3}$$

and

$$\xi(3(1-s)-1) = \xi(3s-1), \tag{1.4}$$

we can easily obtain an analogue of (A) for $\xi_3(s)$. However such expression has many possibilities. Here we notice the functional equations

$$\xi(3(\frac{2}{3}-s)) = \xi(3s-1) \tag{1.5}$$

and

$$\xi(3(\frac{4}{3}-s)-2) = \xi(3s-1). \tag{1.6}$$

These extra functional equations allow us to recast the right-hand side of (1.2) as $\xi_3(s) = Y(s) + Y(1-s)$ so that $Y(s)$ satisfies an analogue of (B). In this process, the extra functional equations (1.5) and (1.6) play a role of (C). At the time, we obtain Theorem 1 by a way similar to the proof of rank 2 case.

Now we put

$$h(s) = \{\frac{3}{2}(\xi(2)-1)s - (\xi(2)-\frac{3}{2})\}\xi(3s). \tag{1.7}$$

This $h(s)$ consists of all terms in the right-hand side of (1.2) containing $\xi(3s)$. We have

$$h(1-s) = \{\frac{3}{2}(\xi(2)-1)(1-s) - (\xi(2) - \frac{3}{2})\}\xi(3s-2). \tag{1.8}$$

This coincides to all terms in the right-hand side of (1.2) containing $\xi(3s-2)$. Further

$$h(\frac{2}{3}-s) + h(s-\frac{1}{3}) = (1-\frac{3}{2}(\xi(2)-1))\xi(3s-1). \tag{1.9}$$

Hence if we take

$$X(s) = h(s) + h((2/3)-s), \tag{1.10}$$

then we have

$$\xi_3(s) = X(s) + X(1-s) + \frac{\pi-3}{2}\xi(3s-1). \tag{1.11}$$

Here we used $\xi(2) = \pi/3$. Since $\xi(3s-1) = \xi(3(1-s)-1)$, this equality also shows the functional equation $\xi_3(s) = \xi_3(1-s)$. By taking

$$Y(s) = X(s) + \frac{\pi-3}{4}\xi(3s-1), \tag{1.12}$$

we obtain

$$\xi_3(s) = Y(s) + Y(1-s). \tag{1.13}$$

As shown in below this expression satisfies properties analogous to (A) and (B).

2 The frame of the proof

To describe the frame of our proof of Theorem 1, we give the proof of the following result which is similar to Theorem 1.

Theorem 2. *Let $X(s)$ be the function defined by* (1.10) *with* (1.7). *If we take $\xi_3^\natural(s) = X(s) + X(1-s)$, then all zeros of $\xi_3^\natural(s)$ lie on the line* $\mathrm{Re}(s) = 1/2$.

To prove Theorem 2, we prepare following two lemmas.

Lemma 1. *Let $F(s)$ be an entire function of genus zero or one, that has the following properties.*

(i) $F(s)$ is real on the real axis and satisfies a functional equation

$$F(s) = \pm F(1-s), \tag{2.1}$$

for some choice of sign.

(ii) There exists $a > 0$ such that all zeros of $F(s)$ lie in the vertical strip

$$|\mathrm{Re}\,(s) - 1/2| < a. \tag{2.2}$$

Then for any real $c \geq a$,

$$\left|\frac{F(s+c)}{F(s-c)}\right| > 1 \quad \textit{for} \quad \mathrm{Re}\,(s) > 1/2, \tag{2.3}$$

and

$$\left|\frac{F(s+c)}{F(s-c)}\right| < 1 \quad \textit{for} \quad \mathrm{Re}\,(s) < 1/2. \tag{2.4}$$

In particular, all zeros of $F(s+c) \pm F(s-c)$ lie on the line $\mathrm{Re}\,(s) = 1/2$.

Proof. We prove the lemma in section 5. □

Lemma 2. *All zeros of $X(s)$ lie on the line* $\mathrm{Re}\,(s) = 1/3$.

Remark. From definition (1.10), $X(s)$ has the functional equation $X(s) = X(2/3 - s)$. The line $\mathrm{Re}\,(s) = 1/3$ is the center of the functional equation.

Proof. We put

$$X^{\natural}(s) = X(s - 1/6).$$

Then Lemma 2 is equivalent to the assertion that all zeros of $X^{\natural}(s)$ lie on the line $\mathrm{Re}\,(s) = 1/2$. Since

$$\begin{aligned} X(s) &= \{(\frac{\pi-3}{2})s - (\frac{\pi}{3} - \frac{3}{2})\}\xi(3s) \\ &\quad + \{(\frac{\pi-3}{2})(\frac{2}{3} - s) - (\frac{\pi}{3} - \frac{3}{2})\}\xi(3s-1), \end{aligned} \tag{2.5}$$

we have

$$X^{\natural}(s) = \{(\frac{\pi-3}{2})(s-\frac{1}{6}) - (\frac{\pi}{3}-\frac{3}{2})\}\cdot\xi(3(s-1/6)) \tag{2.6}$$
$$\times\left\{1+\frac{(\frac{\pi-3}{2})(\frac{2}{3}-(s-\frac{1}{6}))-(\frac{\pi}{3}-\frac{3}{2})}{(\frac{\pi-3}{2})(s-\frac{1}{6})-(\frac{\pi}{3}-\frac{3}{2})}\frac{\xi(3(s-1/6)-1)}{\xi(3(s-1/6))}\right\}.$$

If we take $F(s) = \xi(3s-1)$, then $F(s) = F(1-s)$ and all zeors of $F(s)$ lie in the strip $|\mathrm{Re}\,(s) - 1/2| < 1/6$. Applying Lemma 1 to $F(s)$ with $c = 1/6$, we obtain

$$1 > \left|\frac{F(s-1/6)}{F(s+1/6)}\right| = \left|\frac{\xi(3(s-1/6)-1)}{\xi(3(s-1/6))}\right| \quad \text{for} \quad \mathrm{Re}\,(s) > 1/2. \tag{2.7}$$

On the other hand, we find that

$$\left|\frac{(\frac{\pi-3}{2})(\frac{2}{3}-(s-\frac{1}{6}))-(\frac{\pi}{3}-\frac{3}{2})}{(\frac{\pi-3}{2})(s-\frac{1}{6})-(\frac{\pi}{3}-\frac{3}{2})}\right| < 1 \quad \text{for} \quad \mathrm{Re}\,(s) > 1/2 \tag{2.8}$$

by an elementary calculation. Since

$$\{(\frac{\pi-3}{2})(s-\frac{1}{6}) - (\frac{\pi}{3}-\frac{3}{2})\}\xi(3(s-1/6))$$

has no zeros in the right-half plane $\mathrm{Re}\,(s) > 1/2$, equality (2.6) and (2.7), (2.8) implies $X^{\natural}(s)$ has no zeros in the right-half plane $\mathrm{Re}\,(s) > 1/2$. The functional equation $X^{\natural}(s) = X^{\natural}(1-s)$ shows that $X^{\natural}(s)$ also has no zeros in the left-half plane Re $\mathrm{Re}\,(s) < 1/2$. Thus we obtain the assertion of Lemma 2. □

2.1 Proof of Theorem 2

We have

$$\xi_3^{\natural}(s) = X(s)\left(1+\frac{X(1-s)}{X(s)}\right).$$

We show that

$$\left|\frac{X(1-s)}{X(s)}\right| < 1 \quad \text{for} \quad \mathrm{Re}\,(s) > 1/2. \tag{2.9}$$

This yields the non-existence of the zeros of $\xi_3^\natural(s)$ in the right-half plane $\mathrm{Re}\,(s) > 1/2$, since $X(s)$ has no zeros in the right-half plane $\mathrm{Re}\,(s) > 1/3$ by Lemma 2. We put

$$F(s) = X(s - 1/6).$$

The functional equation $X(s) = X((2/3) - s)$ yields $F(s) = F(1 - s)$. By Lemma 2, all zeros of $F(s)$ lie on the line $\mathrm{Re}\,(s) = 1/2$. These means that $F(s)$ satisfies all conditions of Lemma 1 for any $a > 0$. Applying Lemma 1 to $F(s)$ with $c = 1/6$, we obtain

$$\left|\frac{F(s-1/6)}{F(s+1/6)}\right| < 1 \quad \text{for} \quad \mathrm{Re}\,(s) > 1/2.$$

Using $X(s) = X((2/3) - s)$ we have

$$\frac{F(s-1/6)}{F(s+1/6)} = \frac{X(s-1/3)}{X(s)} = \frac{X(1-s)}{X(s)}.$$

Thus we obtain (2.9). Because of the functional equation $\xi_3^\natural(s) = \xi_3^\natural(1 - s)$, we also obtain the non-existence of the zeros of $\xi_3^\natural(s)$ in the left-half plane $\mathrm{Re}\,(s) < 1/2$. Now we complete the proof of Theorem 2. □

As the above, the proof of Theorem 2 consists of two steps. These two steps correspond to different two symmetries. Lemma 1 is used for each symmetry. In the first step corresponding to $X(s)$, we used Lemma 1 for $h(s)+h((2/3)-s)$ which yields the functional equation $X(s) = X((2/3)-s)$. In the second step corresponding to $\xi_3^\natural(s)$, we used Lemma 1 for $X(s) + X(1 - s)$ which yields the functional equation $\xi_3^\natural(s) = \xi_3^\natural(1 - s)$. This is the rough frame of the proof of Theorem 1.

3 Proof of Theorem 1

Let $Y(s)$ be the function defined by (1.12). Recall equation (1.13);

$$\xi_3(s) = Y(s) + Y(1 - s).$$

Since $Y(s)$ does not have a functional equation, to prove Theorem 1, we need the following modification of Lemma 1.

Lemma 3. *Let $F(s)$ be an entire function of genus zero or one. Suppose that*

(i) $F(s)$ is real on the real axis,

(ii) there exists $\sigma_0 < 1/2$ such that all zeros of $F(s)$ lie in the vertical strip

$$\sigma_0 < \mathrm{Re}\,(s) < 1/2, \tag{3.1}$$

(iii) there exists $C > 0$ such that

$$N(T) \leq CT \log T \quad as \quad T \to \infty, \tag{3.2}$$

where $N(T)$ is the number of zeros ρ of $F(s)$ satisfying $\sigma_0 < \mathrm{Re}\,(\rho) < 1/2$ and $0 \leq \mathrm{Im}\,(\rho) < T$. Further $F(1-\sigma)/F(\sigma) > 0$ for large $\sigma > 1/2$ and

$$\frac{F(1-\sigma)}{F(\sigma)} \to 0 \quad as \quad 1/2 < \sigma \to \infty. \tag{3.3}$$

Then we have

$$\left|\frac{F(1-s)}{F(s)}\right| < 1 \quad for \quad \mathrm{Re}\,(s) > 1/2, \tag{3.4}$$

and

$$\left|\frac{F(1-s)}{F(s)}\right| > 1 \quad for \quad \mathrm{Re}\,(s) < 1/2. \tag{3.5}$$

In particular, all zeros of $F(s) \pm F(1-s)$ lie on the line $\mathrm{Re}\,(s) = 1/2$.

To apply Lemma 3 to $Y(s)$, we prepare the following Lemma 4 which is a modification of Lemma 2.

Lemma 4. *All zeros of $Y(s)$ lie in the strip* $1/3 \leq \mathrm{Re}\,(s) < 1/2$.

3.1 Proof of Theorem 1 under Lemmas 3 and 4

Now we give the proof of Theorem 1 by using Lemma 3 and Lemma 4. We prove these lemmas after the proof of the theorem. By definition of $Y(s)$, it is an entire function which is real on the real axis. By Lemma 4, $Y(s)$ satisfies (ii) of Lemma 3. Hence if (3.2) and (3.3) are shown for $Y(s)$, then we obtain Theorem 1 by Lemma 3.

At first we prove (3.2). It is well known that $|\xi(s)| < \exp(M|s|\log|s|)$ as $|s| \to \infty$ for some constant $M > 0$. Since $Y(s)$ is a linear combination of $\xi(s)$ up to degree one polynomials, we have $|Y(s)| < \exp(M'|s|\log|s|)$ as $|s| \to \infty$ for some constant $M' > 0$. Recall Jensen's formula for entire function $f(s)$ with $f(0) \neq 0$:

$$\int_0^R r^{-1} n(r) dr = \frac{1}{2\pi}\int_0^{2\pi} \log|f(Re^{i\theta})|\, d\theta - \log|f(0)|,$$

where $n(r)$ is the number of zeros in $|s| < r$. We apply Jensen's formula to $Y(s)$. Then the right-hand side is estimated as $\ll R\log R$ for sufficiently large R. Since $\int_R^{2R} r^{-1}n(r)dr \geq n(R)\log 2$, it follows that

$$n(R) = O(R\log R).$$

Hence we obtain (3.2). From the definition of $Y(s)$ we have

$$\frac{Y(1-\sigma)}{Y(\sigma)} = \frac{\xi(3\sigma-1)}{\xi(3\sigma)} \frac{(\frac{\pi-3}{2} - \frac{\pi-5}{4}\sigma^{-1}) - (\frac{\pi-3}{2} - \frac{\pi}{6}\sigma^{-1})\frac{\xi(3\sigma-2)}{\xi(3\sigma-1)}}{(\frac{\pi-3}{2} - (\frac{\pi}{3} - \frac{3}{2})\sigma^{-1}) - (\frac{\pi-3}{2} - \frac{\pi-1}{4}\sigma^{-1})\frac{\xi(3\sigma-1)}{\xi(3\sigma)}}.$$

Since $\xi(s) = s(s-1)\pi^{-s/2}\Gamma(s/2)\zeta(s)$, we have

$$\frac{\xi(3\sigma-1)}{\xi(3\sigma)} = \sqrt{\pi}\Big(1 - \frac{2}{3\sigma}\Big)\frac{\Gamma((3\sigma-1)/2)}{\Gamma(3\sigma/2)}\frac{\zeta(3\sigma-1)}{\zeta(3\sigma)}.$$

Using the Stirling formula and the Dirichlet series expansion of $\zeta(s)$, we have

$$\frac{\Gamma((3\sigma-1)/2)}{\Gamma(3\sigma/2)} = \frac{e^{-\frac{1-\log 2}{2}+O(\sigma^{-1})}}{\sqrt{3\sigma}}\frac{1+O(\sigma^{-1})}{1+O(\sigma^{-1})} \text{ and } \frac{\zeta(3\sigma-1)}{\zeta(3\sigma)} = 1 + O(2^{1-\sigma})$$

for large $\sigma \geq 1$. Hence we obtain $Y(1-\sigma)/Y(\sigma) > 0$ for large $\sigma \geq 1$ and

$$\frac{Y(1-\sigma)}{Y(\sigma)} = O(\sigma^{-1/2}) \quad \text{as} \quad \sigma \to +\infty.$$

Thus (3.3) holds for $Y(s)$. Now we complete the proof of Theorem 1 under Lemma 3 and Lemma 4. □

3.2 Proof of Lemma 3

We prove the lemma only if $F(s)$ has genus one, since if $F(s)$ has genus zero it is easily proved by a way similar to the case of genus one. The genus one assumption is equivalent to the assertion that the Hadamard product factorization

$$F(s) = e^{A+Bs} s^m \prod_{\rho} \Big(1 - \frac{s}{\rho}\Big) e^{\frac{s}{\rho}} \quad (m \in \mathbb{Z}_{\geq 0}) \tag{3.6}$$

converges absolutely and uniformly on any compact subsets of $\mathbb{C}$. This assumption is also equivalent to the bound $\sum_{\rho} |\rho|^{-2} < \infty$. Assumption (i) implies symmetriy of the zeros under $\rho \mapsto \bar{\rho}$. It follows that the set of zeros $\rho = \beta + i\gamma$, counted with multiplicity, can be partitioned into blocks $B(\rho)$ comprising $\{\rho, \bar{\rho}\}$ if $\gamma > 0$ and $\{\rho\}$ if $\beta \neq 0$ and $\gamma = 0$. Each block is labeled with the unique zero in it having $\beta < 1/2$ and $\gamma \geq 0$. Using assumption (ii), we show

$$F(s) = s^m e^{A+B's} \prod_{B(\rho)} \left(\prod_{\rho \in B(\rho)} \Big(1 - \frac{s}{\rho}\Big) \right) \tag{3.7}$$

where the outer product on the right-hand side converges absolutely and uniformly on any compact subsets of $\mathbb{C}$. This assertion holds because the block convergence factors $\exp(c(B(\rho))s)$ are given by $c(B(\rho)) = 2\beta|\rho|^{-2}$ for $\gamma > 0$. Assumption (ii) gives $\sigma_0 < \beta < 1/2$. Hence

$$\sum_{B(\rho)} |c(B(\rho))| \leq \sum_{0 \neq \rho \in (\sigma_0, 1/2)} |\rho|^{-1} + \max\{1,\ 2|\sigma_0|\} \Big(\sum_{\rho} |\rho|^{-2} \Big) < \infty.$$

Thus the convergence factors can be pulled out of the product. Hence we have (3.7) with

$$B' = B + \sum_{B(\rho)} c(B(\rho)). \tag{3.8}$$

To establish (3.4) we proceed block by block in (3.7), using the factorization

$$\left|\frac{F(1-s)}{F(s)}\right| = e^{B'(1-2\mathrm{Re}\,(s))}\left|\frac{1-s}{s}\right|^m \prod_{B(\rho)}\left(\prod_{\rho\in B(\rho)}\left|\frac{1-\frac{1-s}{\rho}}{1-\frac{s}{\rho}}\right|\right). \tag{3.9}$$

Using assumption (iii), we show

$$B' \geq 0. \tag{3.10}$$

We will prove this later. If (3.10) is shown, we have

$$e^{B'(1-2\mathrm{Re}\,(s))}\left|\frac{1-s}{s}\right|^m \leq 1 \quad \text{for} \quad \mathrm{Re}\,(s) > \frac{1}{2}. \tag{3.11}$$

In a single block we can clear denominators to obtain

$$\prod_{\rho\in B(\rho)}\left|\frac{1-\frac{1-s}{\rho}}{1-\frac{s}{\rho}}\right| = \prod_{\rho\in B(\rho)}\left|\frac{s-1+\rho}{s-\rho}\right|.$$

We compare the term in the numerator with ρ against the term in the denominator with $\rho' := 1-\bar{\rho}$. We find that

$$\left|\frac{s-1+\bar{\rho}}{s-\rho}\right|^2 < 1 \quad \text{for} \quad \mathrm{Re}\,(s) > \frac{1}{2}. \tag{3.12}$$

In fact, writing $s = \sigma + it$, we have

$$\left|\frac{s-1+\bar{\rho}}{s-\rho}\right|^2 = \frac{(\sigma-1+\beta)^2+(t-\gamma)^2}{(\sigma-\beta)^2+(t-\gamma)^2} = 1 - \frac{(1-2\beta)(2\sigma-1)}{(\sigma-\beta)^2+(t-\gamma)^2}.$$

This implies (3.12) since $\beta < 1/2$ by assumption (ii). Thus we conclude for $\mathrm{Re}\,(s) > 1/2$ that the absolute value of the product over terms in each block on the right in (3.9) is less than 1. Hence (3.4) holds by (3.11) and (3.12).

Therefore it remains to show (3.10). By (3.3), we have

$$\mathbb{R} \ni \log\left(\frac{F(1-\sigma)}{F(\sigma)}\right) \to -\infty \quad \text{as} \quad 1/2 < \sigma \to +\infty. \tag{3.13}$$

Using (3.3) and (3.7), we have

$$\frac{F(1-\sigma)}{F(\sigma)} = e^{B'(1-2\sigma)}\Big(\frac{\sigma-1}{\sigma}\Big)^m \prod_{\rho=\beta\in\mathbb{R}} \frac{\sigma-1+\beta}{\sigma-\beta} \prod_{\substack{\rho=\beta+i\gamma \\ \gamma>0}} \frac{(\sigma-1+\beta)^2+\gamma^2}{(\sigma-\beta)^2+\gamma^2}.$$

Thus

(3.14)
$$\log\Big(\frac{F(1-\sigma)}{F(\sigma)}\Big) = B'(1-2\sigma) + m\log\Big(1-\frac{1}{\sigma}\Big) + \sum_{\rho=\beta\in\mathbb{R}} \log\Big(1-\frac{1-2\beta}{\sigma-\beta}\Big) + \sum_{\substack{\rho=\beta+i\gamma \\ \gamma>0}} \log\Big(1-\frac{(1-2\beta)(2\sigma-1)}{(\sigma-\beta)^2+\gamma^2}\Big).$$

Here we note that

$$\log\Big(1-\frac{(1-2\beta)(2\sigma-1)}{(\sigma-\beta)^2+\gamma^2}\Big) < 0 \quad \text{for} \quad \sigma > 1/2$$

by assumption (ii). Suppose that $B' < 0$. Then (3.13) and (3.14) claim that

(3.15)
$$\left|\sum_{\substack{\rho=\beta+i\gamma \\ \gamma>0}} \log\Big(1-\frac{(1-2\beta)(2\sigma-1)}{(\sigma-\beta)^2+\gamma^2}\Big)\right| \geq 2|B'|\sigma$$

for sufficiently large $\sigma > 1/2$, because the number of real zeros of $F(s)$ is finite by assumption (ii). On the other hand, for large $\sigma > 1/2$, we have

$$\left|\sum_{\substack{\rho=\beta+i\gamma \\ \gamma>0}} \log\Big(1-\frac{(1-2\beta)(2\sigma-1)}{(\sigma-\beta)^2+\gamma^2}\Big)\right| \leq \left|\sum_{\substack{\rho=\beta+i\gamma \\ \gamma>0}} \log\Big(1-\frac{(1-2\sigma_0)(2\sigma-1)}{(\sigma-1/2)^2+\gamma^2}\Big)\right| \ll (2\sigma-1)\sum_{\substack{\rho=\beta+i\gamma \\ \gamma>0}} \frac{1}{(\sigma-1/2)^2+\gamma^2}.$$

The sum in the right-hand side can be written as the Stieltjes integral

$$\int_{\gamma_0}^{\infty} \frac{dN(t)}{(\sigma-1/2)^2+t^2}.$$

Using (3.2) we have

$$\int_{\gamma_0}^{\infty} \frac{dN(t)}{(\sigma-1/2)^2+t^2} \ll \int_{\gamma_0}^{\infty} \frac{(\log t)\,dt}{(\sigma-1/2)^2+t^2} \ll \frac{\log(\sigma+\gamma_0)}{\sigma-1/2}.$$

Hence we obtain

$$\left|\sum_{\substack{\rho=\beta+i\gamma\\ \gamma>0}} \log\Big(1-\frac{(1-2\beta)(2\sigma-1)}{(\sigma-\beta)^2+\gamma^2}\Big)\right| \ll \log(\sigma+\gamma_0) \tag{3.16}$$

for sufficiently large $\sigma > 1/2$. This contradict (3.15). Thus (3.10) holds.

Inequality (3.5) is proved by a way similar to the proof of (3.4). We complete the proof of Lemma 3. □

3.3 Proof of Lemma 4

Now we complete the proof of Theorem 1 by proving Lemma 4. We put

$$p(s) = (\frac{\pi-3}{2})s - (\frac{\pi}{3}-\frac{3}{2}). \tag{3.17}$$

Then $X(s)$ and $Y(s)$ are written as

$$X(s) = p(s)\,\xi(3s) + p((2/3)-s)\,\xi(3s-1) \tag{3.18}$$

and

$$Y(s) = p(s)\,\xi(3s) + \{\,p(\frac{2}{3}-s) + \frac{\pi-3}{4}\,\}\,\xi(3s-1). \tag{3.19}$$

We find that

(i) the only one zero of $p(s)$ is $\dfrac{2\pi-9}{3\pi-9} \simeq -6.40$,

(ii) the only one zero of $p(\dfrac{2}{3}-s)$ is $\dfrac{1}{\pi-3} \simeq 7.06$,

(iii) the only one zero of $p(\dfrac{2}{3}-s)+\dfrac{\pi-3}{4}$ is $\dfrac{\pi-1}{2\pi-6} \simeq 7.56$.

At the first, we show that $Y(s)$ has no zeros in the left-half plane $\mathrm{Re}(s) < 1/3$. To prove the assertion, we divide $Y(s)$ as

$$Y(s) = \{ p(\frac{2}{3}-s)+\frac{\pi-3}{4} \}\xi(3s-1)\cdot\Big(1+\frac{p(s)}{p((2/3)-s)+(\pi-3)/4}\cdot\frac{\xi(3s)}{\xi(3s-1)}\Big).$$

Here we note that

$$\{ p(\frac{2}{3}-s)+\frac{\pi-3}{4} \}\xi(3s-1)$$

has no zeros in the left-half plane $\mathrm{Re}(s) < 1/3$. Therefore if we show that

$$\left|\frac{p(s)}{p((2/3)-s)+(\pi-3)/4}\cdot\frac{\xi(3s)}{\xi(3s-1)}\right| < 1 \quad \text{for} \quad \mathrm{Re}(s) < \frac{1}{3}, \tag{3.20}$$

then $Y(s)$ has no zeros in $\mathrm{Re}(s) < 1/3$. By a way similar to the proof of Lemma 2,

$$\left|\frac{\xi(3s)}{\xi(3s-1)}\right| < 1 \quad \text{for} \quad \mathrm{Re}(s) < \frac{1}{3}.$$

On the other hand, from (i) and (iii), we have

$$\frac{p(s)}{p((2/3)-s)+(\pi-3)/4} = -\frac{s-(2\pi-9)/(3\pi-9)}{s-(\pi-1)/(2\pi-6)}.$$

Thus

$$\left|\frac{p(s)}{p((2/3)-s)+(\pi-3)/4}\right| < 1 \quad \text{for} \quad \mathrm{Re}(s) < \frac{7\pi-21}{12(\pi-3)} \simeq 0.583.$$

Hence we obtain (3.20).

Next, we prove that $Y(s)$ has no zeros in the right-half plane $\mathrm{Re}(s) \geq 1/2$. To prove the assertion, we divide $Y(s)$ as

$$Y(s) = p(s)\xi(3s)\cdot\Big(1+\frac{p((2/3)-s)+(\pi-3)/4}{p(s)}\cdot\frac{\xi(3s-1)}{\xi(3s)}\Big).$$

Here we note that $p(s)\xi(3s)$ has no zeros in the right-half plane $\mathrm{Re}(s) > 1/3$. Therefore we show that

$$V(s) = 1+\frac{p((2/3)-s)+(\pi-3)/4}{p(s)}\cdot\frac{\xi(3s-1)}{\xi(3s)} \tag{3.21}$$

has no zeros in the right-half plane $\mathrm{Re}(s) \geq 1/2$. By a way similar to the proof of Lemma 2, we obtain

$$\left|\frac{\xi(3s-1)}{\xi(3s)}\right| < 1 \quad \text{for} \quad \mathrm{Re}(s) > \frac{1}{3}.$$

On the other hand, we obtain

$$\left|\frac{p((2/3)-s)+(\pi-3)/4}{p(s)}\right| \leq 1 \quad \text{for} \quad \mathrm{Re}(s) \geq \frac{7\pi-21}{12(\pi-3)} \simeq 0.583$$

by an elementary calculation. Thus $V(s) \neq 0$ for $\mathrm{Re}(s) \geq \dfrac{7\pi-21}{12(\pi-3)}$. Hence to complete the proof of Lemma 4 we should prove that $V(s)$ has no zeros in the strip

$$\frac{1}{2} \leq \mathrm{Re}(s) < \frac{7\pi-21}{12(\pi-3)}. \tag{3.22}$$

Let R be the rectangle $[1/2, (7\pi-21)/(12(\pi-3))] \times [-15, 15]$. Clearly $V(s)$ is holomorphic in R. From a numerical computation (for example by using MATHEMATICA), we obtain

$$\frac{1}{2\pi i}\int_{\partial R}\frac{V'}{V}(s)ds = 0$$

where ∂R is the boundary of R. This means $V(s) \neq 0$ in R by the argument principle.

Now we prove $V(\sigma+it) \neq 0$ for σ in (3.22) and $|t| \geq 15$ by showing that

$$\left|\frac{\xi(3s-1)}{\xi(3s)}\right|\left|\frac{p((2/3)-s)+(\pi-3)/4}{p(s)}\right| < 1 \tag{3.23}$$

holds in this region. In the proof of Lemma 2, we obtained

$$\left|\frac{\xi(3s-1)}{\xi(3s)}\right| < 1 \quad \text{for} \quad \mathrm{Re}(s) > \frac{1}{3} \tag{3.24}$$

by using the method established in [2], which is also explained in section 5. In detail, (3.24) is obtained by showing that this inequality holds term-by-term in the (modified) Hadamard product on a zero-by-zero basis, where the zero $\rho = \beta + i\gamma$ of $\xi(s)$ in the numerator is paired against the zero $\rho' = 1 - \beta + i\gamma$ in the denominator. That is, we established (3.24) using the fact that

$$\left|\frac{3s - 1 - \rho'}{3s - \rho}\right| < 1 \quad \text{for} \quad \mathrm{Re}\,(s) > \frac{1}{3}. \tag{3.25}$$

Hence, to establish (3.23), it suffices to show that for each s in (3.22) with $|t| \geq 15$ there exists a zero ρ of $\xi(s)$ such that

$$\left|\frac{3s - 1 - \rho'}{3s - \rho}\right|\left|\frac{p((2/3) - s) + (\pi - 3)/4}{p(s)}\right| < 1. \tag{3.26}$$

Then (3.25) and (3.26) give (3.23). To prove the existence of a zero ρ of $\zeta(s)$ satisfying (3.26), we need the following lemma which will be proved in section 6.

Lemma 5 (Lagarias). *For any real* $|t| \geq 14$ *there exists a zero* $\rho = \beta + i\gamma$ *of* $\xi(s)$ *such that* $0 < \beta \leq 1/2$ *and* $|t - \gamma| \leq 5$.

For each $s = \sigma + it$ satisfying $|t| \geq 15$, Lemma 5 gives a zero $\rho = \beta + i\gamma$ with $0 < \beta \leq 1/2$ and $|3t - \gamma| \leq 5$. Therefore, to prove (3.26), it is sufficient that

$$\left|\frac{3\sigma - 2 + \beta + it_0}{3\sigma - \beta + it_0}\right| < \left|\frac{p(s)}{p((2/3) - s) + (\pi - 3)/4}\right| \tag{3.27}$$

for any $1/2 \leq \sigma \leq (7\pi - 21)/(12\pi - 36)$, $|t| \geq 15$, $0 < \beta \leq 1/2$ and $|t_0| \leq 5$. Denote by μ and ν the zeros of $p(s)$ and $p((2/3) - s) + (\pi - 3)/4$, respectively. By squaring the inequality, (3.27) is equivalent to

$$\frac{(3\sigma - 2 + \beta)^2 + t_0^2}{(3\sigma - \beta)^2 + t_0^2} < \frac{(\sigma - \mu)^2 + t^2}{(\sigma - \nu)^2 + t^2}. \tag{3.28}$$

For fixed s and β it suffices to prove (3.27) with $t_0 = 5$, for it would then hold for $|t_0| \leq 5$, using the fact that $(3\sigma - \beta)^2 \geq (3\sigma - 2 + \beta)^2$ for $\mathrm{Re}\,(s) > 1/3$. Next, when $t_0 = 5$, it suffices to verify the inequality for $\beta = 1/2$ since if it holds there then the left-hand side of (3.28) with $t_0 = 5$ increases as $\beta > 0$ increases, while the right side is fixed. To establish (3.27) it thus suffices to verify the inequality

$$\frac{(3\sigma - \frac{3}{2})^2 + 25}{(3\sigma - \frac{1}{2})^2 + 25} < \frac{(\sigma - \mu)^2 + t^2}{(\sigma - \nu)^2 + t^2}. \tag{3.29}$$

The left-hand side of (3.29) decreases as σ increases in $1/2 \le \sigma \le (7\pi - 21)/(12\pi - 36)$. On the other hand, for fixed $|t| \ge 15$, the right-hand side of (3.29) increases as σ increases in $1/2 \le \sigma \le (7\pi - 21)/(12\pi - 36)$. Hence to establish (3.29) it suffices to verify the inequality

$$\frac{25}{26} < \frac{((1/2)-\mu)^2 + t^2}{((1/2)-\nu)^2 + t^2}. \tag{3.30}$$

By an elementary calculation, inequality (3.30) is valid for $|t| \ge 3$. Now we obtain (3.26) and complete the proof of Lemma 3. □

4 Proof of Lemma 1

In this section, we give a proof of Lemma 1 according to Lagarias-Suzuki [2].

The genus one assumption is equivalent to that the Hadamard product factorization

$$F(s) = e^{A+Bs} s^m \prod_{\rho} \Big(1 - \frac{s}{\rho}\Big) e^{\frac{s}{\rho}} \tag{4.1}$$

converges absolutely and uniformly on any compact subsets of $\mathbb{C}$. This assumption is also equivalent to the bound $\sum_\rho |\rho|^{-2} < \infty$. Assumption (i) implies symmetries of the zeros under $\rho \mapsto \bar{\rho}$ and $\rho \mapsto 1-\rho$. It follows that the set of zeros $\rho = \beta + i\gamma$, counted with multiplicity, can be partitioned into blocks $B(\rho)$ comprising $\{\rho, 1-\rho, \bar{\rho}, 1-\bar{\rho}\}$ if $\beta \ne 1/2$, $\{\rho, 1-\rho\}$ if $\beta = 1/2$ and $\gamma \ne 0$, and $\{\rho\}$ if $\rho = 1/2$. Each block is labeled with the unique zero in it having $\beta \le 1/2$ and $\gamma \ge 0$. Using assumption (ii), we show

$$F(s) = e^{A+B's} \prod_{B(\rho)} \left(\prod_{\rho \in B(\rho)} \Big(1 - \frac{s}{\rho}\Big) \right), \tag{4.2}$$

where the outer product on the right-hand side converges absolutely and uniformly on any compact subsets of $\mathbb{C}$. This assertion holds because the block convergence factors $\exp(c(B(\rho))s)$ are given by

$$c(B(\rho)) = \begin{cases} \beta|\rho|^{-2} + (1-\beta)|1-\rho|^{-2} & \text{if } \beta \ne 1/2, \\ |\rho|^{-2} & \text{if } \beta = 1/2 \text{ and } \gamma \ne 0, \\ 2 & \text{if } \rho = 1/2. \end{cases}$$

Assumption (ii) gives $-a < \beta - 1/2 < a$. Hence

$$\sum_{B(\rho)} |c(B(\rho))| \le (1+2a)\Big(\sum_{\rho} |\rho|^{-2}\Big) < \infty.$$

Thus the convergence factors can be pulled out of the product. Hence we have (4.2) with $B' = B + \sum_{B(\rho)} c(B(\rho))$.

Using the functional equation (2.1) we infer that $B' = 0$ in (4.2), so that

$$F(s) = e^{A} \prod_{B(\rho)} \bigg(\prod_{\rho \in B(\rho)} \Big(1 - \frac{s}{\rho}\Big) \bigg). \tag{4.3}$$

Indeed the change of variable $s \mapsto 1 - s$ permutes the factors in each block $B(\rho)$, with a possible sign change for $\rho = 1/2$, so it must be that $e^{A+B's} = \pm e^{A+B'(1-s)}$, which forces $B' = 0$.

To establish (2.3) and (2.4) we proceed block by block in (4.3), using the factorization

$$\left| \frac{F(s+c)}{F(s-c)} \right| = \prod_{B(\rho)} \left(\prod_{\rho \in B(\rho)} \left| \frac{1 - \frac{s+c}{\rho}}{1 - \frac{s-c}{\rho}} \right| \right). \tag{4.4}$$

In a single block we can clear denominators to obtain

$$\prod_{\rho \in B(\rho)} \left| \frac{1 - \frac{s+c}{\rho}}{1 - \frac{s-c}{\rho}} \right| = \prod_{\rho \in B(\rho)} \left| \frac{s+c-\rho}{s-c-\rho} \right|.$$

The main point is to compare the term in the numerator with ρ against the term in the denominator with $\rho' := \overline{1-\rho} = 1 - \bar{\rho}$. We show that

$$\left| \frac{s+c-\rho}{s-c-(1-\bar{\rho})} \right|^2 > 1 \qquad \text{for} \qquad \mathrm{Re}\,(s) > \frac{1}{2}, \tag{4.5}$$

and

$$\left| \frac{s+c-\rho}{s-c-(1-\bar{\rho})} \right|^2 < 1 \qquad \text{for} \qquad \mathrm{Re}\,(s) < \frac{1}{2}. \tag{4.6}$$

If (4.5) is shown, then we may conclude for $\mathrm{Re}\,(s) > 1/2$ that the absolute value of the product over terms in each block on the right in (4.4) exceeds 1, and (2.3) follows. Similarly (4.6) implies that for $\mathrm{Re}\,(s) < 1/2$ the product of terms over each block is smaller than 1, and (2.4) follows.

Therefore it remains to show (4.5) and (4.6). Writing $s = \sigma + it$, we have

$$\left|\frac{s+c-\rho}{s-c-(1-\bar{\rho})}\right|^2 = \frac{(\sigma+c-\beta)^2+(t-\gamma)^2}{(\sigma-c-1+\beta)^2+(t-\gamma)^2}.$$

Now (4.5) reduces to the assertion that

$$(\sigma+c-\beta)^2 > (\sigma-c-1+\beta)^2 \quad \text{for} \quad \mathrm{Re}\,(s) > \frac{1}{2}. \tag{4.7}$$

To show this we note that $\mathrm{Re}\,(s) > 1/2$ gives

$$\sigma+c-\beta > \frac{1}{2}+a-\beta > 0,$$

whence (4.7) makes the two assertions

$$\sigma+c-\beta > \sigma-c-1+\beta,$$

$$\sigma+c-\beta > -(\sigma-c-1+\beta).$$

The second of these asserts that $\sigma > 1/2$. While the first asserts that $2c > 2(\beta - 1/2)$. This holds since $c \geq a > \beta - 1/2$. Thus (4.7) holds, whence (4.5) holds.

A similar argument is used to establish (4.6). It reduces to the assertion that

$$(\sigma+c-\beta)^2 < (\sigma-c-1+\beta)^2 \quad \text{for} \quad \mathrm{Re}\,(s) < \frac{1}{2}. \tag{4.8}$$

We have

$$-(\sigma-c-1+\beta) \geq \frac{1}{2}+a-\beta > 0,$$

so that (4.8) is equivalent to the two assertions

$$-\sigma+c+1-\beta > -(\sigma+c-\beta),$$

$$-\sigma+c+1-\beta > \sigma+c-\beta.$$

The second of these is equivalent to $\sigma < 1/2$. While the first of these is equivalent to $c + 1/2 - \beta > 0$. This holds by our choice of c.

The conclusion that all zeros of $F(s+c) \pm F(s-c)$ lie on the line $\mathrm{Re}\,(s) = 1/2$ follows, because two terms $F(s+c)$ and $F(s-c)$ have different absolute values off the line $\mathrm{Re}\,(s) = 1/2$. □

5 Proof of Lemma 5

In this section, we give a proof of Lemma 5 according to Lagarias [1]. Throughout the section, we denote by γ the imaginary part of the non-trivial zeros of $\zeta(s)$. The spacing between consecutive ordinate γ goes to zero as $\gamma \to \infty$. Therefore the result of Lemma 5 holds for $|t|$ exceeding some bound and gives an explicit bound.

First, we prove that the result of the lemma holds for $|t| \geq 14$. Since the zeros are symmetric around the real axis, it suffices to consider the case $t \geq 14$. We verify the lemma directly for $14 \leq t \leq 168\pi + 5 < 525$ by inspection of a table of zeros of $\zeta(s)$. In fact there is no gap of size 5 between any consecutive zeros of $\zeta(s)$ starting with $\gamma_2 \simeq 21.02$, and the smallest zeros have ordinates $\gamma_1 \simeq 14.13$.

For the remaining range we use numerical estimates of Turing [3]. Let $N(T)$ be the number of zeros ρ with $0 < \mathrm{Im}(\rho) < T$ and let $\pi S(T)$ be the argument of $\zeta(1/2 + iT)$ obtained by analytic continuation along a horizontal line from $\infty + iT$. We have

$$N(T) = 2\kappa\Big(\frac{T}{2\pi}\Big) + 1 + S(T), \tag{5.1}$$

where $\kappa(\tau) = (4\pi i)^{-1} \log(\Gamma(1/4 + \pi i\tau)/\Gamma(1/4 - \pi i\tau)) - (1/4)\tau \log \pi$. Theorem 1 of Turing [3] gives

$$\kappa(\tau) = \frac{1}{2}\Big(\tau \log \tau - \tau - \frac{1}{2}\Big) + \varepsilon(\tau) \quad \text{with} \quad |\varepsilon(\tau)| \leq \frac{0.006}{\tau} \quad \text{for} \quad \tau \geq 64. \tag{5.2}$$

For $S_1(t) = \int_0^t S(u)du$, Theorem 4 of Turing [3] assert that if $t_2 > t_1 > 168\pi$ then

$$|S_1(t_2) - S_1(t_1)| \leq 2.30 + 0.128 \log\Big(\frac{t_2}{2\pi}\Big). \tag{5.3}$$

Now we prove Lemma 5 under the following Lemma 6 which is shown in the end of the section.

Lemma 6. *Suppose that there is no zero ρ of $\zeta(s)$ with $168\pi \leq t_1 < \mathrm{Im}(\rho) < t_2$ and that $S(T)$ has one sign over the interval $[t_1, t_2]$. Then*

$$t_2 - t_1 < 10/3. \tag{5.4}$$

Suppose that $t \geq 168\pi + 5$ and there is no zero on $[t-2, t+2]$, that is, $N(T)$ is constant on $[t-2, t+2]$. Because of Lemma 6, inside this interval $S(T)$ must have a zero-crossing in each subinterval of length 10/3. Hence it must have a zero-crossing at some point $t_1 = t + x$ with $|x| \leq 4/3$. Since $N(T)$ is constant, $S(T)$ varies like $-2\kappa(T/2\pi) - N(t_1) - 1$. Therefore (5.2) implies that all other zero-crossing of $S(T)$ in $[t-2, t+2]$ are localized within a distance $\varepsilon = 0.006/(t_1 \log(t_1/2\pi))$ of this one. If there were no zero on $[t-5, t]$, then $N(T)$ is constant there. Hence $S(T)$ varies approximately linearly on the interval. If $t + x$ falls in $[t-5, t]$, zero-crossings of $S(T)$ are located within 0.001 of $t + x$, and otherwise it has no zero-crossings. Since $|x| \leq 4/3$, $S(T)$ has single sign on $[t-5, t-5+10/3]$. This contradicts Lemma 6. Thus there is a zero with the ordinate γ in $[t-5, t-2]$. By a similar argument there is a zero with the ordinate γ in $[t+2, t+5]$.

If $t \geq 168\pi + 5$ and there is a zero with the ordinate γ in $[t-2, t+2]$, this of course implies our assertion. We complete the proof of Lemma 5. □

Proof of Lemma 6. We take $t_2 = t_1 + c$ with $c \geq 10/3$. From the assumption $S(T)$ has single sign on $[t_1, t_2]$. Hence we have

$$|S_1(t_2) - S_1(t_1)| = \left| \int_0^c S(t_1+u)du \right| = \int_0^c |S(t_1+u)|du \geq \int_0^{10/3} |S(t_1+u)|du.$$

Suppose that $S(T)$ is positive. Since $N(t)$ is constant on $[t_1, t_2]$, we obtain

$$S(t_1+u) = N(t_1) - 2\kappa(\frac{t_1+u}{2\pi}) - 1 = 2\kappa(\frac{t_1}{2\pi}) - 2\kappa(\frac{t_1+u}{2\pi}) + S(t_1) > 0 \tag{5.5}$$

for any $0 \leq u \leq 10/3$. Thus

$$S(t_1) > 2\kappa(\frac{t_2}{2\pi}) - 2\kappa(\frac{t_1}{2\pi}). \tag{5.6}$$

Combining (5.5) and (5.6) we obtain

$$S(t_1+u) > 2\kappa(\frac{t_2}{2\pi}) - 2\kappa(\frac{t_1+u}{2\pi})$$

for any $0 \leq u \leq 10/3$. Hence we have

$$|S_1(t_2) - S_1(t_1)| > \int_0^{10/3} \left| 2\kappa(\frac{t_2}{2\pi}) - 2\kappa(\frac{t_1+u}{2\pi}) \right| du. \tag{5.7}$$

Thus (5.2) yields

(5.8)

$$|S_1(t_2) - S_1(t_1)| \geq \int_0^{10/3} \Big(\frac{t_2}{2\pi} - \frac{t_1+u}{2\pi}\Big)\Big(\log\frac{t_2}{2\pi} - 1\Big)du - 3.5\Big(\frac{0.006}{t_1}\Big)$$
$$\geq 0.884\log\Big(\frac{t_2}{2\pi}\Big) - 0.886.$$

Suppose that $S(T)$ is negative. By (5.1) we have $N(t_1) < 2\kappa(t_1/(2\pi)) + 1$. Since $N(T)$ is constant on $[t_1, t_2]$, we obtain

$$-S(t_1+u) = 2\kappa(\frac{t_1+u}{2\pi}) - N(t_1) + 1 > 2\kappa(\frac{t_1+u}{2\pi}) - 2\kappa(\frac{t_1}{2\pi})$$

for $0 \leq u \leq 10/3$. Hence we have

$$(5.9) \qquad |S_1(t_2) - S_1(t_1)| > \int_0^{10/3} \Big|2\kappa(\frac{t_1+u}{2\pi}) - 2\kappa(\frac{t_1}{2\pi})\Big| du.$$

Thus (5.2) yields

$$(5.10) \qquad \begin{aligned} |S_1(t_2) - S_1(t_1)| &\geq \int_0^{10/3} \frac{u}{2\pi}\Big(\log\frac{t_1}{2\pi} - 1\Big)du - 3.5\Big(\frac{0.006}{t_1}\Big) \\ &\geq 0.884\log\Big(\frac{t_1}{2\pi}\Big) - 0.886. \end{aligned}$$

Now $t_1 \geq 168\pi$ gives $\log(t_1/(2\pi)) \geq 4.4$ and $\log(t_1/(2\pi)) \geq \log(t_2/(2\pi)) - 0.01$. Hence (5.8) and (5.10) gives

$$(5.11) \qquad |S_1(t_2) - S_1(t_1)| \geq 0.884\log\Big(\frac{t_2}{2\pi}\Big) - 0.886$$

under the assumptions of Lemma 6. This contradicts (5.3). We complete the proof. □

6 A question

Equality (1.11) can be written as

$$\xi_3(s) = \xi_3^{\natural}(s) + \frac{\pi - 3}{2}\xi(3s-1), \tag{6.1}$$

where $\xi_3^{\natural}(s) = X(s) + X(1-s)$. This yields

$$\frac{\pi - 3}{2}\xi(3s-1) = \xi_3(s) - \xi_3^{\natural}(s). \tag{6.2}$$

As this, the Riemann xi function $\xi(s)$ can be written as the sum of $\xi_3(s)$ and $\xi_3^{\natural}(s)$. Equality (6.2) was first pointed out by Weng. We have already known that all zeros of $\xi_3(s)$ and $\xi_3^{\natural}(s)$ lie on the line $\mathrm{Re}\,(s) = 1/2$. Therefore, equality (6.2) may be a remarkable relation. We conclude the article by the following natural question:

Can one say some nice things for the zeros of $\xi(s)$ by using the properties of $\xi_3(s)$ and $\xi_3^{\natural}(s)$ via relation (6.2)*?*

References

[1] J.C. Lagarias, *On a positivity property of the Riemann ξ-function*, Acta Arithmetica **89** (1999), No. 3, 217–234.

[2] J.C. Lagarias and M. Suzuki, *The Riemann hypothesis for certain integrals of Eisenstein series*, J. Number Theory **118** (2006), no. 1, 98–122.

[3] A.M. Turing, *Some calculations of the Riemann zeta function*, Proc. London Math. Soc. **3** (1953), 99–117. [In: *Collected Works of A. M. Turing, Volume I* (J. L.Britton, Ed.) North-Holland 1992, pp. 79–97. Notes, pp. 254–261.]

[4] L. Weng, *Geometric arithmetic: a program*, Arithmetic Geometry and Number Theory (Series on Number Theory and Its Applications, vol. 1), pp. 211–400, 2006.

[5] L. Weng, *A geometric approach to L-Functions*, this volume.

Masatoshi Suzuki
Graduate School of Mathematics,
Nagoya University,
Chikusa-ku, Nagoya 464-8602,
Japan
Email: m99009t@math.nagoya-u.ac.jp

Elliptic Curves Arising from the Spectral Zeta Function for Non-Commutative Harmonic Oscillators and $\Gamma_0(4)$-Modular Forms

K. KIMOTO[1] & M. WAKAYAMA[2]

The Apéry-like numbers $J_2(n)$ introduced in [6] are indispensable to a description of the special value $\zeta_Q(2)$ of the spectral zeta function $\zeta_Q(s)$ for the non-commutative harmonic oscillator Q. These numbers have remarkable modular form interpretation as the original Apéry numbers possess. In fact, we show that the differential equation satisfied by the generating function $w_2(t)$ of $J_2(n)$ is the Picard-Fuchs equation for the universal family of elliptic curves equipped with rational 4-torsion. The parameter t of this family can be considered as a modular function for the congruent subgroup $\Gamma_0(4)$. Further, we see that the function $w_2(t)$ is regarded as a $\Gamma_0(4)$-modular form of weight 1 in the variable τ by taking t as the classical Legendre modular function $\lambda(\tau)$.

0 Introduction

Let us consider the Apéry numbers a_n and b_n defined by the recurrence relation

$$n^2u_n - (11n^2 - 11n + 3)u_{n-1} - (n-1)^2u_{n-2} = 0 \qquad (0.1)$$

together with the initial conditions $a_0 = 1$, $a_1 = 3$ and $b_0 = 0$, $b_1 = 5$. These numbers were introduced in 1978 by Roger Apéry, who utilized them to prove the irrationality of the special value $\zeta(2) = \frac{\pi^2}{6}$ of the Riemann zeta function $\zeta(s)$. The important point is that a parallel technique allows

[1]Partially supported by Grant-in-Aid for Scientific Research (B) No.16740021.

[2]Partially supported by Grant-in-Aid for Scientific Research (B) No.15340012.

us to provide a proof of the irrationality of $\zeta(3)$ (we refer to [11] for further information on his irrationality proof).

Since then, several people have tried to understand deeply the nature of the Apéry numbers and to generalize the theory to $\zeta(n)$ $(n > 3)$. For instance, there are many works on congruence and/or supercongruence properties, algebro-geometric interpretations, modular properties, etc. For detailed information, we refer to [1], [2], [3] and their references. Among them, in this note, we focus particularly the study on the relation between Apéry numbers and elliptic curves developed by Beukers [1].

Let us hence recall the result in [1] briefly. Consider the generating functions $A(t) = \sum_{n=0}^{\infty} a_n t^n$ and $B(t) = \sum_{n=0}^{\infty} b_n t^n$ of the Apéry numbers a_n and b_n. Then $A(t)$ and $B(t)$ satisfy the differential equation $L_2(A) = 0$ and $L_2(B) = -5$ respectively, where L_2 denotes the Fuchsian differential operator

$$L_2 = t(t^2 + 11t - 1)\frac{d^2}{dt^2} + (3t^2 + 22t - 1)\frac{d}{dt} + (t + 3).$$

The main result of [1] shows that the equation $L_2(Y) = 0$ is the Picard-Fuchs equation associated with the family of curves

$$y^2 = x^3 + \frac{1}{4}(t^2 + 6t + 1)x^2 + \frac{1}{2}t(t + 1)x + \frac{1}{4}t^2. \tag{0.2}$$

We notice that any elliptic curve equipped with rational 5-torsion is birationally equivalent to (0.2) for some value of t [7]. By this fact, the function $A(t)$ is interpreted as the period of the holomorphic 1-form $\frac{\pi i dx}{y}$. Moreover, $A(t)$ is a $\Gamma_1(5)$-modular form of weight 1 in the variable τ ($\operatorname{Im}\tau > 0$), the ratio of the fundamental periods.

In the present note, we deal with analogous objects of the Apéry numbers, say, the "Apéry-like numbers" $J_2(n)$ associated with the special value $\zeta_Q(2)$ introduced in [6]. Here $\zeta_Q(s)$ is the spectral zeta function of a certain differential operator called the non-commutative harmonic oscillator Q. We recall shortly basic properties of the spectral zeta function $\zeta_Q(s)$ and how the numbers $J_2(n)$ arise in connection with the value $\zeta_Q(2)$. Let $Q = Q_{\alpha,\beta}$ be the ordinary differential operator on $L^2(\mathbb{R}) \otimes \mathbb{C}^2$ defined by

$$Q := \begin{pmatrix} \alpha & 0 \\ 0 & \beta \end{pmatrix}\left(-\frac{1}{2}\frac{d^2}{dx^2} + \frac{1}{2}x^2\right) + \begin{pmatrix} 0 & -1 \\ 1 & 0 \end{pmatrix}\left(x\frac{d}{dx} + \frac{1}{2}\right),$$

where the parameters $\alpha, \beta \in \mathbb{R}_{>0}$ satisfy $\alpha\beta > 1$. The system defined by the operator Q is called the *non-commutative harmonic oscillator* [10]. The operator Q is positive and self-adjoint, and has only a discrete spectrum; the eigenvalues are $0 < \lambda_1 \leq \lambda_2 \leq \cdots \leq \lambda_n \leq \cdots \nearrow +\infty$ with uniformly bounded multiplicity. Then, the *spectral zeta function* $\zeta_Q(s)$ of Q is defined as the Dirichlet series

$$\zeta_Q(s) := \sum_{n=1}^{\infty} \lambda_n^{-s}$$

in order to study the structure of the spectrum of Q. This series converges absolutely if $\operatorname{Re} s > 1$ [6]. In [5], it is shown that $\zeta_Q(s)$ is continued to the whole complex plane as a meromorphic function which has a unique simple pole at $s = 1$, and trivial zeros at $s = 0, -2, -4, -6, \ldots$ like the Riemann zeta function $\zeta(s)$. We note that the operator Q is unitarily equivalent to a pair of usual quantum harmonic oscillators when $\alpha = \beta$. In particular, we see that $\zeta_Q(s) = 2(2^s - 1)\zeta(s)$ when $\alpha = \beta = \sqrt{2}$.

The numbers $J_2(n)$ in question are defined by

$$J_2(n) := \int_0^{\infty}\int_0^{\infty} \frac{e^{-(t+s)/2}}{1 - e^{-(t+s)}} \left(\frac{(1 - e^{-2t})(1 - e^{-2s})}{(1 - e^{-(t+s)})^2} \right)^n dtds, \tag{0.3}$$

which arise in the expression of the special value of $\zeta_Q(s)$ at $s = 2$. Actually, we have

$$\zeta_Q(2) = \frac{(\alpha + \beta)^2}{2\alpha\beta(\alpha\beta - 1)} \left(3\zeta(2) + \left(\frac{\alpha - \beta}{\alpha + \beta} \right)^2 \sum_{n=0}^{\infty} \binom{-\frac{1}{2}}{n} J_2(n) \left(\frac{1}{\alpha\beta - 1} \right)^n \right). \tag{0.4}$$

In [6], a recurrence formula for $J_2(n)$ is obtained as

$$4n^2 J_2(n) - (8n^2 - 8n + 3) J_2(n-1) + 4(n-1)^2 J_2(n-2) = 0. \tag{0.5}$$

Introducing the generating function $w_2(t)$ of $J_2(n)$ by

$$w_2(t) = \sum_{n=0}^{\infty} J_2(n) t^n,$$

one finds that the recurrence formula (0.5) is equivalent to the singly confluent Heun differential equation

$$t(1-t)^2 w_2''(t) - (1 - 3t)(1 - t) w_2'(t) + (t - \frac{3}{4}) w_2(t) = 0. \tag{0.6}$$

We show in [8] that the rational numbers $\tilde{J}_2(n) := J_2(n)/J_2(0)$ satisfy the congruence relation $\tilde{J}_2(mp^n) \equiv \tilde{J}_2(mp^{n-1}) \pmod{p^n}$ for any odd prime number p and positive integers m, n, which is quite similar to the one for the Apéry numbers; $a_{mp^n-1} \equiv a_{mp^{n-1}-1} \pmod{p^n}$. Here we note that the Apéry-like numbers $J_3(n)$ are introduced similarly to describe the value $\zeta_Q(3)$, and satisfy the recurrence formula

$$(0.7)\quad 4n^2 J_3(n) - (8n^2 - 8n + 3)J_3(n-1) + 4(n-1)^2 J_3(n-2) = \frac{2^n(n-1)!}{(2n-1)!!}.$$

Remarkably, the only difference between the two recurrence equations (0.5) and (0.7) is the existence of the inhomogeneous term in (0.7). As a result, the generating function $w_3(t)$ for $J_3(n)$ satisfies almost the same differential equation as (0.6) but with inhomogeneous term (see (5.2)). Furthermore, one can obtain a congruence formula for the normalized sequence $\tilde{J}_3(n) := J_3(n) - J_3(0)\tilde{J}_2(n) \in \mathbb{Q}$ too. See [8] for detail.

Based on these similarities between the Apéry numbers and our Apéry-like numbers $\tilde{J}_2(n)$, $\tilde{J}_3(n)$, in the talk at the *Conference on L-functions*, the second author had a chance to state a rather vague conjecture; like the results in [1] (resp. [2]) for the Apéry numbers, the differential equation for $w_2(t)$ (resp. $w_3(t)$) would be understood as a Picard-Fuchs differential equation attached to a certain family of elliptic curves (resp. a certain family of $K3$ surfaces). At that time, he presented some numerical data for $\tilde{J}_2(n)$, from which Zagier immediately showed his interest and indicated that the numbers $\tilde{J}_2(n)$ would be coming from a certain modular form. In fact, on the next day, he kindly suggested us the precise formula from his study [14].

The aim of this note is, therefore, to establish an analogue of the results in [1] for the Apéry-like numbers $J_2(n)$. Precisely, we show that the differential equation satisfied by the generating function $w_2(t)$ of $J_2(n)$ is the Picard-Fuchs equation for the universal family of elliptic curves equipped with rational 4-torsion. The parameter t for the family of such elliptic curves can be considered as a modular function for the congruent subgroup $\Gamma_0(4)$. Moreover, we show that the function $w_2(t)$ is regarded as a $\Gamma_0(4)$-modular form of weight 1 in the variable τ by taking t as the classical Legendre modular function $\lambda(\tau)$. Our strategy is almost parallel to the one in [1].

1 Setting the stage

Changing the variables of integral, we have

$$\begin{aligned}
J_2(n) &= \int_0^{\infty}\int_0^{\infty} \frac{e^{-(t+s)/2}}{1-e^{-(t+s)}}\left(\frac{(1-e^{-2t})(1-e^{-2s})}{(1-e^{-(t+s)})^2}\right)^n dtds \\
&= 4\int_0^1\int_0^1 \left(\frac{(1-X^4)(1-Y^4)}{(1-X^2Y^2)^2}\right)^n \frac{dXdY}{1-X^2Y^2}.
\end{aligned}$$

Thus we obtain the expression
(1.1)

$$w_2(t) = \sum_{n=0}^{\infty} J_2(n)t^n = 4\int_0^1\int_0^1 \frac{1-X^2Y^2}{(1-X^2Y^2)^2 - t(1-X^4)(1-Y^4)}dXdY$$

of $w_2(t)$ as an integration of a rational function. This integral expression is calculated as

$$\begin{aligned}
\frac{1}{4}w_2(t) &= \int_0^1\int_0^1 \frac{1-X^2Y^2}{(1-t)(1-X^2Y^2)^2 + t(X^2-Y^2)^2}dXdY \\
&= \frac{1}{2(1-t)}\Big(\int_0^1\int_0^1 \frac{dXdY}{(1-X^2Y^2)+T(X^2-Y^2)} \\
&\qquad + \int_0^1\int_0^1 \frac{dXdY}{(1-X^2Y^2)-T(X^2-Y^2)}\Big) \\
&= \frac{1}{1-t}\int_0^1\int_0^1 \frac{dXdY}{(1-X^2Y^2)+T(X^2-Y^2)},
\end{aligned}$$

where we put $T = \sqrt{\frac{t}{t-1}}$ or $t = \frac{T^2}{T^2-1}$. Set

$$Q_T(X,Y) = (1-X^2Y^2) + T(X^2-Y^2). \tag{1.2}$$

Then we define $\mathcal{W}_2(T)$ by

$$\mathcal{W}_2(T) = \frac{1-t}{4}w_2(t) = \int_{\square} \frac{dXdY}{Q_T(X,Y)}, \tag{1.3}$$

where $\square$ denotes the domain $[0,1]\times[0,1]$. The denominator $Q_T(X,Y)$ of the integrand in $\mathcal{W}_2(T)$ defines an algebraic curve $Q_T : Q_T(X,Y) = 0$ in $\mathbb{C}^2$. From (0.6), it is easy to see that the function $\mathcal{W}_2(T)$ satisfies the differential equation $\mathcal{L}(\mathcal{W}_2) = 0$. Here the differential operator $\mathcal{L}$ is given by

$$\mathcal{L} = (T^3 - T)\frac{d^2}{dT^2} + (3T^2 - 1)\frac{d}{dT} + T. \tag{1.4}$$

In the sequel, we explain actually that the differential equation $\mathcal{L}(\mathcal{W}_2) = 0$ has a geometric origin, that is,

(1) the algebraic curve Q_T is birationally equivalent to a certain elliptic curve C_T for all but finite values of T, and $\{C_T\}_{T\in\sqrt{-1}\mathbb{Q}^\times}$ gives the family of elliptic curves having rational 4-torsion.

(2) the differential equation $\mathcal{L}(\mathcal{W}_2) = 0$ is the Picard-Fuchs equation corresponding to the family $\{C_T\}_T$.

Remark 1.1. In [9], the differential equation (0.6) is solved so that we have

$$w_2(t) = \frac{J_2(0)}{1-t}\,{}_2F_1\left(\frac{1}{2}, \frac{1}{2}; 1; \frac{t}{t-1}\right), \tag{1.5}$$

where ${}_2F_1(a, b; c; z)$ denotes the Gaussian hypergeometric function. As a corollary, we obtain the hypergeometric expression

$$\mathcal{W}_2(T) = \frac{\pi^2}{8}\,{}_2F_1\left(\frac{1}{2}, \frac{1}{2}; 1; T^2\right)$$

for $\mathcal{W}_2(T)$. This yields the binomial expression for $J_2(n)$ as

$$J_2(n) = \frac{\pi^2}{2}\sum_{k=0}^{n}(-1)^k\binom{-\frac{1}{2}}{k}^2\binom{n}{k}.$$

2 Elliptic curves associated with $J_2(n)$

Let us consider the series of birational transformations

$$X = \frac{1}{1+x_1}, \quad Y = \frac{x_1+y_1+1}{\sqrt{T}(x_1+y_1)+1}, \tag{2.1a}$$

$$x_1 = \frac{(1+\sqrt{T})y_2}{(1+T)x_2-(1+\sqrt{T})y_2-1}, \tag{2.1b}$$

$$y_1 = \frac{(1+\sqrt{T})(1+\sqrt{T}(1-\sqrt{T})y_2)}{(1+T)((1+T)x_2-(1+\sqrt{T})y_2-1)},$$

$$x_2 = x_3, \quad y_2 = \frac{y_3}{x_3}, \tag{2.1c}$$

$$x_3 = \frac{-1+6\sqrt{T}-6T+6\sqrt{T}T-T^2+12x}{12\sqrt{T}(1+T)^2}, \tag{2.1d}$$

$$y_3 = \frac{1-5T-5T^2+T^3-12(1+T)x+24y}{24T(1+\sqrt{T})(1+T)^2}.$$

The first transformation (2.1a) reduces the curve Q_T defined in (1.2) to the cubic curve

$$\begin{aligned} &2(1+T)(1-\sqrt{T})y_1+(1+T)(1-T)y_1^2-2\sqrt{T}(1-\sqrt{T})^2x_1 \\ &+2(1-\sqrt{T})(1-\sqrt{T}+T+T\sqrt{T})x_1y_1+\sqrt{T}(\sqrt{T}-1)(T+\sqrt{T}-4)x_1^2 \\ &\qquad -2\sqrt{T}(1-\sqrt{T})x_1^2y_1-2\sqrt{T}(1-\sqrt{T})x_1^3=0, \end{aligned}$$

and the remaining three steps (2.1b), (2.1c), (2.1d) are the regular procedure to obtaining the standard form of an elliptic curve (see, e.g. [12]).

By a tedious step-by-step calculation, we obtain the following key lemma.

Lemma 2.1. *By the transformation $X = X(x,y)$ and $Y = Y(x,y)$ given in (2.1), the curve Q_T is birationally equivalent to the elliptic curve C_T : $C_T(x,y)=0$ defined by*

$$C_T(x,y) = \left(x^3 - \frac{T^4+14T^2+1}{48}x + \frac{T^6-33T^4-33T^2+1}{864}\right) - y^2. \tag{2.2}$$

Furthermore, the equality

$$\frac{dXdY}{Q_T(X,Y)} = \frac{dxdy}{2C_T(x,y)} \tag{2.3}$$

holds. □

Let us look at the resulting elliptic curve

$$\begin{aligned} C_T\,:\,y^2 &= x^3 - \frac{T^4+14T^2+1}{48}x + \frac{T^6-33T^4-33T^2+1}{864} \\ &= \frac{1}{864}(1-6T+T^2-12x)(1+6T+T^2-12x)(1+T^2+6x). \end{aligned}$$

The points of order 2 on C_T are

$$\left(\frac{1-6T+T^2}{12},0\right),\ \left(\frac{1+6T+T^2}{12},0\right),\ \left(-\frac{1+T^2}{6},0\right).$$

For any $T \in \mathbb{C}$, these three points together with the point of infinity (the identity element of the group $C_T(\mathbb{C})$) form a finite subgroup of $C_T(\mathbb{C})$ isomorphic to $\mathbb{Z}/2\mathbb{Z} \times \mathbb{Z}/2\mathbb{Z}$. We also conclude that the curve C_T is singular if and only if $T = 0, \pm 1, \infty$ since the discriminant of C_T is equal to $\frac{1}{256}T^2(1-T^2)^4$.

When $T = \sqrt{-1}u \in \sqrt{-1}\mathbb{Q}$, C_T has only one *rational* 2-torsion $(\frac{u^2-1}{6}, 0)$. In this case, there are just two rational points

$$\left(\frac{1+5u^2}{12}, \pm\frac{u(1+u^2)}{4}\right)$$

on C_T of order 4, and hence the torsion part of $C_T(\mathbb{Q})$ is isomorphic to $\mathbb{Z}/4\mathbb{Z}$ by Mazur's theorem on the structure of torsion subgroups.

In [7], it is shown that the universal family of elliptic curves with rational 4-torsion is given by

$$E_b\,:\,y^2+xy-by = x^3-bx^2, \qquad \Delta = b^4(1+16b) \neq 0.$$

In this form, the origin $(0,0)$ is a generator of the 4-torsion. If we set

$$b = \frac{\sqrt{T}(T+1)}{2(\sqrt{T}-1)^4},$$

then we see that E_b and C_T are birationally equivalent, and the family $\{C_T\}$ is the universal family of the elliptic curves equipped with rational 4-torsion described by another form.

3 Geometric interpretation of the differential equation for $\mathcal{W}_2(T)$

We follow the discussion expanded in [1] to see that the differential equation (0.6) is the Picard-Fuchs equation associated with the family of elliptic curves $\{C_T\}$.

We recall that the function $\mathcal{W}_2(T)$ is a unique (up to constant) holomorphic solution of the differential equation $\mathcal{L}(\mathcal{W}_2) = 0$ ($\mathcal{L}$ is given in (1.4)) around $T = 0$.

Fix a point $T_0 \in \mathbb{C}$ near the origin and take a closed curve γ through T_0 such that γ does not contain any singularity of the form $\frac{dXdY}{Q_T(X,Y)}$. Then, following the idea in [1], we consider the analytic continuation of $\mathcal{W}_2(T)$ along γ, which is given by

$$\widetilde{\mathcal{W}}_2(T) := \int_D \frac{dXdY}{Q_T(X,Y)}$$

where $D \subset \mathbb{C}^2$ is a suitable domain such that $\partial D = \partial\square$ (the same boundary). Notice that $\widetilde{\mathcal{W}}_2(T)$ also satisfies the differential equation $\mathcal{L}(\widetilde{\mathcal{W}}_2) = 0$, and the integral

$$\widetilde{\mathcal{W}}_2(T) - \mathcal{W}_2(T) = \int_{D-\square} \frac{dXdY}{Q_T(X,Y)}$$

is over the closed and oriented surface (2-cycle) $D-\square$. Namely, this integral can be written as the form

$$I_\Gamma(T) := \int_\Gamma \frac{dXdY}{Q_T(X,Y)}$$

for a certain 2-cycle Γ in $\mathbb{C}^2 \setminus Q_T$. By Lemma 2.1, the integral $I_\Gamma(T)$ becomes

$$I_S(T) := \frac{1}{2}\int_S \frac{dxdy}{C_T(x,y)}$$

for some $S \in H_2(\mathbb{C}^2 \setminus C_T, \mathbb{Z})$. By the same 'topological reduction' discussion as in [1], it follows that

$$I_S(T) = \frac{\pi i}{2}\int_\gamma \frac{dx}{y}$$

for a certain 1-cycle $\gamma \in H_1(\mathbb{C}^2 \setminus C_T, \mathbb{Z})$. Thus the difference $\widetilde{\mathcal{W}}_2(T) - \mathcal{W}_2(T)$ is a period $I_S(T)$ of the holomorphic 1-form $\frac{\pi i dx}{2y}$. Hence $\mathcal{W}_2(T)$ (and $\widetilde{\mathcal{W}}_2(T)$) and the period $I_S(T)$ of $\frac{\pi i dx}{2y}$ satisfy the same differential equation (see also Remark 3.1).

It is well known that the integral

$$I_\gamma(T) := \pi i \int_\gamma \frac{dx}{y} \qquad (\gamma \in H_1(\mathbb{C}^2 \setminus C_T, \mathbb{Z}))$$

satisfies a second order Fuchsian differential equation (the *Picard-Fuchs differential equation* for $\{C_T\}$). The local exponents of $I_\gamma(T)$ at the singularities $0, 1, -1, \infty$ are respectively given by $0, 0, 0, 1$. Therefore, the differential equation for $I_\gamma(T)$ is of the form

$$(T^3 - T)F''(T) + (3T^2 - 1)F'(T) + (T + A)F(T) = 0 \tag{3.1}$$

for a certain constant A.

To determine this constant A, we take γ as a closed path in the x-plane surrounding just two of the singularities $(\frac{1 \pm 6T + T^2}{12}, 0)$ of the 1-form $\frac{\pi i dx}{y}$, which go to $\frac{1}{12}$ as $T \to 0$, so that this integral $I_\gamma(T)$ gives the *holomorphic* solution of this differential equation (3.1) around the origin $T = 0$. Calculating the integral $I_\gamma(T)$ directly using the residue theorem, we notice that the Taylor expansion of $I_\gamma(T)$ for sufficient small T is given by

$$I_\gamma(T) = 1 + \frac{1}{4}T^2 + \frac{9}{64}T^4 + \cdots. \tag{3.2}$$

From this expansion (3.2), it follows that $A = 0$.

Hence, putting the discussions above together, we now conclude the

Theorem 3.1. *The differential equation $\mathcal{L}(\mathcal{W}_2) = 0$ is the Picard-Fuchs equation associated with the family*

$$C_T \, : \, y^2 = x^3 - \frac{T^4 + 14T^2 + 1}{48}x + \frac{T^6 - 33T^4 - 33T^2 + 1}{864} \tag{3.3}$$

of elliptic curves equipped with rational 4-torsion. □

Remark 3.1. Let $\widetilde{\mathcal{L}}(I_\gamma) = 0$ be the second order linear differential equation for the period $I_\gamma(T)$ of the holomorphic 1-form $\frac{dx}{2y}$ associated to the elliptic curves $\{C_T\}$. From the equation $\widetilde{\mathcal{W}}_2(T) - \mathcal{W}_2(T) = \frac{1}{2}I_\gamma(T)$, we have

$\widetilde{\mathcal{L}}(\widetilde{\mathcal{W}_2}) = \widetilde{\mathcal{L}}(\mathcal{W}_2)$. Hence we see that the monodromy group of $\widetilde{\mathcal{L}}(\mathcal{W}_2)$ is trivial so that it is a rational function. In our case, we know a priori that the function $\mathcal{W}_2(T)$ itself satisfy the second order linear differential equation of the form $\mathcal{L}(\mathcal{W}_2) = 0$, and it follows that the operators $\mathcal{L}$ and $\widetilde{\mathcal{L}}$ are identical, up to a constant multiple.

4 Modular properties

Let $\lambda(\tau)$ be the classical Legendre modular function

$$(4.1)\quad \lambda(\tau) = 16q\prod_{n=1}^{\infty}\left(\frac{1+q^{2n}}{1+q^{2n-1}}\right)^8 = 16\frac{\eta(\tau)^8\eta(4\tau)^{16}}{\eta(2\tau)^{24}} = 16q-128q^2+\cdots,$$

where $\operatorname{Im}\tau > 0$, $q = e^{\tau\pi i}$ and $\eta(\tau) = q^{\frac{1}{24}}\prod_{n=1}^{\infty}(1-q^n)$, the Dedekind eta-function. This $\lambda(\tau)$ is a modular function for the congruent subgroup

$$\Gamma_0(4) = \left\{\begin{pmatrix} a & b \\ c & d\end{pmatrix} \in SL_2(\mathbb{Z})\,;\; c \equiv 0 \,(\mathrm{mod}\, 4)\right\}.$$

Further, it is known that the modular function field for $\Gamma_0(4)$ is equal to $\mathbb{C}(\lambda(\tau))$. Based on the study [14] (see also Remark 4.1), Zagier pointed out the following theorem from the list therein.

Theorem 4.1. *The equality*

$$(4.2)\qquad w_2(\lambda(\tau)) = \frac{\vartheta_0(\tau)^2}{1-\lambda(\tau)} = \frac{\eta(2\tau)^{22}}{\eta(\tau)^{12}\eta(4\tau)^8}$$

holds. Here $\vartheta_0(\tau)$ is the elliptic theta function (which is a $\Gamma_0(4)$-modular form of weight $\frac{1}{2}$)

$$(4.3)\qquad \vartheta_0(\tau) = \frac{\eta(\tau)^2}{\eta(2\tau)} = \sum_{n\in\mathbb{Z}}(-1)^n q^{n^2} = 1-2q+2q^4+\cdots.$$

Notice that $w_2(\lambda)$ is not holomorphic. We show that the theorem is deduced by the same mechanism as the discussion in [1].

Proof. Let $\omega(\tau)$ be a $\Gamma_0(4)$-modular form of weight 1 and put $\tilde{\omega}(\tau) = \tau\omega(\tau)$. Then

$$\omega\left(\frac{a\tau+b}{c\tau+d}\right) = c\,\tilde{\omega}(\tau) + d\,\omega(\tau), \qquad \tilde{\omega}\left(\frac{a\tau+b}{c\tau+d}\right) = a\,\tilde{\omega}(\tau) + b\,\omega(\tau)$$

for $\begin{pmatrix} a & b \\ c & d \end{pmatrix} \in \Gamma_0(4)$. Hence, if we regard ω and $\tilde{\omega}$ as a multivalued functions in λ, then the analytic continuation $\omega_\gamma(\lambda)$ and $\tilde{\omega}_\gamma(\lambda)$ along a closed path γ in the λ-plane corresponding to an element $\begin{pmatrix} a & b \\ c & d \end{pmatrix} \in \Gamma_0(4)$ is given by

$$\tilde{\omega}_\gamma(\lambda) = a\,\tilde{\omega}(\lambda) + b\,\omega(\lambda), \qquad \omega_\gamma(\lambda) = c\,\tilde{\omega}(\lambda) + d\,\omega(\lambda). \tag{4.4}$$

It is immediate that both $\omega(\lambda)$ and $\tilde{\omega}(\lambda)$ satisfy the differential equation

$$\begin{vmatrix} \omega & \omega' \\ \tilde{\omega} & \tilde{\omega}' \end{vmatrix} F'' - \begin{vmatrix} \omega & \omega'' \\ \tilde{\omega} & \tilde{\omega}'' \end{vmatrix} F' + \begin{vmatrix} \omega' & \omega'' \\ \tilde{\omega}' & \tilde{\omega}'' \end{vmatrix} F = 0 \tag{4.5}$$

where the prime denotes the differential with respect to λ. The coefficients in the differential equation are calculated as

$$\begin{vmatrix} \omega & \omega' \\ \tilde{\omega} & \tilde{\omega}' \end{vmatrix} = \frac{\omega^2}{\frac{d\lambda}{d\tau}}, \quad \begin{vmatrix} \omega & \omega'' \\ \tilde{\omega} & \tilde{\omega}'' \end{vmatrix} = \left(\frac{\omega^2}{\frac{d\lambda}{d\tau}}\right)', \quad \begin{vmatrix} \omega' & \omega'' \\ \tilde{\omega}' & \tilde{\omega}'' \end{vmatrix} = \frac{2(\frac{d\omega}{d\tau})^2 - \omega\frac{d^2\omega}{d\tau^2}}{(\frac{d\lambda}{d\tau})^3}. \tag{4.6}$$

Notice that the monodromy groups of these determinants are trivial due to (4.4), so that they are rational functions in λ.

Now we take $\omega(\tau) = \frac{\vartheta_0(\tau)^2}{1-\lambda(\tau)}$. Using well-known formulas among the values of elliptic functions and theta functions, we see that the equations

$$\begin{aligned} \begin{vmatrix} \omega & \omega' \\ \tilde{\omega} & \tilde{\omega}' \end{vmatrix} &= \frac{1}{\pi i}\frac{1}{\lambda(1-\lambda)^2}, \begin{vmatrix} \omega & \omega'' \\ \tilde{\omega} & \tilde{\omega}'' \end{vmatrix} = -\frac{1}{\pi i}\frac{(1-3\lambda)(1-\lambda)}{\lambda^2(1-\lambda)^4}, \\ \begin{vmatrix} \omega' & \omega'' \\ \tilde{\omega}' & \tilde{\omega}'' \end{vmatrix} &= \frac{1}{\pi i}\frac{\lambda - \frac{3}{4}}{\lambda^2(1-\lambda)^4} \end{aligned} \tag{4.7}$$

holds (we postpone the calculation; see Lemma 4.2 below). Thus we find that ω satisfies

$$\lambda(1-\lambda)^2\omega''(\lambda) - (1-3\lambda)(1-\lambda)\omega'(\lambda) + (\lambda - \frac{3}{4})\omega(\lambda) = 0. \tag{4.8}$$

Comparing this differential equation with (0.6), we have proved the lemma.

□

Lemma 4.2. *The equations of determinants in* (4.7) *hold.*

Proof. We use the convention for elliptic and/or theta functions in [4]. Denote by ω_1, ω_2 the fundamental periods so that $\tau = \frac{\omega_2}{\omega_1}$, and by $\vartheta_j(v, \tau)$ $(j = 0, 1, 2, 3)$ the elliptic theta functions. We put $\vartheta_j = \vartheta_j(0, \tau)$ $(j = 0, 2, 3)$ and $\vartheta_1' = \frac{\partial \vartheta_1}{\partial v}(0, \tau)$. We also put $e_1 = \wp\left(\frac{\omega_1}{2}\right)$ and $e_2 = \wp\left(\frac{\omega_1+\omega_2}{2}\right)$, where $\wp(z) = \wp(z; \omega_1, \omega_2)$ is the Weierstrass $\wp$-function. Then we have

$$e_1 = \frac{4\pi i}{\omega_1^2}\left(\frac{1}{3}\frac{d\log\vartheta_1'}{d\tau} - \frac{d\log\vartheta_2}{d\tau}\right), e_2 = \frac{4\pi i}{\omega_1^2}\left(\frac{1}{3}\frac{d\log\vartheta_1'}{d\tau} - \frac{d\log\vartheta_3}{d\tau}\right),$$

and it follows that

$$\vartheta_0^4 = \frac{\omega_1^2}{\pi^2}(e_1 - e_2) = \frac{4}{\pi i}\left(\frac{d\log\vartheta_2}{d\tau} - \frac{d\log\vartheta_3}{d\tau}\right).$$

On the other hand, since $\lambda(\tau) = \left(\frac{\vartheta_2}{\vartheta_3}\right)^4$, we have

$$\frac{d\log\lambda}{d\tau} = 4\left(\frac{d\log\vartheta_2}{d\tau} - \frac{d\log\vartheta_3}{d\tau}\right) = \pi i \vartheta_0^4.$$

Thus we obtain

$$\begin{vmatrix} \omega & \omega' \\ \tilde{\omega} & \tilde{\omega}' \end{vmatrix} = \frac{\omega^2}{\frac{d\lambda}{d\tau}} = \frac{\vartheta_0^4}{\lambda(1-\lambda)^2\frac{d\log\lambda}{d\tau}} = \frac{1}{\pi i}\frac{1}{\lambda(1-\lambda)^2}.$$

The third determinant $\begin{vmatrix} \omega & \omega'' \\ \tilde{\omega} & \tilde{\omega}'' \end{vmatrix}$ is also calculated in the same way, but the calculation is rather complicated. The second determinant $\begin{vmatrix} \omega & \omega'' \\ \tilde{\omega} & \tilde{\omega}'' \end{vmatrix} = -\frac{1}{\pi i}\frac{(1-3\lambda)(1-\lambda)}{\lambda^2(1-\lambda)^4}$ is readily obtained by differentiate $\frac{1}{\pi i}\frac{1}{\lambda(1-\lambda)^2}$ by λ. □

Remark 4.1. Consider the differential equation

$$\left((t^3 + at^2 + bt)F'(t)\right)' + (t - \lambda)F(t) = 0 \tag{4.9}$$

with rational parameters a, b, λ, which is due to Beukers. When $a = 11$, $b = -1$ and $\lambda = -3$, this equation (4.9) is exactly the one for the generating function $A(t)$ of the Apéry numbers a_n. In [14], Zagier searches the triplets

(a, b, λ) of integers within a certain domain such that (4.9) has an integral solution (i.e. solutions in $\mathbb{Z}[[t]]$), and presents a list of 36 such solutions. It is shown that the twelve solutions of them have parametrizations in terms of modular functions. Zagier noticed and pointed out that our generating function $w_2(t)$ of $J_2(n)$ is #19 in his list.

5 Closing remarks

In the final position of the note, we give several remarks for the future study.

5.1

In the discussions above, we see that the numbers $J_2(n)$ arising from the value $\zeta_Q(2)$ acquire the $\Gamma_0(4)$-modularity associated with the family of elliptic curves equipped with rational 4-torsion. However, at this moment, we have no intrinsic explanation from the level of non-commutative harmonic oscillators why such things hold.

5.2

The Apéry-like numbers $J_2(n)$ are arising in the expression (0.4) for $\zeta_Q(2)$ via the generating function

$$g_2(z) = \sum_{n=0}^{\infty} \binom{-\frac{1}{2}}{n} J_2(n) z^n.$$

This function satisfies the differential equation $\mathcal{D}(g_2) = 0$ (see [8]). Here $\mathcal{D}$ is given by

$$\mathcal{D} = 8z^2(1+z)^2 \frac{d^3}{dz^3} + 24z(1+z)(1+2z)\frac{d^2}{dz^2} + 2(4+27z+27z^2)\frac{d}{dz} + 3(1+2z).$$

Does the function $g_2(z)$ have a modular form interpretation like $w_2(t)$? If this is true, then $g_2(z)$ should be a modular form of weight 2 from the result in [13]. We also note that $g_2(z)$ has the following explicit expressions by hypergeometric functions [9]:

$$g_2(z) = \frac{1}{\sqrt{1+z^2}}\, {}_3F_2\left(\frac{1}{2}, \frac{1}{2}, \frac{1}{2}; 1, 1; \frac{z^2}{1+z^2}\right) = {}_2F_1\left(\frac{1}{4}, \frac{3}{4}; 1; -z^2\right)^2.$$

5.3

The Apéry numbers a_n and b_n are corresponding to $\zeta(2)$, satisfying the same recurrence formula. Their generating functions satisfy the differential equations $L_2(A) = 0$ and $L_2(B) = -5$ for the same operator L_2 but one is homogeneous and the other is inhomogeneous.

Denote by α_n and β_n the Apéry numbers corresponding to $\zeta(3)$. They satisfy the same recurrence formula

$$n^3 u_n - (34n^3 - 51n^2 + 27n - 5)u_{n-1} + (n-1)^3 u_{n-2} = 0$$

with initial conditions $\alpha_0 = 0, \alpha_1 = 6$ and $\beta_0 = 1, \beta_1 = 5$. Their generating functions $\mathbb{A}(t) = \sum_{n=0}^{\infty} \alpha_n t^n$ and $\mathbb{B}(t) = \sum_{n=0}^{\infty} \beta_n t^n$ satisfy the differential equations $L_3(\mathbb{A}) = 0$ and $L_3(\mathbb{B}) = 5$ for the same operator

$$L_3 = t^2(t^2 - 34t + 1)\frac{d^3}{dt^3} + 3t(2t^2 - 51t + 1)\frac{d^2}{dt^2} + (7t^2 - 112t + 1)\frac{d}{dt} + (t - 5),$$

but one is homogeneous and the other is inhomogeneous. Thus the situation is exactly the same for the Apéry numbers corresponding to $\zeta(2)$. It was proved in [2] that the differential equation $L_3(\mathbb{A}) = 0$ is the Picard-Fuchs equation associated to a certain family of $K3$ surfaces.

In our spectral case, the situation seems similar at a glance, but it is different. The Apéry-like numbers $J_2(n)$ are corresponding to the special value $\zeta_Q(2)$, while

$$J_3(n) := 8 \int_0^1 \int_0^1 \int_0^1 \left(\frac{(1 - X^4)(1 - Y^4 Z^4)}{(1 - X^2 Y^2 Z^2)^2} \right)^n \frac{dZ dY dZ}{1 - X^2 Y^2 Z^2} \tag{5.1}$$

are corresponding to the special value $\zeta_Q(3)$ (see [8]). Let $w_3(t)$ be the generating function for $J_3(n)$ defined by $w_3(t) = \sum_{n=0}^{\infty} J_3(n) t^n$. The generating functions $w_2(t)$ and $w_3(t)$ satisfy the differential equations

$$D(w_2) = 0 \qquad \text{and} \qquad D(w_3) = \frac{1}{2}\, {}_2F_1\left(1, 1; \frac{3}{2}; t\right) \tag{5.2}$$

respectively for the same operator

$$D = t(1 - t^2)\frac{d^2}{dt^2} - (1 - 3t)(1 - t)\frac{d}{dt} + (t - \frac{3}{4}).$$

Notice that the former equation is homogeneous, but the latter is inhomogeneous. This fact shows that the structure of 'pairing' in our theory is different from the case of the Apéry numbers.

5.4

From the expressions (0.3) and (5.1) of the Apéry-like numbers associated to $\zeta_Q(2)$ and $\zeta_Q(3)$, it is quite natural to consider the numbers $J_k(n)$ defined as

$$\begin{aligned}&\int_0^\infty\int_0^\infty\cdots\int_0^\infty \frac{e^{-(t_1+\cdots+t_k)/2}}{1-e^{-(t_1+\cdots+t_k)}}\left(\frac{(1-e^{-2t_1})(1-e^{-2(t_2+\cdots+t_k)})}{(1-e^{-(t_1+\cdots+t_k)})^2}\right)^n dt_1dt_2\cdots dt_k\\&=2^k\int_0^1\int_0^1\cdots\int_0^1\frac{1}{1-x_1^2\cdots x_k^2}\left(\frac{(1-x_1^4)(1-x_2^4\cdots x_k^4)}{(1-x_1^2\cdots x_k^2)^2}\right)^n dx_1dx_2\cdots dx_k.\end{aligned}$$

Note that the numbers $J_k(n)$ are all positive. Their generating functions are

$$\begin{aligned}w_k(z)&:=\sum_{n=0}^\infty J_k(n)z^n\\&=2^k\int_0^1\int_0^1\cdots\int_0^1\frac{1-x_1^2\cdots x_k^2}{(1-x_1^2\cdots x_k^2)^2-(1-x_1^4)(1-x_2^4\cdots x_k^4)z}dx_1dx_2\cdots dx_k\\&=2^k\int_0^1\int_0^1\cdots\int_0^1\frac{1-x_1^2\cdots x_k^2}{(1-z)(1-x_1^2\cdots x_k^2)^2+z(x_1^2-x_2^2\cdots x_k^2)^2}dx_1dx_2\cdots dx_k\\&=\frac{2^{k-1}}{1-z}\left(\int_0^1\int_0^1\cdots\int_0^1\frac{dx_1dx_2\cdots dx_k}{(1-x_1^2x_2^2\cdots x_k^2)+Z(x_1^2-x_2^2\cdots x_k^2)}\right.\\&\qquad\left.+\int_0^1\int_0^1\cdots\int_0^1\frac{dx_1dx_2\cdots dx_k}{(1-x_1^2x_2^2\cdots x_k^2)-Z(x_1^2-x_2^2\cdots x_k^2)}\right)\end{aligned}$$

where $Z=\sqrt{\frac{z}{z-1}}$. It is not hard to see the formula $w_k(0)=(2^k-1)\zeta(k)$. Then, we can ask the question; is it true that the function $w_k(z)$ $(k\geq 3)$ satisfies a Picard-Fuchs type differential equation coming from certain family of algebraic varieties?

We remark that we have the double integral expression for $J_k(n)$

$$J_k(n)=\frac{2^k}{\Gamma(k-1)}\int_0^1\int_0^1\frac{(-\log X)^{k-2}}{1-X^2Y^2}\left(\frac{(1-X^4)(1-Y^4)}{(1-X^2Y^2)^2}\right)^n dXdY,$$

as well as for $w_k(z)$

$$w_k(z)=\frac{(-1)^k}{1-z}\frac{2^{k-1}}{\Gamma(k-1)}\int_0^1\int_0^1\frac{(\log X)^{k-2}+(\log Y)^{k-2}}{(1-X^2Y^2)+Z(X^2-Y^2)}dXdY.$$

Acknowledgement

The second author would like to thank Lin Weng and Masanobu Kaneko, the organizers of the *Conference on L-functions*, for presenting him the precious occasion to give a talk on the subject. We are also very grateful to Don Zagier, who gave us an accurate suggestion and his strong interest on our Apéry-like numbers, and kindly sent us his unpublished note [14].

References

[1] Beukers, F.: Irrationality of π^2, periods of an elliptic curve and $\Gamma_1(5)$. Diophantine approximations and transcendental numbers (Luminy, 1982), 47–66, Progr. Math., **31**, Birkhäuser, Boston, Mass., 1983.

[2] Beukers, F. and Peters, C. A. M.: A family of $K3$ surfaces and $\zeta(3)$. *J. Reine Angew. Math.* **351** (1984), 42–54.

[3] Deutsch, E. and Sagan, B. E.: Congruences for Catalan and Motzkin numbers and related sequences. *J. Number Theory* **117** (2006), no. 1, 191–215.

[4] Hurwitz, A. and Courant, R.: "Vorlesungen über allgemeine Funktionentheorie und elliptische Funktionen." Interscience Publishers, Inc., New York, (1944).

[5] Ichinose, T. and Wakayama, M.: Zeta functions for the spectrum of the non-commutative harmonic oscillators. *Commun. Math. Phys.* **258** (2005), 697–739.

[6] Ichinose, T. and Wakayama, M.: Special values of the spectral zeta function of the non-commutative harmonic oscillator and confluent Heun equations. *Kyushu J. Math.* **59** (2005), 39–100.

[7] Kubert, D. S.: Universal bounds on the torsion of elliptic curves. *Proc. London Math. Soc.* **33** (1976), 193–237.

[8] Kimoto, K. and Wakayama, M.: Apéry-like numbers arising from special values of spectral zeta functions for non-commutative harmonic oscillators. *Kyushu J. Math.* **60** (2006), 383–404.

[9] Ochiai, H.: A special value of the spectral zeta function of the non-commutative harmonic oscillators. To appear in *Ramanujan J.* (ArXiv: `math.RT/0507111`).

[10] Parmeggiani, A. and Wakayama, M.: Oscillator representations and systems of ordinary differential equations. *Proc. Natl. Acad. Sci. USA* **98** (2001), 26–30.

[11] Van der Poorten, A. J.: A proof that Euler missed... Apéry's proof of the irrationality of $\zeta(3)$. *Math. Intelligencer* **1** (1978/1979), 195–203.

[12] Silverman, J. H. and Tate, J.: "Rational points on elliptic curves." Undergraduate Texts in Mathematics. Springer-Verlag, New York, 1992.

[13] Zagier, D.: Modular forms and differential operators. *Proc. Indian Acad. Sci. Math. Sci.* **104** (1994), no 1, 57–75.

[14] Zagier, D.: Integral solutions of Apéry-like recurrence equations. Preprint.

Kazufumi KIMOTO
Department of Mathematical Science,
University of the Ryukyus,
Senbaru, Nishihara,
Okinawa 903-0231, Japan
Email: kimoto@math.u-ryukyu.ac.jp

Maasato WAKAYAMA
Faculty of Mathematics,
Kyushu University,
Hakozaki,
Fukuoka 812-8518, Japan
Email: wakayama@math.kyushu-u.ac.jp

A Geometric Approach to L-Functions

L. WENG

In this paper, we initiate a geometric approach to the theory of L-functions. It consists of two parts: Chapters 1-4, used to be called 'Non-Abelian L-Functions for Number Fields', reveal a general theory, and Chapters 5-9 give a detailed account of ranks two and three zetas.

The starting point is a stability for lattices in terms of intersection. This is followed by a new yet genuine geo-arithmetical cohomology, which satisfies Serre duality and Riemann-Roch. Based on them, then we introduce high rank zeta functions for number fields and establish their basic properties. At it turns out, high rank zetas are canonical: When rank is one, our zetas give the well-known Dedekind zetas; moreover, they can be meromorphic extended, satisfy functional equations, and admit only two singularities with the same residues given by volumes of certain compact moduli spaces of stable lattices. All this is done in Chapter 1.

By singling out the volume part, using Mellin transform, we find that high rank zetas can also be understood as integrations of certain Epstein type zetas over the above mentioned compact moduli spaces. Note that, despite we call them zetas, Epstein type zetas are indeed a special kind of Eisenstein series. This then leads to our definition of non-abelian L-functions (as integrations of relative Eisenstein series associated to L^2-automorphic forms on some far-reaching generalizations of moduli spaces).

As such, basic properties such as meromorphic continuations, functional equations and singularities of these non-abelian L-functions are also studied based on Langlands' fundamental theory on Eisenstein series associated L^2-automorphic forms. This is given in Chapter 2.

With our primitive objects introduced and basic properties being established, our next aim is to study them quantitatively. This then leads to the discussion in Chapter 3, in which we successfully relate our geo-arithmetical truncation with Arthur's analytic truncation, a result already appeared in Lafforgue's work on Langlands conjecture for curves defined over finite fields. Built on it, in Chapter 4, we develop an advance version of Rankin-Selberg & Zagier method, by generalizing a result of Jacquet-Lapid-Rogowski. As a direct application of this, we then give a general formula for what we call high rank L-functions associated to cusp forms.

Surely, the point here is that unlike in Arthur's theory of analytic truncations, parameters used in our natural geo-arithmetical truncation are just normal, or better, are not sufficiently regular. So additional cares have to be taken. After this settled, at the end of Chapter 4, we give an explicit formula for the volume of mentioned moduli space based on an earlier work of Henry Kim and the author. This then reveals an intrinsic relation between special values of Dedekind zetas and that of our zetas, a result rooted back to Siegel and Langlands.

From Chapter 5, we concentrate our attention to the study of our zeta functions in ranks two and three. In particular, in Chapter 6, we show that a lattice of rank two is semi-stable if and only if the generalized Siegel type distances of the corresponding modular point in the upper half space model to all cusps are bigger or equal to 1. This is a beautiful fact. Based on this, in Chapter 7, we then give an explicit expression of rank two zetas in terms of Dedekind zetas, by further developing Rankin-Selberg & Zagier method in a classical fashion with the help of our study on Fourier expansions for Epstain type zeta functions involved. Meeting everyone's expectation, the fact exposed in Chapter 8 then is that all zeros of rank two zetas lie on the central line Re $= 1/2$, an analogue of the standard Riemann Hypothesis for Dedekind zetas, based on a method of Suzuki and Lagarias who first prove the result for the field of rationals.

Finally, in Chapter 9, we give a detailed account of rank three zeta for the field of rationals. Our approach uses classical Eisenstein series such as Koecher zeta functions and Siegel's Eisenstein series and the powerful formula in Chapter 4 mentioned above. The up-shot is a precise formula for the associated rank three zeta. We end our presentation by the following recent result of Suzuki: All zeros of this rank three zeta lie on the central line Re $= 1/2$ as well. As such, then it appears that the time is ripe to openly make the following generalization of the Riemann Hypothesis:

The RH. *All zeros of zetas, rank one or higher, lie on the central line* $\mathrm{Re}\, s = \frac{1}{2}$.

Contents

1 High Rank Zetas for Number Fields

1.1 Iwasawa's ICM Talk on Dedekind Zetas

We start with a discussion on Dedekind zeta functions. However, we will not adopt the classical approach, rather we would like to recall Iwasawa's interpretation.

Let F be a number field. Denote by S the collection of all (inequivalent) normalized places of F. Set S_∞ be the collection of all Archimedean places of F and $S_{\text{fin}} := S \backslash S_\infty$.

Denote by $\mathbf{I}$ the idele group of F, $N : \mathbf{I} \to \mathbf{R}_{\geq 0}$ and $\deg : \mathbf{I} \to \mathbf{R}$ the norm map and the degree map on ideles respectively. Also introduce the following subgroups of $\mathbf{I}$:

$$\begin{aligned} \mathbf{I}^0 :=&\{a = (a_v) \in \mathbf{I} : \deg(a) = 0\}, \\ F^* :=&\{a = (a_v) \in \mathbf{I}^0 : a_v = \alpha \in F\backslash\{0\}, \forall v \in S\}, \\ U :=&\{a = (a_v) \in \mathbf{I}^0 : |a_v|_v = 1 \forall v \in S\}, \\ \mathbf{I}_{\text{fin}} :=&\{a = (a_v) \in \mathbf{I} : a_v = 1 \forall v \in S_\infty\}, \\ \mathbf{I}_\infty :=&\{a = (a_v) \in \mathbf{I} : a_v = 1 \forall v \in S_{\text{fin}}\}. \end{aligned}$$

Set $U_{\text{fin}} := U \cap \mathbf{I}_{\text{fin}}$. Then, with respect to the natural topology on $\mathbf{I}$, we have

(1) $F^* \hookrightarrow \mathbf{I}$ is discrete and $\mathbf{I}^0/F^*$ is compact. Write $\mathrm{Pic}(F) = \mathbf{I}/F^*$;

(2) $U \hookrightarrow \mathbf{I}$ is compact;

(3) $U_{\text{fin}} \hookrightarrow \mathbf{I}_{\text{fin}}$ is both open and compact. Moreover, the morphism I : $[a = (a_v)] \mapsto I(a) := \prod_{v \in S_{\text{fin}}} P_v^{\mathrm{ord}_v(a_v)}$ induces an isomorphism between $\mathbf{I}_{\text{fin}}/U_{\text{fin}}$ and the ideal group of F, where P_v denotes the maximal ideal of the ring of integers O_{F_v} corresponding to the place v; and, $N(a) = \prod_{v \in S} |a_v|_v^{N_v := [F_v : \mathbf{Q}_p]} = N(I(a))^{-1}$ with $N(I(a))$ the norm of the ideal $I(a)$;

(4) $\mathbf{I} = \mathbf{I}_{\text{fin}} \times \mathbf{I}_\infty$. Hence we may write an idele a as $a = a_{\text{fin}} \cdot a_\infty$ with $a_{\text{fin}} \in \mathbf{I}_{\text{fin}}$ and $a_\infty \in \mathbf{I}_\infty$ respectively. In particular, if $d\mu(a)$ denotes the normalized Haar measure on $\mathbf{I}$ as say in Weil's Basic Number Theory, we have

$$d\mu(a) = d\mu(a_{\text{fin}}) \cdot d\mu(a_\infty)$$

corresponding to the decomposition $\mathbf{I} = \mathbf{I}_{\text{fin}} \times \mathbf{I}_\infty$.

Set

$$e(a_{\rm fin}) := \begin{cases} 1, & \text{if } I(a_{\rm fin}) \subset O_F; \\ 0, & \text{otherwise}, \end{cases}$$

$$e(a_\infty) = \exp\Big(-\pi \sum_{v:\mathbf{R}} a_v^2 - 2\pi \sum_{v:\mathbf{C}} |a_v|^2 \Big)$$

and

$$e(a_{\rm fin} \cdot a_\infty) = e(a_{\rm fin}) \cdot e(a_\infty),$$

for $a_{\rm fin} \in \mathbf{I}_{\rm fin}$ and $a_\infty \in \mathbf{I}_\infty$, Denote by Δ_F (the absolute values of) the discriminant of F, r_1 and r_2 the number of real and complex places in S_∞ as usual.

With this preparation in notation, next we recall Iwasawa's interpretation of Dedekind zeta function for F in the form suitable for our later study. This goes as follows.

For $s \in \mathbf{C}, \mathrm{Re}(s) > 1$,

$$\begin{aligned}
\xi_F(s) &:= \Delta_F^{\frac{s}{2}} \Big(2\pi^{-\frac{s}{2}} \Gamma\Big(\frac{s}{2}\Big)\Big)^{r_1} \Big((2\pi)^{-s} \Gamma(s) \Big)^{r_2} \sum_{0 \neq I \subset O_F} N(I)^{-s} \\
&= \Delta_F^{\frac{s}{2}} \cdot \sum_{0 \neq I \subset O_F} N(I)^{-s} \\
&\quad \times \int_{t_v \in F_v, v \in S_\infty} \prod_v |t_v|_v^s \exp\Big(-\pi \sum_{v:\mathbf{R}} t_v^2 - 2\pi \sum_{v:\mathbf{C}} |t_v|^2 \Big) \prod_v d^* t_v \\
&= \Delta_F^{\frac{s}{2}} \cdot \sum_{0 \neq I \subset O_F} N(I)^{-s} \cdot \int_{\mathbf{I}_\infty} N(a_\infty)^s e(a_\infty) d\mu(a_\infty) \\
&= \Delta_F^{\frac{s}{2}} \cdot \frac{1}{\mathrm{vol}(U_{\rm fin})} \\
&\quad \times \Big(\int_{\mathbf{I}_{\rm fin}} N(a_{\rm fin})^s e(a_{\rm fin}) d\mu(a_{\rm fin}) \cdot \int_{\mathbf{I}_\infty} N(a_\infty)^s e(a_\infty) d\mu(a_\infty) \Big) \\
&= \frac{1}{\mathrm{vol}(U_{\rm fin})} \cdot \Delta_F^{\frac{s}{2}} \\
&\quad \times \int_{\mathbf{I}_{\rm fin} \times \mathbf{I}_\infty} \Big(N(a_{\rm fin}) N(a_\infty) \Big)^s \Big(e(a_{\rm fin}) e(a_\infty) \Big) \Big(d\mu(a_{\rm fin}) d\mu(a_\infty) \Big) \\
&= \frac{1}{\mathrm{vol}(U_{\rm fin})} \cdot \Delta_F^{\frac{s}{2}} \cdot \int_{\mathbf{I}} N(a)^s e(a) d\mu(a).
\end{aligned}$$

Now denote by $d\mu([a])$ the induced Haar measure on the Picard group $\mathrm{Pic}(F) := \mathbf{I}/F^*$. Note that

$$e(a_{\mathrm{fin}}a_\infty) := \begin{cases} e(a_\infty), & \text{if } I(a_{\mathrm{fin}}) \subset O_F, \\ 0, & \text{otherwise,} \end{cases}$$

and that by (1) above, $F^* \hookrightarrow \mathbf{I}$ is discrete, (hence taking integration over F^* means taking summation), we get

$$\begin{aligned}
&\Big(\mathrm{vol}(U_{\mathrm{fin}}) \cdot \Delta_F^{\frac{-s}{2}}\Big) \cdot \xi_F(s) \\
&= \int_{\mathbf{I}/F^*} \Big(\int_{F^*} N(\alpha a)^s e(\alpha a) d\alpha \Big) d\mu([a]) \\
&= \int_{\mathbf{I}/F^*} \Big(\sum_{\alpha \in F^*} e(\alpha a) \Big) \cdot N([a])^s d\mu([a]) \\
&\qquad \text{(by the product formula)} \\
&= \int_{\mathbf{I}/F^*} N([a])^s d\mu([a]) \cdot \Big(\sum_{\alpha \in F^*} e(\alpha a_{\mathrm{fin}}) e(\alpha a_\infty) \Big) \\
&= \int_{\mathbf{I}/F^*} N([a])^s d\mu([a]) \cdot \Big(\sum_{\alpha \in F^*, I(\alpha a_{\mathrm{fin}}) \subset O_F} e(\alpha a_\infty) \Big) \\
&= \int_{\mathbf{I}/F^*} N([a])^s d\mu([a]) \cdot \Big(\sum_{\alpha \in F^*, \alpha a_v \subset O_v, \forall v \in S_{\mathrm{fin}}} e(\alpha a_\infty) \Big).
\end{aligned}$$

For an idele $L = (a_v)$, define the 0-th algebraic cohomology group of the idele L^{-1} by

$$H^0(F, L^{-1}) := \{\alpha \in F^*, \alpha a_v \subset O_v, \forall v \in S_{\mathrm{fin}}\} = I(L^{-1});$$

moreover for the associated idele class, introduce its associated geometric 0-th cohomology via

$$\begin{aligned}
&h^0(F, L^{-1}) \\
&:= \log\Big(\sum_{\alpha \in H^0(F, L^{-1}) \setminus \{0\}} \exp\Big(-\pi \sum_{v:\mathbf{R}} |g_v \alpha|^2 - 2\pi \sum_{v:\mathbf{C}} |g_v \alpha|^2 \Big)\Big).
\end{aligned}$$

Then easily, we have

$$\begin{aligned}
&\Big(\mathrm{vol}(U_{\mathrm{fin}})\cdot\Delta_F^{-\frac{s}{2}}\Big)\cdot\xi_F(s)\\
&=\int_{\mathrm{Pic}(F)}\frac{1}{w_F}\sum_{\alpha\in H^0(F,L^{-1})\setminus\{0\}}\exp\Big(-\pi\sum_{v:\mathbf{R}}|g_v\alpha|^2-2\pi\sum_{v:\mathbf{C}}|g_v\alpha|^2\Big)\\
&\qquad\cdot N(L)^s d\mu(L)\\
&=\int_{\mathrm{Pic}(F)}\frac{1}{w_F}\Big(e^{h^0(F,L^{-1})}-1\Big)N(L)^s d\mu(L)\\
&=\frac{1}{w_F}\cdot\int_{\mathrm{Pic}(F)}\Big(e^{h^0(F,L)}-1\Big)N(L)^{-s} d\mu(L)\\
&=\int_{\mathrm{Pic}(F)}\frac{e^{h^0(F,L)}-1}{\#\mathrm{Aut}(L)}\cdot\Big(e^{\deg(L)}\Big)^{-s}d\mu(L).
\end{aligned}$$

Here w_F denotes the number of units of F, and $\#\mathrm{Aut}(L)(=w_F)$ denotes the number of automorphisms of L.

With this said, it is quite clear how a high rank zeta for number field should be introduced. The key points are as follows:
(1) We should develop a geo-ari cohomology for which Serre type duality and Riemann-Roch type results must hold;
(2) We should use moduli spaces of high rank lattices, instead of the Picard, which works only for rank 1.

However, by doing this, then a serious problem appears: In the case of rank 1, the Picard can be factorized as the product of Picard zero, which is compact, with the volume part. For higher ranks, even we have a similar decomposition, the analogue of Picard zero is no longer compact. Hence, convergence problem does arise. It is for this, borrowing a powerful idea from algebraic geometry,
(2)′ We should introduce a suitable stability in terms of intersection for hermitian bundles over number fields, or better, for lattices.

1.2 A Geo-Arithmetical Cohomology

1.2.1 Cohomology Groups

For number field F, denote by $\mathbb{A}=\mathbb{A}_F$ its ring of adeles, by $\mathrm{GL}_r(\mathbb{A})$ the rank r general linear group over $\mathbb{A}$, and write $\mathbb{A}:=\mathbb{A}_{\mathrm{fin}}\oplus\mathbb{A}_\infty$, $\mathrm{GL}_r(\mathbb{A}):=$

$\mathrm{GL}_r(\mathbb{A}_{\mathrm{fin}}) \times \mathrm{GL}_r(\mathbb{A}_\infty)$ according to finite and infinite partition. Also for simplicity, we often use v (resp. σ) to denote elements in S_{fin} (resp. S_∞).

For any $g = (g_{\mathrm{fin}} : g_\infty) = (g_v; g_\sigma) \in \mathrm{GL}_r(\mathbb{A})$, define an injective morphism $i(g) := i(g_\infty) : F^r \to \mathbb{A}^r$ by $f \mapsto (f; g_\sigma \cdot f)$. Let $F^r(g) := \mathrm{Im}(i(g))$ and set

$$\mathbb{A}^r(g) := \left\{ (a_v; a_\sigma) \in \mathbb{A}^r : \begin{array}{c} \text{(i) } g_v(a_v) \in O_v^r,\ \forall v; \\ \text{(ii) } \exists f \in F^r \text{ such that} \\ \text{(ii.a) } g_v(f) \in O_v^r,\ \forall v; \quad \text{(ii.b) } (f; a_\sigma) = i(g_\infty)(f) \end{array} \right\}$$

Then we have the following 9-diagram with exact columns and rows:

$$\begin{array}{ccccc} 0 & & 0 & & 0 \\ \downarrow & & \downarrow & & \downarrow \\ \mathbb{A}^r(g) \cap F^r(g) & \rightarrowtail & \mathbb{A}^r(g) & \twoheadrightarrow & \mathbb{A}^r(g)/\mathbb{A}^r(g) \cap F^r(g) \\ \downarrow & & \downarrow & & \downarrow \\ F^r(g) & \rightarrowtail & \mathbb{A}^r & \twoheadrightarrow & \mathbb{A}^r/F^r(g) \\ \downarrow & & \downarrow & & \downarrow \\ F^r(g)/\mathbb{A}^r(g) \cap F^r(g) & \rightarrowtail & \mathbb{A}^r/\mathbb{A}^r(g) & \twoheadrightarrow & \mathbb{A}^r/\mathbb{A}^r(g) + F^r(g) \\ \downarrow & & \downarrow & & \downarrow \\ 0 & & 0 & & 0. \end{array}$$

Guided by Chevelley's adelic cohomology theory for divisors over algebraic curves, (see e.g, [S]), we introduce the following

Definition. *For any $g \in \mathrm{GL}_r(\mathbb{A})$, define its 0-th and 1-st geo-arithmetical cohomology groups by*

$$H^0(\mathbb{A}_F, g) := \mathbb{A}^r(g) \cap F^r(g), \text{ and } H^1(\mathbb{A}_F, g) := \mathbb{A}^r/\mathbb{A}^r(g) + F^r(g).$$

Proposition 1. (Serre Duality = Pontryagin Duality) *As locally compact groups,*

$$H^1(\mathbb{A}_F, g) \simeq \widehat{H^0(\mathbb{A}_F, \kappa_F \otimes g^{-1})}.$$

Here κ_F denotes an idelic dualizing element of F, and $\hat{\ }$ the Pontryagin dual. In particular, H^0 is discrete and H^1 is compact.

Remark. For $v \in S_{\text{fin}}$, denote by ∂_v the local different of F_v, the v-completion of F, and by O_v the valuation ring with π_v a local parameter. Then $\partial_v =: \pi_v^{\text{ord}_v(\partial_v)} \cdot O_v$. We call $\kappa_F := (\partial_v^{\text{ord}_v(\partial_v)}; 1) \in \mathbb{I}_F := \text{GL}_1(\mathbb{A})$ an idelic dualizing element of F.

Proof. As usual, fix a basic character χ on $\mathbb{A}$ by $\chi =: (\chi_v^{(r)}; \chi_\sigma^{(r)})$ where $\chi_v := \lambda_v \circ \text{Tr}_{\mathbb{Q}_v}^{F_v}$ with $\lambda_v : \mathbb{Q}_v \to \mathbb{Q}_v/\mathbb{Z}_v \hookrightarrow \mathbb{Q}/\mathbb{Z} \hookrightarrow \mathbb{R}/\mathbb{Z}$, and $\chi_\sigma := \lambda_\infty \circ \text{Tr}_{\mathbb{R}}^{F_\sigma}$ with $\lambda_\infty : \mathbb{R} \to \mathbb{R}/\mathbb{Z}$. Then the pairing $(x, y) \mapsto e^{2\pi i \chi(x \cdot y)}$ induces natural isomorphisms $\widehat{\mathbb{A}^r} \simeq \mathbb{A}^r$ (as locally compact groups) and $(F^r)^\perp \simeq F^r$ (as discrete subgroups). With this, a direct local calculation, see e.g, [T], shows that $(\mathbb{A}^r(g))^\perp = \mathbb{A}^r(\kappa_F \otimes g^{-1})$ and $(F^r(g))^\perp \simeq F^r(\kappa_F \otimes g^{-1})$. This completes the proof since

$$\left(\mathbb{A}^r(g) \cap F^r(g)\right)^\perp = (\mathbb{A}^r(g))^\perp + (F^r(g))^\perp.$$

□

1.2.2 Geo-Arithmetical Counts

Motivated by the Pontryagin duality and that the dimensions of a vector space and its dual are the same, one basic principle we adopt in counting locally compact groups is the following:

Counting Axiom. *If $\#_{\text{ga}}$ counts a certain class of locally compact groups G, then $\#_{\text{ga}}(G) = \#_{\text{ga}}(\hat{G})$.*

Practically, our counts of arithmetic cohomology groups are based on Fourier inverse formula, or more accurately, the Plancherel formula in Fourier analysis over locally compact groups. (See e.g, [Fo].)

While any reasonable test function on $\mathbb{A}^r$ would do, as a continuation of a more traditional mathematics and also for simplicity, we set $f := \prod_v f_v \cdot \prod_\sigma f_\sigma$, where
(i) f_v is the characteristic function of O_v^r;
(ii.a) $f_\sigma(x_\sigma) := e^{-\pi|x_\sigma|^2/2}$ if σ is real; and
(ii.b) $f_\sigma(x_\sigma) := e^{-\pi|x_\sigma|^2}$ if σ is complex.
Moreover, we take the following normalization for the Haar measure dx, which we call *standard*, on $\mathbb{A}$: locally

for v, dx is the measure for which O_v gets the measure $N(\partial_v)^{-1/2}$;

while for σ real (resp. complex), dx is the ordinary Lebesgue measure (resp. twice the ordinary Lebesgue measure).

Definition. (1) *The geo-arithmetical counts of the 0-th and the 1-st cohomology groups for* $g \in \mathrm{GL}_r(\mathbb{A})$ *are*

$$\#_{\mathrm{ga}}(H^0(\mathbb{A}_F, g)) :=\#_{\mathrm{ga}}\Big(H^0(\mathbb{A}_F, g); f, dx\Big) := \int_{H^0(\mathbb{A}_F,g)} |f(x)|^2 dx;$$
$$\#_{\mathrm{ga}}(H^1(\mathbb{A}_F, g)) :=\#_{\mathrm{ga}}\Big(H^1(\mathbb{A}_F, g); \hat{f}, d\xi\Big) := \int_{H^1(\mathbb{A}_F,g)} |\hat{f}(\xi)|^2 d\xi.$$

Here dx denotes (the restriction of) the standard Haar measure on $\mathbb{A}$, $d\xi$ (the induced quotient measure from) the dual measure (with respect to χ), and $\hat{f}$ the corresponding Fourier transform of f;
(2) *The 0-th and the 1-st geo-arithmetical cohomologies of* $g \in \mathrm{GL}_r(\mathbb{A})$ *are*

$$h^0(\mathbb{A}_F, g) := \log\Big(\#_{\mathrm{ga}}(H^0(\mathbb{A}_F, g))\Big);$$
$$h^1(\mathbb{A}_F, g) := \log\Big(\#_{\mathrm{ga}}(H^1(\mathbb{A}_F, g))\Big).$$

Remark. Even groups $H^0(\mathbb{A}_F, g)$ and $H^1(\mathbb{A}_F, g)$ do depend on $g \in \mathrm{GL}_r(\mathbb{A})$, from definition, it is easy to see that their counts $h^0(\mathbb{A}_F, g)$ and $h^1(\mathbb{A}_F, g)$ depend only on the class of $g \in \mathrm{GL}_r(F)\backslash\mathrm{GL}_r(\mathbb{A})$.

1.2.3 Serre Duality and Riemann-Roch

For arithmetic cohomologies just introduced, we have the following

Theorem. *For any* $g \in \mathrm{GL}_r(F)\backslash\mathrm{GL}_r(\mathbb{A})$,
(1) (Serre Duality) $h^1(\mathbb{A}_F, g) = h^0(\mathbb{A}_F, \kappa_F \otimes g^{-1})$;
(2) (Riemann-Roch Theorem)

$$h^0(\mathbb{A}_F, g) - h^1(\mathbb{A}_F, g) = \deg(g) - \frac{r}{2} \cdot \log|\Delta_F|.$$

Proof. By the choice of our f in the definition, (1) is a direct consequence of the topological Serre duality, i.e, Prop. 1 in 1.2.1, and the Plancherel Formula; while (2) is a direct consequence of the Serre duality just proved and Tate's Riemann-Roch theorem ([T, Thm. 4.2.1] and/or [Neu, XIV, §6]), i.e, the Poisson summation formula, by the fact that $\Big(H^0(\mathbb{A}_F, g)\Big)^\perp = H^0(\mathbb{A}_F, \kappa_F \otimes g^{-1})$. □

Often, we also write $e^{h^i(\mathbb{A}_F,g)} := H^i_{\rm ga}(F,g),\ i = 1,2.$

Remarks. (1) Our work here is motivated by the works of Weil [W2], Tate [T], Iwasawa [Iw1,2], van der Geer-Schoof [GS], and Li [Li], as well as the works of Lang [L1,2], Arakelov [L3], Szpiro [L3], Parshin [Pa], Moreno [Mo], Neukirch [Ne], Deninger [De 1,2], Connes [Co], and Borisov [Bor]. Also, it would be extremely interesting if one could relate the work here with that of Connes [Co] and Deninger [De1,2,3].
(2) One may apply the discussion here to wider classes of (multiplicative) characters and test functions so as to obtain a more general class of L-functions as done in Tate's Thesis.
(3) Instead of working on S_F, the set of equivalence classes of valuations of F, or better, normalized valuations over F, one may also work on the set of *all* valuations of F, which in particular is parametrized by $S_F \times \mathbb{R}$ to get a much more refined geo-arithmetical cohomology theory.

To understand the above theory, we suggest the reader to use language of lattices (whose details will be recall in the following section): Denote by O_F the ring of integers of F. Then for an O_F-lattice Λ of rank r, its geo-arithmetical cohomology groups are defined to be locally compact groups

$$H^0(F,\Lambda) := \Lambda \qquad \text{and} \qquad H^1(F,\Lambda) := (\mathbb{R}^{r_1} \times \mathbb{C}^{r_2})^r/\Lambda.$$

Consequently, then
(i) $H^0(F,\Lambda)$ is discrete while $H^1(F,\Lambda)$ is compact;
(ii) The Serre duality coincides with the Pontryagin duality, say in the case $F = \mathbb{Q}$,

$$\Big((\mathbb{R}^{r_1} \times \mathbb{C}^{r_2})^r/\Lambda\Big)^\vee \simeq \Lambda;$$

(iii) the 0-th cohomology is then simply given by

$$h^0(F,\Lambda) := \log\Big(\sum_{\mathbf{x}\in\Lambda} \exp\Big(-\pi\sqrt{-1}\sum_{\sigma:\mathbb{R}} \|\mathbf{x}\|_\sigma - 2\pi\sqrt{-1}\sum_{\tau:\mathbb{C}} \|\mathbf{x}\|_\tau\Big)\Big),$$

a definition due to van der Geer and Schoof [GS].

1.3 Geometric Truncations

1.3.1 Projective O_K-modules

Let K be an algebraic number field, i.e., a finite extension of the field of rationals $\mathbb{Q}$. Denote by O_K the ring of integers of K. Then by definition, an

O_K-module M is called *projective* if there exists an O_K-module N such that $M \oplus N$ is a free O_K-module. For projective O_K-modules, it is well-known that

(i) If M is a projective O_K-module and is a part of a short exact sequence (of O_K-modules) $0 \to M_0 \to M_1 \to M \to 0$, then $M_1 \simeq M_0 \oplus M$.
(ii) All fractional O_K-ideals are projective; and
(iii) Rank 1 projective O_K-submodules in K are simply fractional O_K-ideals.

Thus, by finiteness of the ideal class group of K, up to equivalence relation (defined by isomorphisms as O_K-modules), there are only finitely many rank 1 projective O_K-modules in K. Therefore, we may choose integral O_K-ideals $\mathfrak{a}_i, i = 1, \ldots, h$ with $h = h(K)$, the class number of K, such that

(a) Any rank 1 projective O_K-module is isomorphic to one of the $\mathfrak{a}_i$; while
(b) None of the $\mathfrak{a}_i$ and $\mathfrak{a}_j$ are isomorphic to each other if $i \neq j$, $i, j = 1, \ldots, h$.

We will fix a choice of $\mathfrak{a}_i, i = 1, \ldots, h$ satisfying (a) and (b) above for the rest of this paper, and sometimes use $\mathfrak{a}$ as a running symbol for them.

With this discussion of the rank 1 projective O_K-modules, now let us consider high rank O_K-modules. Clearly for a fractional ideal $\mathfrak{a}$,

$$P_{\mathfrak{a}} := P_{r;\mathfrak{a}} := O_K^{r-1} \oplus \mathfrak{a}$$

is a rank r projective O_K-module. The nice thing is that such types of projective O_K-modules, up to isomorphism, give all rank r projective O_K-modules. Indeed, we have the following structural

Proposition. (1) *For fractional ideals* $\mathfrak{a}$ *and* $\mathfrak{b}$, $P_{r;\mathfrak{a}} \simeq P_{r;\mathfrak{b}}$ *if and only if* $\mathfrak{a} \simeq \mathfrak{b}$*;*
(2) *For a rank r projective* O_K*-module P, there exists a fractional ideal* $\mathfrak{a}$ *such that* $P \simeq P_{\mathfrak{a}}$.

The proof of this proposition is based on an induction on the rank and the following basic relation about fractional ideals: For two fractional ideals $\mathfrak{a}$ and $\mathfrak{b}$, as O_K-modules, $\mathfrak{a} \oplus \mathfrak{b} \simeq O_K \oplus \mathfrak{a}\mathfrak{b}$. (See e.g., [FT].) So we omit the details.

Now use the natural inclusion of fractional ideals in K to embed $P_{r;\mathfrak{a}}$ into K^r. View an element in K^r as a column vector. In such a way, any O_K-module morphism $A : P_{r;\mathfrak{a}} \to P_{r;\mathfrak{b}}$ may be written down as an element in $A \in M_{r\times r}(K)$ so that the image $A(x)$ of x under A becomes simply the matrix multiplication $A \cdot x$. In particular, one checks easily the following

Lemma. *If $A \in GL(r, O_K)$ defines an O_K-isomorphism $A : P_{r;\mathfrak{a}} \to P_{r;\mathfrak{b}}$, then*

$$\mathfrak{b} \simeq (\det A) \cdot \mathfrak{a}.$$

In particular, in the case when both $\mathfrak{a}$ and $\mathfrak{b}$ are integral O_K-ideals,
(i) $\det A \in U_K$, *the group of units of K;*
(ii) $\mathfrak{a} = \mathfrak{b}$; *and*
(iii) $A \in \mathrm{Aut}_{O_K}(P_{r;\mathfrak{a}})$.

1.3.2 O_K-Lattices: First Level

We start our first level discussion on lattices in the next three sections, mainly follow [Gr] for the expositions.

Let σ be an Archimedean place of K, and K_σ be the σ-completion of K. It is well-known that the $\mathbb{R}$-algebra K_σ is either equal to $\mathbb{R}$, or equal to $\mathbb{C}$. Accordingly, we call σ (to be) real or complex, write sometimes in terms of $\sigma : \mathbb{R}$ or $\sigma : \mathbb{C}$ accordingly.

If V_σ is a finite dimensional K_σ-vector space, then an inner product on V_σ is a positive definite bilinear form $V_\sigma \times V_\sigma \to K_\sigma$ which is symmetric if σ is real and is hermitian if σ is complex. When equipped with an inner product, V_σ is called a *metrized space*.

By definition, an *O_K-lattice* Λ consists of
(1) a projective O_K-module $P = P(\Lambda)$ of finite rank; and
(2) an inner product on the vector space $V_\sigma := P \otimes_{O_K} K_\sigma$ for each of the Archimedean place σ of K.
Set $V = P \otimes_{\mathbb{Z}} \mathbb{R}$. Then $V = \prod_{\sigma\in S_\infty} V_\sigma$. Indeed, this is a direct consequence of the fact that as a $\mathbb{Z}$-module, an O_K-ideal is of rank $n = r_1 + 2r_2$ where $n = [K : \mathbb{Q}]$, r_1 denotes the number of real places and r_2 denotes the number of complex places (in S_∞).

1.3.3 Space of O_K-Lattices: First Level

Let P be a rank r projective O_K-module. Denote by $GL(P) := \mathrm{Aut}_{O_K}(P)$. Let $\widetilde{\mathbf{\Lambda}} := \widetilde{\mathbf{\Lambda}}(P)$ be the space of (O_K-)lattices Λ whose underlying O_K-module is P. For $\sigma \in S_\infty$, let $\widetilde{\mathbf{\Lambda}}_\sigma$ be the space of inner products on V_σ; if a basis is chosen for V_σ as a real or a complex vector space according to whether σ is real or complex, $\widetilde{\mathbf{\Lambda}}_\sigma$ may be realized as an open set of a real or complex vector space. (See 5.2 below for details.) We have $\widetilde{\mathbf{\Lambda}} = \prod_{\sigma \in S_\infty} \widetilde{\mathbf{\Lambda}}_\sigma$ and this provides us a natural topology on $\widetilde{\mathbf{\Lambda}}$.

As above, consider $GL(P)$ to act on P from the left. With the chosen basis, elements of $\widetilde{\mathbf{\Lambda}}$ will be thought of as column vectors, and matrices of linear maps will be written on the left.

Given $\Lambda \in \widetilde{\mathbf{\Lambda}}$ and $u, w \in V_\sigma$, let $\langle u, w\rangle_{\Lambda,\sigma}$ or $\langle u, w\rangle_{\rho_\Lambda(\sigma)}$ denote the value of the inner product on the vectors u and w associated to the lattice Λ.

As such, if $A \in GL(P)$, we may define a new lattice $A \cdot \Lambda$ in $\widetilde{\mathbf{\Lambda}}$ by the following formula

$$\langle u, w\rangle_{A\cdot\Lambda,\sigma} := \langle A^{-1} \cdot u, A^{-1} \cdot w\rangle_{\Lambda,\sigma}.$$

This defines an action of $GL(P)$ on $\widetilde{\mathbf{\Lambda}}$. Clearly, then the map $v \mapsto Av$ gives an isometry $\Lambda \cong A \cdot \Lambda$ of the lattices. (By an isometry here, we mean an isomorphism of O_K-modules for the underlying O_K-modules subjecting the condition that the isomorphism also keeps the inner product unchanged.) Conversely, suppose that $A : \Lambda_1 \cong \Lambda_2$ is an isometry of O_K-lattices, each of which is in $\widetilde{\mathbf{\Lambda}}$. Then, A defines an element, also denoted by A, of $GL(P)$. Clearly $\Lambda_2 \cong A \cdot \Lambda_1$. (Here we follow a nice notation initiated by Miyaoka; an isomorphism of corresponding algebraic structures is only denoted by $\simeq$, while an isometry is denoted by $\cong$.)

Therefore, the orbit set $GL(P)\backslash\widetilde{\mathbf{\Lambda}}(P)$ can be regarded as the set of isometry classes of O_K-lattices whose underlying O_K-modules are isomorphic to P.

We end this section by introducing an operation among the lattices in $\widetilde{\mathbf{\Lambda}}$. If T is a positive real number, then from Λ, we can produce a new O_K-lattice called $\Lambda[T]$ by multiplying each of the inner products on Λ, or better, on Λ_σ for $\sigma \in S_\infty$, by T^2. Let then $\mathbf{\Lambda} = \mathbf{\Lambda}(P)$ be the quotient of $\widetilde{\mathbf{\Lambda}}$ by the equivalence relation $\Lambda \sim \Lambda[T]$. As such, $\mathbf{\Lambda}$ admits a natural topological structure as well. Furthermore, as it becomes clear later, the construction of Λ from $\widetilde{\Lambda}$ plays a key role when we want to get the compactness statement for our moduli spaces. (Indeed, the $[T]$-construction naturally fixes a specific volume for a certain family of lattices, while does not really change the 'essential' structures of lattices involved. Consequently, by reduction theory, semi-stable lattices of a fixed volume form a compact family.)

1.3.4 Semi-Stable Lattices: First Level

Let Λ be an O_K-lattice with underlying O_K-module P. Then any submodule $P_1 \subset P$ can be made into an O_K-lattice by restricting the inner product on each V_σ to the subspace $V_{1,\sigma} := P_1 \otimes_K K_\sigma$. Call the resulting O_K-lattice $\Lambda_1 := \Lambda \cap P_1$ and write $\Lambda_1 \subset \Lambda$. If moreover, P/P_1 is projective, we say that Λ_1 is a *sublattice* of Λ.

The orthogonal projections $\pi_\sigma : V_\sigma \to V_{1,\sigma}^\perp$ to the orthogonal complement $V_{1,\sigma}^\perp$ of $V_{1,\sigma}$ in V_σ provide isomorphisms $(P/P_1) \otimes_{O_K} K_\sigma \simeq V_{1,\sigma}^\perp$, which can be used to make P/P_1 into an O_K-lattice. We call this resulting lattice the *quotient lattice* of Λ by Λ_1, and denote it by Λ/Λ_1.

There is a procedure called *restriction of scalars* which makes an O_K-lattice into a standard $\mathbb{Z}$-lattice. Recall that $V = \Lambda \otimes_{\mathbb{Z}} \mathbb{R} = \prod_{\sigma \in S_\infty} V_\sigma$. Define an inner product on the real vector space V by

$$\langle u, w \rangle_\infty := \sum_{\sigma:\mathbb{R}} \langle u_\sigma, w_\sigma \rangle_\sigma + \sum_{\sigma:\mathbb{C}} \mathrm{Re}\, \langle u_\sigma, w_\sigma \rangle_\sigma .$$

Denote $\mathrm{Res}_{K/\mathbb{Q}}\Lambda$ the $\mathbb{Z}$-lattice obtained by equipped P, regarding as a $\mathbb{Z}$-module, with this inner product (at the unique infinite place ∞ of $\mathbb{Q}$).

We let $\mathrm{rk}(\Lambda)$ denote the O_K-module rank of P (or of Λ) and let $\dim(\Lambda)$ denote the rank of P as $\mathbb{Z}$-module. Clearly,

$$\dim(\Lambda) = \mathrm{rk}(\Lambda) \cdot \dim(O_K) = \mathrm{rk}(\Lambda) \cdot [F : \mathbb{Q}].$$

We define the *Lebesgue volume* of Λ, denoted by $\mathrm{Vol}_{\mathrm{Leb}}(\Lambda)$, to be the (co)volume of the lattice $\mathrm{Res}_{K/\mathbb{Q}}\Lambda$ inside its inner product space V. As such, this volume may be computed as $\left|\det \langle l_i, e_j \rangle\right|$, where $\{l_i\}$ is a $\mathbb{Z}$-basis of $\mathrm{Res}_{K/\mathbb{Q}}\Lambda$ and $\{e_j\}$ is an orthonormal basis of Λ with respect to $\langle \cdot, \cdot \rangle_\infty$. For examples,

(a) If $\dim\Lambda = 0$, then $\mathrm{Vol}_{\mathrm{Leb}}(\Lambda) = 1$;

(b) If $\dim\Lambda = 1$, then $\mathrm{Vol}_{\mathrm{Leb}}(\Lambda)$ is the length of a generator of Λ;
(c) If $\dim\Lambda = 2$, then $\mathrm{Vol}_{\mathrm{Leb}}(\Lambda)$ is the area of a fundamental parallelopiped, and so on.

Clearly, if P' is a submodule of finite index in P, then

$$\mathrm{Vol}_{\mathrm{Leb}}(\Lambda') = [P : P']\mathrm{Vol}_{\mathrm{Leb}}(\Lambda),$$

where $\Lambda' = \Lambda \cap P$ is the lattice induced from P'.

Examples: 1) Take $P = O_K$ and for each place σ, let $\{1\}$ be an orthonormal basis of $V_\sigma = K_\sigma$, i.e., equipped $V_\sigma = \mathbb{R}$ or $\mathbb{C}$ with the standard Lebesgue measure. This makes O_K into an O_K-lattice $\overline{O_K} = (O_K, \mathbf{1})$ in a natural way. It is a well-known fact, see e.g., [L1], that

$$\mathrm{Vol}_{\mathrm{Leb}}\Big(\overline{O_K}\Big) = 2^{-r_2} \cdot \sqrt{\Delta_F},$$

where Δ_F denotes the absolute value of the discriminant of K.

More generally, take $P = \mathfrak{a}$ an fractional idea of K and equip the same inner product as above on V_σ. Then $\mathfrak{a}$ becomes an O_K-lattice $\overline{\mathfrak{a}} = (\mathfrak{a}, \mathbf{1})$ in a natural way with $\mathrm{rk}(\mathfrak{a}) = 1$. It is a well-known fact, see e.g., [Ne], that

$$\mathrm{Vol}_{\mathrm{Leb}}\Big(\overline{\mathfrak{a}}\Big) = 2^{-r_2} \cdot \Big(N(\mathfrak{a}) \cdot \sqrt{\Delta_K}\Big),$$

where $N(\mathfrak{a})$ denote the norm of $\mathfrak{a}$.

Due to the factor 2^{-r_2}, we also define the *canonical volume* of Λ, denoted by $\mathrm{Vol}_{\mathrm{can}}(\Lambda)$ or simply by $\mathrm{Vol}(\Lambda)$, to be $2^{r_2 \mathrm{rk}(\Lambda)}\mathrm{Vol}_{\mathrm{Leb}}(\Lambda)$. (This canonical volume is theoretically correct. In fact, in Arakelov theory, where the minus of the log of the canonical vomule is defined to be the Arakelov-Euler characteristic $\chi(\Lambda)$ of Λ. That is,

$$\chi(\Lambda) := -\log\Big(\mathrm{Vol}(\Lambda)\Big).$$

For details, see the justification given in 1.3.5 below.) So in particular,

$$\mathrm{Vol}\Big(\overline{\mathfrak{a}}\Big) = N(\mathfrak{a}) \cdot \sqrt{\Delta_K},$$

with

$$\mathrm{Vol}\Big(\overline{O_K}\Big) = \sqrt{\Delta_K}$$

as its special case.

2) The $[T]$-construction changes volumes of lattices in the following way

$$\mathrm{Vol}(\Lambda[T]) = T^{\dim(\Lambda)} \cdot \mathrm{Vol}(\Lambda).$$

Now we are ready to introduce our first key definition.

Definition. An O_K lattice Λ is called *semi-stable* (resp. *stable*) if for any proper sublattice Λ_1 of Λ,

$$\mathrm{Vol}(\Lambda_1)^{\mathrm{rk}(\Lambda)} \geq \text{ (resp. } >) \, \mathrm{Vol}(\Lambda)^{\mathrm{Vol}(\Lambda_1)}.$$

Clearly the last inequality is equivalent to

$$\mathrm{Vol}_{\mathrm{Leb}}(\Lambda_1)^{\mathrm{rk}(\Lambda)} \geq \mathrm{Vol}_{\mathrm{Leb}}(\Lambda)^{\mathrm{Vol}(\Lambda_1)}.$$

So it does not matter which volume, the canonical one or the Lebesgue one, we use.

Remark. Despite the fact that we introduce the stability for lattices independently, many others, notably Stuhler, introduced the stability earlier. (See e.g. [Gr], [St].)

1.3.5 Minkowski Metric versus Canonical Metric

This section is specially introduced for the reader who wants to see clearly the relation between Lebesgue and canonical volumes. We mainly follow [Neu] for the presentation. For the first reading, one can skip it, except perhaps the Arakelov Riemann-Roch.

Minkowski's basic idea on using Geometry of Numbers to study algebraic number field $K/\mathbb{Q}$ of degree n is to interpret its numbers as points in an n-dimensional space. For this, let us consider the canonical map

$$j : K \to K_{\mathbb{C}} := \prod_{\tau} \mathbb{C}, \qquad a \mapsto j(a) := (\tau a)$$

induced from the n complex embeddings $\tau : K \to \mathbb{C}$. (Here, if $\sigma : K \to K_\sigma = \mathbb{R}$ is a real Archimedean place, we use the natural embedding $\mathbb{R} \hookrightarrow \mathbb{C} = \mathbb{R} + i\mathbb{R}$ so as to get a natural map $\tau : K \to \mathbb{C}$ ending with $\mathbb{C}$.) The $\mathbb{C}$-vector space $K_{\mathbb{C}}$ is equipped with the hermitian scalar product

$$\langle x, y \rangle = \sum_{\tau} x_\tau \bar{y}_\tau.$$

In the sequel, we always view $K_{\mathbb{C}}$ as the hermitian space with respect to this standard metric.

Let $G(\mathbb{C}|\mathbb{R})$ be the Galois group generated by complex conjugation $F: z \mapsto \bar{z}$. Then F acts on both the factors $\mathbb{C}$ of the product $\prod_\tau \mathbb{C}$ and on the index set of τ's at the same time; that is to say, $a \mapsto \bar{a}$ for $a \in \mathbb{C}$, while $\tau \mapsto \bar{\tau}$ for each embedding $\tau : K \to \mathbb{C}$ with associated complex conjugate $\bar{\tau} : K \to \mathbb{C}$. Altogether, this defines an involution $F : K_{\mathbb{C}} \to K_{\mathbb{C}}$, which, in terms of points $z = (z_\tau) \in K_{\mathbb{C}}$, is given by $(Fz)_\tau = \bar{z}_{\bar{\tau}}$. Clearly, the scalar product $\langle \cdot, \cdot \rangle$ is equivariant under F. That is to say, $\langle Fx, Fy \rangle = \langle x, y \rangle$.

We now concentrate on the $\mathbb{R}$-vector space

$$K_{\mathbb{R}} := K_{\mathbb{C}}^{+} := \Big[\prod_\tau \mathbb{C}\Big]^{+}$$

consisting of the $G(\mathbb{C}|\mathbb{R})$-invariant, i.e., F-invariant, points of $K_{\mathbb{C}}$. Easily, the F-invariant points of $K_{\mathbb{C}}$ are exactly these points (z_τ) such that $z_{\bar{\tau}} = \bar{z}_\tau$. In particular, then we have

$$K_{\mathbb{R}} \simeq \mathbb{R}^{r_1} \times \mathbb{C}^{r_2}.$$

via the map $(x_{\sigma_1}, \ldots, x_{\sigma_{r_1}}; z_{\sigma_{r_1+1}}, \bar{z}_{\sigma_{r_1+1}}, \ldots, z_{\sigma_{r_1+r_2}}, \bar{z}_{\sigma_{r_1+r_2}})$

$$\mapsto (x_{\sigma_1}, \ldots, x_{\sigma_{r_1}}; z_{\sigma_{r_1+1}}, \ldots, z_{\sigma_{r_1+r_2}}).$$

Since $\bar{\tau}(a) = \overline{\tau a}$ for $a \in K$, we have $F(j(a)) = F(a)$ for all $a \in K$. This then further induces a map $j : K \to K_{\mathbb{R}}$.

The restriction of hermitian scalar product $\langle \cdot, \cdot \rangle$ from $K_{\mathbb{C}}$ to $K_{\mathbb{R}}$ gives a scalar product

$$\langle \cdot, \cdot \rangle : K_{\mathbb{R}} \times K_{\mathbb{R}} \to \mathbb{R}$$

on the $\mathbb{R}$-vector space $K_{\mathbb{R}}$. We call the Euclidean vector space $K_{\mathbb{R}} = [\prod_\tau \mathbb{C}]^+$ the *Minkowski space*, its scalar product $\langle \cdot, \cdot \rangle$ the *canonical metric*, and the association Haar measure the canonical measure.

Moreover, the map $j : K \to K_{\mathbb{R}}$ identifies the vector space $K_{\mathbb{R}}$ with the tensor product $K \otimes_{\mathbb{Q}} \mathbb{R}$. That is, we have a natural identification

$$K \otimes_{\mathbb{Q}} \mathbb{R} \simeq K_{\mathbb{R}}, \qquad a \otimes x \mapsto (j(a)) \cdot x.$$

Likewise, $K \otimes_{\mathbb{Q}} \mathbb{C} \simeq K_{\mathbb{C}}$. With this said, then the inclusion $K_{\mathbb{R}} \subset K_{\mathbb{C}}$ corresponds exactly to the canonical map $K \otimes_{\mathbb{Q}} \mathbb{R} \to K \otimes_{\mathbb{Q}} \mathbb{C}$ induced from the natural inclusion $\mathbb{R} \hookrightarrow \mathbb{C}$. And simply, F corresponds to $a \otimes z \mapsto a \otimes \bar{z}$.

On the other hand, the Lebesgue volume is associated with the following alternative explicit descripion of the Minkowski space $K_{\mathbb{R}}$. When embeddings $\tau : K \to \mathbb{C}$ are real, their images land already in $\mathbb{R}$. As for complexes, i.e., these which are not real, by writing them in pairs, we may list (all real embeddings as $\sigma_1, \ldots, \sigma_{r_1} : K \to \mathbb{R}$, and) all complex (conjugate) embeddings as $\tau_1, \bar{\tau}_1, \ldots, \tau_{r_2}, \bar{\tau}_{r_2} : K \to \mathbb{C}$. (Here $[K : \mathbb{Q}] = n = r_1 + 2r_2$.) Choose from each pair τ_j, $\bar{\tau}_j$ a certain fixed complex embedding τ_j. With this done, then let σ vary over the family of real embeddings and τ over the family of chosen complex embeddings. Write $\mathfrak{p}$ as a (common) running symbol for σ_i and $\{\tau_j, \bar{\tau}_j\}$. Since F leaves the σ invariant, but exchanges τ and $\bar{\tau}$, we have

$$K_{\mathbb{R}} = \Big\{(z_{\mathfrak{p}}) \in \prod_{\mathfrak{p}} \mathbb{C} : z_\sigma \in \mathbb{R}, z_{\bar{\tau}} = \overline{z_\tau} \in \mathbb{C}\Big\}.$$

As a direct consequence, we then obtain a natural isomorphism

$$f : K_{\mathbb{R}} \to \prod_{\mathfrak{p}} \mathbb{R} = \mathbb{R}^{r_1+2r_2=n=[K:\mathbb{Q}]}, \qquad (z_{\mathfrak{p}}) \mapsto (x_{\mathfrak{p}}),$$

where for the reals $x_\sigma = z_\sigma$ while for the complexes $x_\tau = \mathrm{Re}\,(z_\tau), x_{\bar{\tau}} = \mathrm{Im}\,(z_\tau)$. (So in particular for complex τ,

$$\begin{aligned}(z_\tau, z_{\bar{\tau}}) = (x_\tau + iy_\tau, x_\tau - iy_\tau) &\mapsto (x_\tau, y_\tau),\\ (x_\tau^2 + y_\tau^2) + (x_\tau^2 + y_\tau^2) &\mapsto 2(x_\tau^2 + y_\tau^2).)\end{aligned}$$

This isomorphism transforms the canonical metric $\langle\cdot,\cdot\rangle$ on $K_{\mathbb{R}}$ into the scalar product $(x, y) = \sum_{\mathfrak{p}} N_{\mathfrak{p}} x_{\mathfrak{p}} y_{\mathfrak{p}}$ where $N_{\mathfrak{p}} = 1$ resp. $N_{\mathfrak{p}} = 2$ if $\mathfrak{p}$ is real, resp. complex. The scalar product $(x, y) = \sum_{\mathfrak{p}} N_{\mathfrak{p}} x_{\mathfrak{p}} y_{\mathfrak{p}}$ transforms the canonical measure from $K_{\mathbb{R}}$ to a measure on $\mathbb{R}^{r_1+2r_2}$. It obviously differes from the standard Lebesgue measure by

$$\mathrm{Vol}_{\mathrm{can}}(X) = 2^{r_2}\mathrm{Vol}_{\mathrm{Leb}}(f(X)).$$

Minkowski himself worked with the Lebesgue measure on $\mathbb{R}^{r_1+2r_2}$, and afterwards, most works, e.g., research papers and textbooks, follow suit. As we said before, we will write $\mathrm{Vol}_{\mathrm{can}}$ simply as Vol. Also in the calculations below, we use both canonical and Lebesgue measures without any clear indication on which one we really use for our convenience.

As said too, the canonical measure has an advantage theoretically. For example, we have the following
Arakelov-Riemann-Roch Formula: For an O_K-lattice Λ of rank r,

$$-\log\Big(\mathrm{Vol}(\Lambda)\Big) = \deg(\Lambda) - \frac{r}{2}\log \Delta_K.$$

For a proof, see e.g., [L2]. (For the reader who does not know the definition of the Arakelov degree, he or she may simply take this relation as the definition.)

1.3.6 Canonical Filtration

Based on stability, we may introduce a more general geometric truncation for space of lattices. Let us start with the following well-known

Lemma. *For a fixed O-lattice Λ, $\Big\{\mathrm{Vol}(\Lambda_1) : \Lambda_1 \subset \Lambda\Big\} \subset \mathbb{R}_{\geq 0}$ is discrete and bounded from below.*

Proof. It is based on the follows:
1) If a lattice Λ_1 induced from a submodule of the lattice Λ has the minimal volume among all lattices induced from submodules of Λ of the same rank, then Λ_1 is a sublattice;
2) Taking wedge product, we may further assume that the rank is 1. It is then clear. □

As a direct consequence, we have the following
Proposition. *Let Λ be an O-lattice. Then*
(1) (**Canonical Filtration**) *There exists a unique filtration of proper sub-lattices*

$$0 = \Lambda_0 \subset \Lambda_1 \subset \cdots \subset \Lambda_s = \Lambda$$

such that

(i) *for all $i = 1, \ldots, s$, Λ_i/Λ_{i-1} is semi-stable*; *and*

(ii) *for all $j = 1, \ldots, s-1$,*

$$\Big(\mathrm{Vol}(\Lambda_{j+1}/\Lambda_j)\Big)^{\mathrm{rk}(\Lambda_j/\Lambda_{j-1})} > \Big(\mathrm{Vol}(\Lambda_j/\Lambda_{j-1})\Big)^{\mathrm{rk}(\Lambda_{j+1}/\Lambda_j)};$$

(2) (**Jordan-Hölder Filtration**) *If moreover,* Λ *is semi-stable, then there exists a filtration of proper sublattices,*

$$0 = \Lambda^{t+1} \subset \Lambda^t \subset \cdots \subset \Lambda^0 = \Lambda$$

such that

(i) *for all* $k = 0, \ldots, t$, Λ^k/Λ^{k+1} *is stable; and*

(ii) *for all* $l = 1, \ldots, t$,

$$\Big(\mathrm{Vol}(\Lambda^l/\Lambda^{l+1})\Big)^{\mathrm{rk}(\Lambda^{l-1}/\Lambda^l)} = \Big(\mathrm{Vol}(\Lambda^{l-1}/\Lambda^l)\Big)^{\mathrm{rk}(\Lambda^l/\Lambda^{l+1})}.$$

Moremore, up to isometry, the graded lattice $\mathrm{Gr}(\Lambda) := \oplus_{k=0}^t \Lambda^k/\Lambda^{k+1}$ *is uniquely determined by* Λ.

Proof. Existence is clear and the uniqueness of Jordan-Hölder graded lattice is fairly standard. Let us look at the uniqueness of canonical filtration, an analogue of the well-known Harder-Narasimhan filtration for vector bundles: Existence of two such filtrations will lead to a contradiction by applying the fact that for any two sublattices Λ_1, Λ_2 of Λ,

$$\mathrm{Vol}\Big(\Lambda_1/(\Lambda_1 \cap \Lambda_2)\Big) \geq \mathrm{Vol}\Big((\Lambda_1 + \Lambda_2)/\Lambda_2\Big).$$

□

1.3.7 Compactness

For a lattice $\Lambda \in \widetilde{\mathbf{\Lambda}}$ with the associated canonical filtration

$$0 = \overline{\Lambda}_0 \subset \overline{\Lambda}_1 \subset \cdots \subset \overline{\Lambda}_s = \Lambda$$

define the associated *canonical polygone* $\overline{p}_\Lambda : [0, r] \to \mathbb{R}$ by the following conditions:

(1) $\overline{p}_\Lambda(0) = \overline{p}_\Lambda(r) = 0$;

(2) $\overline{p}_\Lambda$ is affine over the closed interval $[\mathrm{rk}\overline{\Lambda}_i, \mathrm{rk}\overline{\Lambda}_{i+1}]$; and

(3) $\overline{p}_\Lambda(\mathrm{rk}\overline{\Lambda}_i) = \deg(\overline{\Lambda}_i) - \mathrm{rk}(\overline{\Lambda}_i) \cdot \frac{\deg(\overline{\Lambda})}{r}$.

Clearly, the canonical polygon is well-defined on the space $\widetilde{\mathbf{\Lambda}}$.

To go further, let us introduce an operation among the lattices in $\widetilde{\mathbf{\Lambda}}$. If T is a positive real number, then from Λ, we can produce a new O_K-lattice called $\Lambda[T]$ by multiplying each of the inner products on Λ, or better, on Λ_σ for $\sigma \in S_\infty$, by T^2. Obviously, the $[T]$-construction changes volumes of lattices in the following way

$$\mathrm{Vol}(\Lambda[T]) = T^{\mathrm{rk}(\Lambda)\cdot[K:\mathbb{Q}]} \cdot \mathrm{Vol}(\Lambda).$$

That is to say, the $[T]$-construction naturally fixes a specific volume for a certain family of lattices, while does not really change the 'essential' structures of lattices involved.

Accordingly, let $\mathbf{\Lambda} = \mathbf{\Lambda}(P)$ be the quotient topological space of $\widetilde{\mathbf{\Lambda}}$ modulo the equivalence relation $\Lambda \sim \Lambda[T]$. Since canonical polygon is invariant under the scaling operation and hence descends to this quotient space $\mathbf{\Lambda}$.

With all this, we are ready to state the following fundamental

Theorem. *For any fixed convex polygon $p : [0, r] \to \mathbb{R}$, the subset $\Big\{[\Lambda] \in \mathbf{\Lambda} : \overline{p}_\Lambda \leq p\Big\}$ is compact in $\mathbf{\Lambda}$.*

Proof. This is based on the follows:
(1) Dirichlet's Unit Theorem; and
(2) Minkowski's Reduction Theory: Due to the fact that volumes of lattices involved are fixed, semi-stability condition implies that the first Minkowski successive minimums of these lattices admit a natural lower bound away from 0 (depending only on r). Hence by the standard reduction theory, see e.g., Borel [Bo1,2], the subset $\Big\{[\Lambda] \in \mathbf{\Lambda} : \overline{p}_\Lambda \leq p\Big\}$ is compact. □

1.3.8 Geometric Truncation: Adelic Version

Let $X = \mathrm{Spec}\, O_F$. If E is a vector sheaf of rank r over X, i.e, a locally free O_F-sheaf of rank r, denote by E_F the fiber of E at the generic point $\mathrm{Spec}(F) \hookrightarrow \mathrm{Spec}\, O_F$ of X (E_F is an F-vector space of dimension r), and for each $v \in S_f := S_{\mathrm{fin}}$, set $E_{O_v} := H^0(\mathrm{Spec}\, O_{F_v}, E)$ a free O_v-module of rank r. In particular, we have a canonical isomorphism:

$$\mathrm{can}_v : F_v \otimes_{O_v} E_{O_v} \simeq F_v \otimes_F E_F.$$

Now if E is a vector sheaf of rank r over X equipped with a basis $\alpha_F : F^r \simeq E_F$ of its generic fiber and a basis $\alpha_{O_v} : O_v^r \simeq E_{O_v}$ for any $v \in S_f$, the elements $g_v := (F_v \otimes_F \alpha_F)^{-1} \circ \mathrm{can}_v \circ (F_v \otimes_{O_v} \alpha_{O_v}) \in GL_r(F_v)$ for all $v \in S_f$ define an element $g_{\mathbb{A}} := (g_v)_{v \in S_f}$ of $GL_r(\mathbb{A}_f)$, i.e, for almost every v we have $g_v \in GL_r(O_v)$. By this construction, we obtain a bijection from the set of isomorphism classes of triples $(E; \alpha_F; (\alpha_{O_v})_{v \in S_f})$ as above onto $GL_r(\mathbb{A}_f)$. Moreover, if $\gamma \in GL_r(F), \kappa \in GL_r(O_F)$ and if this bijection maps the triple $(E; \alpha_F; (\alpha_{O_v})_{v \in S_f})$ onto $g_{\mathbb{A}}$, the same map maps the triple $(E; \alpha_F \circ \gamma^{-1}; (\alpha_{O_v} \circ \kappa_v)_{v \in S_f})$ onto $\gamma g_{\mathbb{A}} \kappa$. Therefore the above bijection induces a bijection between the set of isomorphism classes of vector sheaves of rank r on $\mathrm{Spec}\, O_F$ and the double coset space $GL_r(F) \backslash GL_r(\mathbb{A}_f)/GL_r(O_F)$.

More generally, let $r = r_1 + \cdots + r_s$ be a partition $I = (r_1, \ldots, r_s)$ of r and let P_I be the corresponding standard parabolic subgroup of GL_r. Then we have a natural bijection from the set of isomorphism classes of triple $(E_*; \alpha_{*,F} : (\alpha_{*,O_v})_{v \in S_f})$ onto $P_I(\mathbb{A}_f)$, where $E_* := ((0) = E_0 \subset E_1 \subset \cdots \subset E_s)$ is a filtration of vector sheaves of rank $(r_1, r_1 + r_2, \ldots, r_1 + r_2 + \cdots + r_s = r)$ over X, (i.e, each E_j is a vector sheaf of rank $r_1 + r_2 + \cdots + r_j$ over X and each quotient E_j/E_{j-1} is torsion free,) which is compactible with an isomorphism of filtrations of F-vector spaces

$$\alpha_{*,F} : ((0) = F_0 \subset F^{r_1} \subset \cdots \subset F^{r_1 + r_2 + \cdots + r_s = r}) \simeq (E_*)_F,$$

and with an isomorphism of filtrations of free O_v-modules

$$\alpha_{*,O_v} : ((0) \subset O_v^{r_1} \subset \cdots \subset O_v^{r_1 + r_2 + \cdots + r_s = r}) \simeq (E_*)_{O_v},$$

for every $v \in S_f$. Clearly, this bijection induces a bijection between the set of isomorphism classes of filtrations of vector sheaves of rank $(r_1, r_1 + r_2, \ldots, r_1 + r_2 + \cdots + r_s = r)$ over X and the double coset space $P_I(F) \backslash P_I(\mathbb{A}_f)/P_I(O_F)$. The natural embedding $P_I(\mathbb{A}_f) \hookrightarrow P_I(\mathbb{A})$ (resp. canonical projections $P_I(\mathbb{A}_f) \to M_I(\mathbb{A}_f) \to GL_{r_j}(\mathbb{A}_f)$ for $j = 1, \ldots, s$, where M_I denotes the standard Levi of P_I,) admits a modular interpretation

$$(E_*; \alpha_{*,F} : (\alpha_{*,O_v})_{v \in S_f}) \mapsto (E_s; \alpha_{s,F} : (\alpha_{s,O_v})_{v \in S_f})$$

(resp.

$$(E_*; \alpha_{*,F} : (\alpha_{*,O_v})_{v \in S_f}) \mapsto (\mathrm{gr}_j(E_*); \mathrm{gr}_j(\alpha_{*,F}), \mathrm{gr}_j(\alpha_{*,O_v})_{v \in S_f}),$$

where $\mathrm{gr}_j(E_*) := E_j/E_{j-1}$, $\mathrm{gr}_j(\alpha_{*,F}) : F^{r_j} \simeq \mathrm{gr}_j(E_*)_F$ and $\mathrm{gr}_j(\alpha_{*,O_v}) : O_v{}^{r_j} \simeq \mathrm{gr}_j(E_*)_{O_v}$, $v \in S_f$ are induced by $\alpha_{*,F}$ and α_{*,O_v} respectively.)

Moreover, any $g = (g_f; g_\infty) \in GL_r(\mathbb{A}_f) \times GL_r(\mathbb{A}_\infty) = GL_r(\mathbb{A})$ gives first a rank r vector sheaf E_g on $\mathrm{Spec}\, O_F$, which via the embedding $E_F \hookrightarrow (\mathbb{R}^{r_1} \times \mathbb{C}^{r_2})^r$ gives a discrete subgroup, a free rank r O_F-module. In particular, $g_\infty = (g_\sigma)$ then induces a natural metric on E_g by twisting the standard one on $(\mathbb{R}^{r_1} \times \mathbb{C}^{r_2})^r$ via the linear transformation induced from g_∞. As a direct consequence, see e.g, [L3],

$$\deg(E_g, \rho_g) = -\log\left(N(\det g)\right)$$

with $N : GL_1(\mathbb{A}_F) = \mathbb{I}_F \to \mathbb{R}_{>0}$ the standard norm map of the idelic group of F.

With this, for $g = (g_f; g_\infty) \in GL_r(\mathbb{A})$ and a parabolic subgroup Q of GL_r, denote by $E_*^{g;Q}$ the filtration of the vector sheaf E_{g_f} induced by the parabolic subgroup Q. Then we have a filtration of hermitian vector sheaves $(E_*^{g;Q}, \rho_*^{g;Q})$ with the hermitian metrics $\rho_j^{g;Q}$ on $E_j^{g;Q}$ obtained via the restrictions of ρ_{g_∞}.

Now introduce an associated polygon $p_Q^g : [0, r] \to \mathbb{R}$ by the following conditions:
(i) $p_Q^g(0) = p_Q^g(r) = 0$;
(ii) p_Q^g is affine on the interval $[\mathrm{rk}E_{i-1}^{g;Q}, \mathrm{rk}E_i^{g;Q}]$; and
(iii) for all indices i,

$$p_Q^g(\mathrm{rk}E_i^{g;Q}) = \deg(E_i^{g;Q}, \rho_i^{g;Q}) - \mathrm{rk}E_i^{g;Q} \cdot \frac{\deg(E_g, \rho_g)}{r}.$$

Then, by Prop. 1.3.6, there is a unique convex polygon $\bar{p}^g$, the canonical polygon associated with g, which bounds all p_Q^g from above for all parabolic subgroups Q for GL_r. Moreover there exists a parabolic subgroup $\bar{Q}^g$ such that $p_{\bar{Q}^g}^g = \bar{p}^g$.

Hence we have the following fundamental

Main Lemma. *For any fixed polygon $p : [0, r] \to \mathbb{R}$ and any $d \in \mathbb{R}$, the subset*

$$\{g \in GL_r(F)\backslash GL_r(\mathbb{A}) : \deg g = d, \bar{p}^g \le p\}$$

is compact.

Proof. This is a restatement of the classical reduction theory, i.e., Thm. 1.3.7. More precisely, it consists of two parts: In terms of O_F-lattices, by fixing the degree, we get a fixed volume for the free O_F-lattice of rank r corresponding to (E_g, ρ_g), by the Arakelov Riemann-Roch formula. Thus the condition $\overline{p}^g \leq p$ gives an upper bound for the volumes of all the sub-lattices of (E_g, ρ_g) and a lower bound of the Minkowski successive minimums. That is to say, at the lattice level, the corresponding space is compact.

On the other hand, a general theory says that fibers of the natural map from $GL_r(F)\backslash GL_r(\mathbb{A})/GL_r(O_F)$ to isometry classes of O_F-lattices are all compact: This is in fact generally true for all reductive groups, a result due to Borel [Bo1,2]. In fact, our current case for which only the group GL_r is involved is rather obvious: As to be seen in 5.2, essentially, the fibers are the compact subgroup $SO_r(\mathbb{R})^{r_1} \times SU_r(\mathbb{C})^{r_2}$. □

Similarly yet more generally, for a fixed parabolic subgroup P of GL_r and $g \in GL_r(\mathbb{A})$, there is a unique maximal element $\bar{p}_P^g$ among all p_Q^g, where Q runs over all parabolic subgroups of GL_r which are contained in P. And we have

Main Lemma′. *For any fixed polygon $p : [0, r] \to \mathbb{R}$, $d \in \mathbb{R}$ and any standard parabolic subgroup P of GL_r, the subset*

$$\{g \in GL_r(F)\backslash GL_r(\mathbb{A}) : \deg g = d, \bar{p}_P^g \leq p, p_P^g \geq -p\}$$

is compact.

1.4 High Rank Zeta Functions for Number Fields

1.4.1 Definition

For a number field F with (absolute value of) discriminant Δ_F, denote by $\mathcal{M}_{F,r}$ the moduli space of semi-stable O_F-lattices of rank r. Then naturally $\mathcal{M}_{F,r} := \cup_{T \in \mathbb{R}_+} \mathcal{M}_{F,r}[T]$ where $\mathcal{M}_{F,r}[T] := \mathcal{M}_{F,r}(\log T)$ denotes the volume T part of the moduli. For an adele element $g \in \mathrm{GL}_r(\mathbb{A})$, denote its associated (isometric class of) O_F-lattice of rank r by $\Pi(g)$. Clearly, Π factors through $\mathrm{GL}_r(F)\backslash \mathrm{GL}_r(\mathbb{A})$. Denote by $\mathcal{M}_{\mathbb{A}_F,r} \subset \mathrm{GL}_r(F)\backslash \mathrm{GL}_r(\mathbb{A})$ the inverse image of $\mathcal{M}_{F,r}$. It is well known that fibers of $\Pi : \mathcal{M}_{\mathbb{A}_F,r} \to \mathcal{M}_{F,r}$

are all compact. Hence so is $\mathcal{M}_{\mathbb{A}_F,r}[T] := \Pi^{-1}\big(\mathcal{M}_{F,r}[T]\big)$. Consequently, we obtain natural measures $d\mu$ on $\mathcal{M}_{F,r}$ and $\mathcal{M}_{\mathbb{A}_F,r}[T]$ induced from that on $\mathrm{GL}_r(\mathbb{A})$. In particular, being compact, $\mathcal{M}_{\mathbb{A}_F,r}[\Delta_F^{\frac{r}{2}}]$ is of finite volume.

For any $\Lambda \in \mathcal{M}_{\mathbb{A}_F,r}$, set $h^i(F,\Lambda) := h^i(\mathbb{A}_F, g)$ where g is an element of $\mathrm{GL}_r(\mathbb{A})$ whose associated O_F-lattice coincides with Λ. This is well-defined since for any $a \in \mathrm{GL}_r(F)$, $h^i(\mathbb{A}_F, a\cdot g) = h^i(\mathbb{A}_F, g)$ by definition. Indeed, as said earlier,

$$h^0(F,\Lambda) = \log\Big(\sum_{\mathbf{x}\in\Lambda}\exp\Big(-\pi\sqrt{-1}\sum_{\sigma:\mathbb{R}}\|\mathbf{x}\|_\sigma - 2\pi\sqrt{-1}\sum_{\tau:\mathbb{C}}\|\mathbf{x}\|_\tau\Big)\Big).$$

Definition. The (completed) rank r zeta function $\xi_{F,r}(s)$ of F is defined by

$$\xi_{F,r}(s) := \big(\Delta_F^{\frac{r}{2}}\big)^s\int_{\Lambda\in\mathcal{M}_{F,r}}\big(e^{h^0(F,\Lambda)}-1\big)(e^{-s})^{\deg(\Lambda)}\cdot d\mu,\quad \mathrm{Re}(s)>1.$$

Following 1.1, i.e., Iwasawa's interpretation of Dedekind zeta functions, we see that $\xi_{F,1}(s)$ is essentially the completed Dedekind zeta function $\xi_F(s)$ for F.

1.4.2 Basic Properties

Just as Dedekind zeta functions, high rank zetas are canonical as well.

Theorem. (1) *Rank r zeta function*

$$\xi_{F,r}(s) := \big(\Delta_F^{\frac{r}{2}}\big)^s\int_{\Lambda\in\mathcal{M}_{F,r}}\big(e^{h^0(F,\Lambda)}-1\big)(e^{-s})^{\deg(\Lambda)}\cdot d\mu$$

converges absolutely and uniformly when $\mathrm{Re}(s) \geq 1+\delta$ *for any* $\delta > 0$. *Moreover,* $\xi_{F,r}(s)$ *admits a unique meromorphic continuation to the whole complex s-plane;*
(2) *The extended* $\xi_{F,r}(s)$ *satisfies the functional equation*

$$\xi_{F,r}(s) = \xi_{F,r}(1-s);$$

(3) *The extended* $\xi_{F,r}(s)$ *has two singularities only, both are simple poles, at* $s = 0, 1$ *with*

$$\mathrm{Res}_{s=1}\,\xi_{F,r}(s) \;=\; \mathrm{Res}_{s=0}\,\xi_{F,r}(s) \;=\; \mathrm{Vol}\big(\mathcal{M}_{F,r}[\Delta_F^{\frac{r}{2}}]\big).$$

Remarks. The global non-abelian invariant $\mathrm{Vol}\big(\mathcal{M}_{F,r}[\Delta_F^{\frac{r}{2}}]\big)$ for F can be calculated in terms of special values $\xi_F(2), \xi_F(3), \ldots, \xi_F(r)$ of the Dedekind zeta. For details, see 4.8.

Proof. Motivated by Weil's treatment of functional equation in [W2], we first give a formal discussion to show the key points of functional equation, but under a more general context.

For the number field F with discriminant Δ_F, denote by $\mathcal{M}_{\mathbb{A}_F,r}$ the moduli space of semi-stable adelic bundles of rank r, that is, $\mathcal{M}_{\mathbb{A}_F,r} := \dot{\cup}_{N\in\mathbb{R}_+} \mathcal{M}_{\mathbb{A}_F,r}[N]$ where $\mathcal{M}_{\mathbb{A}_F,r}[N] := \mathcal{M}_{\mathbb{A}_F,r}(\log N)$. As topological spaces, $\mathcal{M}_{\mathbb{A}_F,r} \simeq \mathcal{M}_{\mathbb{A}_F,r}[|\Delta_F|^{\frac{r}{2}}] \times \mathbb{R}_+$. Hence we obtain a natural measure $d\mu$ on $\mathcal{M}_{\mathbb{A}_F,r}$ from the Tamagawa measures on $\mathcal{M}_{\mathbb{A}_F,r}[N]$ and $\frac{dT}{T}$ on $\mathbb{R}_+$.

Recall that for $E \in \mathcal{M}_{\mathbb{A}_F,r}$, $H^i_{\mathrm{ga}}(F,E) := H^i_{\mathrm{ga}}(F,g) = e^{h^i(\mathbb{A}_F,g)}$ where $g \in \mathrm{GL}_r(\mathbb{A})$ is such that $E = [g]$. Since for any $a \in \mathrm{GL}_r(F)$, $H^i_{\mathrm{ga}}(F, a\cdot g) = H^i_{\mathrm{ga}}(F,g)$ by definition. $H^i_{\mathrm{ga}}(F,E)$ is well-defined for $i = 0, 1$.

With respect to fixed real constants A, B, C, α and β, introduce the formal integration

$$Z_{F,r;A,B,C;\alpha,\beta}(s) := \\ (|\Delta_F|^{-\frac{rB}{2}})^s \int_{E\in\mathcal{M}_{\mathbb{A}_F,r}} \Big(H^0_{\mathrm{ga}}(F,E)^A \cdot N(E)^{Bs+C} - N(E)^{\alpha s+\beta}\Big) d\mu(E).$$

Then formally, $Z_{F,r;A,B,C;\alpha,\beta}(s) = I(s) - II(s) + III(s)$, where

$$I(s) := (|\Delta_F|^{-\frac{rB}{2}})^s \int_{E\in\mathcal{M}_{\mathbb{A}_F,r}, N(E)\le|\Delta_F|^{\frac{r}{2}}} \Big(H^0_{\mathrm{ga}}(F,E)^A \cdot N(E)^{Bs+C} - N(E)^{\alpha s+\beta}\Big) d\mu(E);$$

$$II(s) := (|\Delta_F|^{-\frac{rB}{2}})^s \int_{E\in\mathcal{M}_{\mathbb{A}_F,r}, N(E)\ge|\Delta_F|^{\frac{r}{2}}} N(E)^{\alpha s+\beta} d\mu(E);$$

$$III(s) := (|\Delta_F|^{-\frac{rB}{2}})^s \int_{E\in\mathcal{M}_{\mathbb{A}_F,r}, N(E)\ge|\Delta_F|^{\frac{r}{2}}} H^0_{\mathrm{ga}}(F,E)^A \cdot N(E)^{Bs+C} d\mu(E).$$

By Theorem 1.2.3, i.e., the Serre duality and Riemann-Roch theorem, we have

$$III(s) = (|\Delta_F|^{-\frac{rB}{2}})^{-s-\frac{A+2C}{B}} \times \int_{E\in\mathcal{M}_{\mathbb{A}_F,r},N(E)\leq|\Delta_F|^{\frac{r}{2}}} H^0_{\mathrm{ga}}(F,E)^A\cdot N(E)^{B(-s-\frac{A+2C}{B})+C}d\mu(E),$$

by the fact that $N(E_1\otimes E_2^\vee) = N(E_1)^{\mathrm{rank}(E_2)}.N(E_2)^{-\mathrm{rank}(E_1)}$. Hence, formally,

$$Z_{F,r;A,B,C;\alpha,\beta}(s) = I(s) + I(-s-\frac{A+2C}{B}) - II(s) + IV(s),$$

where

$$IV(s) := (|\Delta_F|^{-\frac{rB}{2}})^{-s-\frac{A+2C}{B}} \times \int_{E\in\mathcal{M}_{\mathbb{A}_F,r},N(E)\leq|\Delta_F|^{\frac{r}{2}}} N(E)^{\alpha\cdot(-s-\frac{A+2C}{B})+\beta}d\mu(E).$$

Moreover, by definition,

$$\begin{aligned}-II(s) &= -\int_{E\in\mathcal{M}_{\mathbb{A}_F,r},N(E)\geq\Delta_F^{\frac{r}{2}}} N(E)^{\alpha s+\beta}d\mu(E)\\ &= -\int_{\mathcal{M}_{\mathbb{A}_F,r}[|\Delta_F|^{\frac{r}{2}}]} d\mu(E)\cdot\int_1^\infty t^{\alpha t+\beta}\frac{dt}{t}\\ &= -W_F(r)\cdot\frac{t^{\alpha s+\beta}}{\alpha s+\beta}\Big|_1^\infty = W_F(r)\cdot\frac{1}{\alpha s+\beta},\end{aligned}$$

provided that $\alpha s+\beta<0$.

Similarly,

$$\begin{aligned}IV(s) &= \int_{\mathcal{M}_{\mathbb{A}_F,r}[|\Delta_F|^{\frac{r}{2}}]} d\mu(E)\cdot\int_0^1 t^{\alpha(-s-\frac{A+2C}{B})+\beta}\frac{dt}{t}\\ &= W_F(r)\cdot\frac{1}{\alpha(-s-\frac{A+2C}{B})+\beta}\end{aligned}$$

provided that $\alpha(-s-\frac{A+2C}{B})+\beta>0$.

Therefore, formally,

$$\begin{aligned}Z_{F,r;A,B,C;\alpha,\beta}(s) =& I(s) + I(-s-\frac{A+2C}{B})\\ &+ W_F(r)\cdot\Big(\frac{1}{\alpha s+\beta}+\frac{1}{\alpha(-s-\frac{A+2C}{B})+\beta}\Big).\end{aligned}$$

As a direct consequence, we have

$$Z_{F,r;A,B,C;\alpha,\beta}(s) = Z_{F,r;A,B,C;\alpha,\beta}(-s - \frac{A+2C}{B}).$$

This is the functional equation.

Next let us justify the arguments above, we consider two types of convergence problems.

Type 1. Convergence for $II(s)$ and $IV(s)$, where

$$\begin{aligned} II(s) &= (|\Delta_F|^{-\frac{rB}{2}})^s \int_{E \in \mathcal{M}_{\mathbb{A}_F,r}, N(E) \geq |\Delta_F|^{\frac{r}{2}}} N(E)^{\alpha s+\beta} d\mu(E); \\ IV(s) &= (|\Delta_F|^{-\frac{rB}{2}})^{-s-\frac{A+2C}{B}} \\ &\quad \times \int_{E \in \mathcal{M}_{\mathbb{A}_F,r}, N(E) \leq |\Delta_F|^{\frac{r}{2}}} N(E)^{\alpha\cdot(-s-\frac{A+2C}{B})+\beta} d\mu(E). \end{aligned}$$

From above calculation, when $\mathrm{Re}(\alpha\cdot s+\beta) < 0$ and $\mathrm{Re}(\alpha\cdot(-s-\frac{A+2C}{B})+\beta) > 0$, being holomorphic functions,

$$II(s) = -W_F(r)\cdot\frac{1}{\alpha\cdot s+\beta},\quad IV(s) = W_E(r)\cdot\frac{1}{\alpha\cdot(-s-\frac{A+2C}{B})+\beta}.$$

Type 2. Convergence for $I(s)$ and $I(-s-\frac{A+2C}{B})$ where

$$\begin{aligned} I(s) = (|\Delta_F|^{-\frac{rB}{2}})^s &\int_{E \in \mathcal{M}_{\mathbb{A}_F,r}, N(E) \leq |\Delta_F|^{\frac{r}{2}}} \\ &\Big(H^0_{\mathrm{ga}}(F,E)^A \cdot N(E)^{Bs+C} - N(E)^{\alpha s+\beta}\Big) d\mu(E). \end{aligned}$$

By the discussion above about $II(s)$, unless $B=\alpha$, $I(s)$ and $I(-s-\frac{A+2C}{B})$, and hence $Z_{F,r;A,B,C;\alpha,\beta}(s)$ cannot be meromorphically extended as a meromorphic function to the whole s-plane. (See also the discussion below.) Thus, we introduce the following

Compatibility Conditions: $\alpha = B$ and $\beta = C$.

With this, by a change of variables, we may afterwards assume that $B=1$ and $C=0$ without loss of generality so as to obtain the following integration:

$$\begin{aligned} Z_{F,r;A}(s) &:= Z_{F,r;A,-1,0;-1,0}(s) \\ &:= (|\Delta_F|^{\frac{r}{2}})^s \cdot \int_{E \in \mathcal{M}_{\mathbb{A}_F,r}} \Big(H^0_{\text{ga}}(F,E)^A - 1\Big) \cdot N(E)^{-s} d\mu(E). \end{aligned}$$

(Note that then $Z_{F,r;1}(s) = \xi_{F,r}(s)$.) Furthermore, for any $g \in \mathrm{GL}_r(\mathbb{A})$, note that $H^0(\mathbb{A}_F, g)$ is discrete, so from the definition, the associated measure is simply the standard counting measure. Thus by writing down precisely,

$$\begin{aligned} &H^0_{\text{ga}}(F,g) \\ &= 1 + \sum_{\alpha \in H^0(\mathrm{Spec}O_F, \mathcal{E}(g)) \setminus \{0\}} \exp\Big(-\pi \sum_{\sigma:\mathbb{R}} |g_\sigma \cdot \alpha|^2 - 2\pi \sum_{\sigma:\mathbb{C}} |g_\sigma \cdot \alpha|^2\Big) \\ &=: 1 + H'_{\text{ga}}(F,g). \end{aligned}$$

In this expression, the first term is simply the constant function 1 on the moduli space, while each term in the second decays exponentially. Hence, by a standard argument about convergence of an integration of theta series for higher rank lattices, see. e.g, [Ne], and the fact that 1 in the first term $H^0_{\text{ga}}(F,E)^A$ cancels with 1 appeared as the second term in the combination $H^0_{\text{ga}}(F,E)^A - 1$, we conclude that $II(s)$ and $II(-s-A)$ are all holomorphic functions, provided $A > 0$. □

1.5 First Example

1.5.1 Epstein Zeta and High Rank Zeta

Accordingly, the rank r zeta function $\xi_{\mathbb{Q},r}(s)$ of $\mathbb{Q}$ is given by

$$\xi_{\mathbb{Q},r}(s) = \int_{\mathcal{M}_{\mathbb{Q},r}} \Big(e^{h^0(\mathbb{Q},\Lambda)} - 1\Big) \cdot (e^{-s})^{\deg(\Lambda)} d\mu(\Lambda), \qquad \mathrm{Re}(s) > 1,$$

where

$$h^0(\mathbb{Q},\Lambda) := \log\Bigg(\sum_{x \in \Lambda} \exp\big(-\pi|x|^2\big)\Bigg), \qquad \deg(\Lambda) = -\log \mathrm{Vol}(\mathbb{R}^r/\Lambda).$$

And we have

Theorem′. (i) $\xi_{\mathbb{Q},1}(s)$ coincides with the (completed) Riemann-zeta;
(ii) $\xi_{\mathbb{Q},r}(s)$ can be meromorphically extended to the whole plane;
(iii) $\xi_{\mathbb{Q},r}(s)$ satisfies the functional equation $\xi_{\mathbb{Q},r}(s) = \xi_{\mathbb{Q},r}(1-s)$;
(iv) $\xi_{\mathbb{Q},r}(s)$ has two singularities at $s = 0, 1$ only, all simple poles with the same residues $\mathrm{Vol}\left(\mathcal{M}_{\mathbb{Q},r}[1]\right)$.

Decompose according to their volumes, $\mathcal{M}_{\mathbb{Q},r} = \cup_{T>0}\mathcal{M}_{\mathbb{Q},r}[T]$, and there is a natural morphism $\mathcal{M}_{\mathbb{Q},r}[T] \to \mathcal{M}_{\mathbb{Q},r}[1]$, $\Lambda \mapsto T^{\frac{1}{r}} \cdot \Lambda$. Consequently,

$$\begin{aligned}\xi_{\mathbb{Q},r}(s) &= \int_{\cup_{T>0}\mathcal{M}_{\mathbb{Q},r}[T]} \left(e^{h^0(\mathbb{Q},\Lambda)} - 1\right) \cdot (e^{-s})^{\deg(\Lambda)}\, d\mu(\Lambda)\\ &= \int_0^\infty T^s \frac{dT}{T} \int_{\mathcal{M}_{\mathbb{Q},r}[1]} \left(e^{h^0(\mathbb{Q},T^{\frac{1}{r}}\cdot\Lambda)} - 1\right) d\mu(\Lambda).\end{aligned}$$

But $h^0(\mathbb{Q}, T^{\frac{1}{r}} \cdot \Lambda) = \log\left(\sum_{x\in\Lambda} \exp\left(-\pi|x|^2 \cdot T^{\frac{2}{r}}\right)\right)$ and

$$\int_0^\infty e^{-AT^B} T^s \frac{dT}{T} = \frac{1}{B} \cdot A^{-\frac{s}{B}} \cdot \Gamma\left(\frac{s}{B}\right) \qquad \text{for } B \neq 0,$$

we have

$$\xi_{\mathbb{Q},r}(s) = \frac{r}{2} \cdot \pi^{-\frac{r}{2}s}\Gamma\left(\frac{r}{2}s\right) \cdot \int_{\mathcal{M}_{\mathbb{Q},r}[1]} \left(\sum_{x\in\Lambda\setminus\{0\}} |x|^{-rs}\right) d\mu_1(\Lambda).$$

Accordingly, introduce the completed Epstein zeta function for Λ by

$$\hat{E}(\Lambda; s) := \pi^{-s}\Gamma(s) \cdot \sum_{x\in\Lambda\setminus\{0\}} |x|^{-2s}.$$

We then arrive at

Proposition. (Eisenstein Zeta and High Rank Zeta)

$$\xi_{\mathbb{Q},r}(s) = \frac{r}{2} \int_{\mathcal{M}_{\mathbb{Q},r}[1]} \hat{E}\left(\Lambda, \frac{r}{2}s\right) d\mu_1(\Lambda).$$

1.5.2 Rankin-Selberg & Zagier Method

Consider the action of $\mathrm{SL}(2,\mathbb{Z})$ on the upper half plane $\mathcal{H}$. Then a standard 'fundamental domain' is given by $D = \{z = x + iy \in \mathcal{H} : |x| \leq \frac{1}{2}, y > 0, x^2 + y^2 \geq 1\}$. Recall also the completed standard Eisenstein series

$$\hat{E}(z; s) := \pi^{-s}\Gamma(s) \cdot \sum_{(m,n)\in\mathbb{Z}^2\setminus\{(0,0)\}} \frac{y^s}{|mz+n|^{2s}}.$$

Naturally, we are led to consider the integral $\int_D \hat{E}(z,s)\frac{dx\,dy}{y^2}$. However, this integration diverges. Indeed, near the only cusp $y = \infty$, by the Chowla-Selberg formula, $\hat{E}(z,s)$ has the Fourier expansion

$$\hat{E}(z; s) = \sum_{n=-\infty}^{\infty} a_n(y,s)e^{2\pi inx}$$

with

$$a_n(y,s) = \begin{cases} \xi(2s)y^s + \xi(2-2s)y^{1-s}, & \text{if } n = 0; \\ 2|n|^{s-\frac{1}{2}}\sigma_{1-2s}(|n|)\,\sqrt{y}K_{s-\frac{1}{2}}(2\pi|n|y), & \text{if } n \neq 0, \end{cases}$$

where $\xi(s)$ is the completed Riemann zeta function, $\sigma_s(n) := \sum_{d|n} d^s$, and $K_s(y) := \frac{1}{2}\int_0^\infty e^{-y(t+\frac{1}{t})/2}t^s\frac{dt}{t}$ is the K-Bessel function. Moreover,

$$|K_s(y)| \leq e^{-y/2}K_{\mathrm{Re}(s)}(2), \text{ if } y > 4, \qquad \text{and} \qquad K_s = K_{-s}.$$

So $a_{n\neq 0}(y,s)$ decay exponentially, and the problematic term comes from $a_0(y,s)$, which is of slow growth.

Therefore, to make the original integration meaningful, we need to cut-off the slow growth part. There are two ways to do so: one is geometric and hence rather direct and simple; while the other is analytic, and hence rather technical and traditional, dated back to Rankin-Selberg.

(a) **Geometric Truncation**

Draw a horizontal line $y = T \geq 1$ and set

$$D_T = \{z = x + iy \in D : y \leq T\}, \qquad D^T = \{z = x + iy \in D : y \geq T\}.$$

Then $D = D_T \cup D^T$. Introduce a well-defined integration

$$I_T^{\mathrm{Geo}}(s) := \int_{D_T} \hat{E}(z,s)\,\frac{dx\,dy}{y^2}.$$

(b) **Analytic Truncation**

Define a truncated Eisenstein series $\hat{E}_T(z;s)$ by

$$\hat{E}_T(z;s) := \begin{cases} \hat{E}(z;s), & \text{if } y \le T; \\ \hat{E}(z,s) - a_0(y;s), & \text{if } y > T. \end{cases}$$

Introduce a well-defined integration

$$I_T^{\mathrm{Ana}}(s) := \int_D \hat{E}_T(z;s)\,\frac{dx\,dy}{y^2}.$$

With this, from the Rankin-Selberg method, one checks that we have the following:

Proposition. (Analytic Truncation = Geometric Truncation in Rank 2)

$$I_T^{\mathrm{Geo}}(s) = \frac{\xi(2s)}{s-1}\cdot T^{s-1} - \frac{\xi(2s-1)}{s}\cdot T^{-s} = I_T^{\mathrm{Ana}}(s).$$

For a proof, please see Ch. 7.

Each of the above two integrations has its own merit: for the geometric one, we keep the Eisenstein series unchanged, while for the analytic one, we keep the original fundamental domain of $\mathcal{H}$ under $\mathrm{SL}(2,\mathbb{Z})$ as it is.

Note that the nice point about the fundamental domain is that it admits a modular interpretation. Thus it would be very nice if we could keep the Eisenstein series unchanged, while offer some integration domains which appear naturally in certain moduli problems. Guided by this, in Ch. 2, we will introduce non-abelian L-functions using integrations of Eisenstein series over generalized moduli spaces.

(c) **Arithmetic Truncation**

Now we explain why the above discussion and Rankin-Selberg method have anything to do with our high rank zeta functions. For this, we introduce yet another truncation, the stability one.

So back to the moduli space of rank 2 lattices of volume 1 over $\mathbb{Q}$. Then classical reduction theory gives a natural map from this moduli space to the fundamental domain D above: For any lattice Λ, fix $\mathbf{x}_1 \in \Lambda$ such that its length gives the first Minkowski minimum λ_1 of Λ. Then via rotation, we may assume that $\mathbf{x}_1 = (\lambda_1, 0)$. Further, from the reduction theory $\frac{1}{\lambda_1}\Lambda$ may be viewed as the lattice of the volume $\lambda_1^{-2} = y_0$ which is generated by $(1,0)$ and $\omega = x_0 + iy_0 \in D$. That is to say, the points in D_T constructed above are in one-to-one corresponding to the rank two lattices of volume one whose first Minkowski minimum $\lambda_1^{-2} \leq T$, i.e, $\lambda_1 \geq T^{-\frac{1}{2}}$. Set $\mathcal{M}_{\mathbb{Q},2}^{\leq\frac{1}{2}\log T}[1]$ be the moduli space of rank 2 lattices Λ of volume 1 over $\mathbb{Q}$ all of whose sublattices Λ_1 of rank 1 have degrees $\leq \frac{1}{2}\log T$. As a direct consequence, we have the following

Proposition. (Geometric Truncation = Arithmetic Truncation) *There is a natural one-to-one, onto morphism*

$$\mathcal{M}_{\mathbb{Q},2}^{\leq\frac{1}{2}\log T}[1] \simeq D_T.$$

In particular,

$$\mathcal{M}_{\mathbb{Q},2}^{\leq 0}[1] = \mathcal{M}_{\mathbb{Q},2}[1] \simeq D_1.$$

Consequently, we have the following

Example in Rank 2. $\quad \xi_{\mathbb{Q},2}(s) = \dfrac{\xi(2s)}{s-1} - \dfrac{\xi(2s-1)}{s}.$

2 Non-Abelian L-Functions

2.1 Automorphic Forms and Eisenstein Series

To facilitate our ensuing discussion, we make the following preparation. For details, see e.g., [MW] and [We7].

Fix a connected reductive group G defined over F, denote by Z_G its center. Fix a minimal parabolic subgroup P_0 of G. Then $P_0 = M_0U_0$, where as usual we fix once and for all the Levi M_0 and the unipotent radical U_0. A parabolic subgroup P is G is called standard if $P \supset P_0$. For such groups write $P = MU$ with $M_0 \subset M$ the standard Levi and U the unipotent radical. Denote by $\mathrm{Rat}(M)$ the group of rational characters of M, i.e, the morphism $M \to \mathbb{G}_m$ where $\mathbb{G}_m$ denotes the multiplicative group. Set

$$\mathfrak{a}_M^* := \mathrm{Rat}(M) \otimes_{\mathbb{Z}} \mathbb{C}, \qquad \mathfrak{a}_M := \mathrm{Hom}_{\mathbb{Z}}(\mathrm{Rat}(M), \mathbb{C}),$$

and

$$\mathrm{Re}\mathfrak{a}_M^* := \mathrm{Rat}(M) \otimes_{\mathbb{Z}} \mathbb{R}, \qquad \mathrm{Re}\mathfrak{a}_M := \mathrm{Hom}_{\mathbb{Z}}(\mathrm{Rat}(M), \mathbb{R}).$$

For any $\chi \in \mathrm{Rat}(M)$, we obtain a (real) character $|\chi| : M(\mathbb{A}) \to \mathbb{R}^*$ defined by $m = (m_v) \mapsto m^{|\chi|} := \prod_{v \in S} |m_v|_v^{\chi_v}$ with $|\cdot|_v$ the v-absolute values. Set then $M(\mathbb{A})^1 := \cap_{\chi \in \mathrm{Rat}(M)} \mathrm{Ker}|\chi|$, which is a normal subgroup of $M(\mathbb{A})$. Set X_M be the group of complex characters which are trivial on $M(\mathbb{A})^1$. Denote by $H_M := \log_M : M(\mathbb{A}) \to \mathfrak{a}_M$ the map such that $\forall \chi \in \mathrm{Rat}(M) \subset \mathfrak{a}_M^*, \langle \chi, \log_M(m) \rangle := \log(m^{|\chi|})$. Clearly,

$$M(\mathbb{A})^1 = \mathrm{Ker}(\log_M); \qquad \log_M(M(\mathbb{A})/M(\mathbb{A})^1) \simeq \mathrm{Re}\mathfrak{a}_M.$$

Hence in particular there is a natural isomorphism $\kappa : \mathfrak{a}_M^* \simeq X_M$. Set

$$\mathrm{Re}X_M := \kappa(\mathrm{Re}\mathfrak{a}_M^*), \qquad \mathrm{Im}X_M := \kappa(i \cdot \mathrm{Re}\mathfrak{a}_M^*).$$

Moreover define our working space X_M^G to be the subgroup of X_M consisting of complex characters of $M(\mathbb{A})/M(\mathbb{A})^1$ which are trivial on $Z_{G(\mathbb{A})}$.

Fix a maximal compact subgroup $\mathbb{K}$ such that for all standard parabolic subgroups $P = MU$ as above, $P(\mathbb{A}) \cap \mathbb{K} = \big(M(\mathbb{A}) \cap \mathbb{K}\big) \cdot \big(U(\mathbb{A}) \cap \mathbb{K}\big)$. Hence we get the Langlands decomposition $G(\mathbb{A}) = M(\mathbb{A}) \cdot U(\mathbb{A}) \cdot \mathbb{K}$. Denote by $m_P : G(\mathbb{A}) \to M(\mathbb{A})/M(\mathbb{A})^1$ the map $g = m \cdot n \cdot k \mapsto M(\mathbb{A})^1 \cdot m$ where $g \in G(\mathbb{A}), m \in M(\mathbb{A}), n \in U(\mathbb{A})$ and $k \in \mathbb{K}$.

Fix Haar measures on $M_0(\mathbb{A}), U_0(\mathbb{A}), \mathbb{K}$ respectively such that
(1) the induced measure on $M(F)$ is the counting measure and the volume of the induced measure on $M(F)\backslash M(\mathbb{A})^1$ is 1. (Recall that it is a fundamental fact that $M(F)\backslash M(\mathbb{A})^1$ is of finite volume.)
(2) the induced measure on $U_0(F)$ is the counting measure and the volume of $U(F)\backslash U_0(\mathbb{A})$ is 1. (Recall that being unipotent radical, $U(F)\backslash U_0(\mathbb{A})$ is compact.)
(3) the volume of $\mathbb{K}$ is 1.

Such measures then also induce Haar measures via $\log_M$ to the spaces $\mathfrak{a}_{M_0}, \mathfrak{a}^*_{M_0}$, etc. Furthermore, if we denote by ρ_0 the half of the sum of the positive roots of the maximal split torus T_0 of the central Z_{M_0} of M_0, then

$$f \mapsto \int_{M_0(\mathbb{A})\cdot U_0(\mathbb{A})\cdot\mathbb{K}} f(mnk)\,dk\,dn\,m^{-2\rho_0}dm$$

defined for continuous functions with compact supports on $G(\mathbb{A})$ defines a Haar measure dg on $G(\mathbb{A})$. This in turn gives measures on $M(\mathbb{A}), U(\mathbb{A})$ and hence on $\mathfrak{a}_M, \mathfrak{a}^*_M$, $P(\mathbb{A})$, etc, for all parabolic subgroups P. In particular, one checks that the following compatibility condition holds

$$\begin{aligned}&\int_{M_0(\mathbb{A})\cdot U_0(\mathbb{A})\cdot\mathbb{K}} f(mnk)\,dk\,dn\,m^{-2\rho_0}dm\\ &= \int_{M(\mathbb{A})\cdot U(\mathbb{A})\cdot\mathbb{K}} f(mnk)\,dk\,dn\,m^{-2\rho_P}dm\end{aligned}$$

for all continuous functions f with compact supports on $G(\mathbb{A})$, where ρ_P denotes one half of the sum of all positive roots of the maximal split torus T_P of the central Z_M of M. For later use, denote also by Δ_P the set of positive roots determined by (P, T_P) and $\Delta_0 = \Delta_{P_0}$.

Fix an isomorphism $T_0 \simeq \mathbb{G}_m^R$. Embed $\mathbb{R}^*_+$ by the map $t \mapsto (1;t)$. Then we obtain a natural injection $(\mathbb{R}^*_+)^R \hookrightarrow T_0(\mathbb{A})$ which splits. Denote by $A_{M_0(\mathbb{A})}$ the unique connected subgroup of $T_0(\mathbb{A})$ which projects onto $(\mathbb{R}^*_+)^R$. More generally, for a standard parabolic subgroup $P = MU$, set $A_{M(\mathbb{A})} := A_{M_0(\mathbb{A})} \cap Z_{M(\mathbb{A})}$ where as used above Z_* denotes the center of the group $*$. Clearly, $M(\mathbb{A}) = A_{M(\mathbb{A})} \cdot M(\mathbb{A})^1$. For later use, set also $A^G_{M(\mathbb{A})} := \{a \in A_{M(\mathbb{A})} : \log_G a = 0\}$. Then $A_{M(\mathbb{A})} = A_{G(\mathbb{A})} \oplus A^G_{M(\mathbb{A})}$.

Note that $\mathbb{K}$ and $U(F)\backslash U(\mathbb{A})$ are all compact, and $M(F)\backslash M(\mathbb{A})^1$ is of finite volume. With the Langlands decomposition $G(\mathbb{A}) = U(\mathbb{A})M(\mathbb{A})\mathbb{K}$ in mind, the reduction theory for $G(F)\backslash G(\mathbb{A})$ or more generally $P(F)\backslash G(\mathbb{A})$ is reduced to that for $A_{M(\mathbb{A})}$ since $Z_G(F)\cap Z_{G(\mathbb{A})}\backslash Z_{G(\mathbb{A})}\cap G(\mathbb{A})^1$ is compact as well. (The reader should know that one of the chief purposes to have a nice reduction theory is to make sure that the integrations of moderate growth functions are controlled. Thus with $M(F)\backslash M(\mathbb{A})^1$ of finite volume, a good control is not a big problem. This then leads us to the affine part A.) As such, for $t_0\in M_0(\mathbb{A})$ set

$$A_{M_0(\mathbb{A})}(t_0) := \{a\in A_{M_0(\mathbb{A})} : a^\alpha > t_0^\alpha \forall\alpha\in\Delta_0\}.$$

Then, for a fixed compact subset $\omega\subset P_0(\mathbb{A})$, we have the corresponding Siegel set

$$S(\omega;t_0) := \{p\cdot a\cdot k : p\in\omega, a\in A_{M_0(\mathbb{A})}(t_0), k\in\mathbb{K}\}.$$

In particular, for big enough ω and small enough t_0, i.e, t_0^α is very close to 0 for all $\alpha\in\Delta_0$, the classical reduction theory may be restated as $G(\mathbb{A}) = G(F)\cdot S(\omega;t_0)$. More generally set

$$A^P_{M_0(\mathbb{A})}(t_0) := \{a\in A_{M_0(\mathbb{A})} : a^\alpha > t_0^\alpha \forall\alpha\in\Delta_0^P\},$$

and

$$S^P(\omega;t_0) := \{p\cdot a\cdot k : p\in\omega, a\in A^P_{M_0(\mathbb{A})}(t_0), k\in\mathbb{K}\}.$$

Then similarly as above for big enough ω and small enough t_0, $G(\mathbb{A}) = P(F)\cdot S^P(\omega;t_0)$. (Here Δ_0^P denotes the set of positive roots for $(P_0\cap M, T_0)$.)

Fix an embedding $i_G : G \hookrightarrow SL_n$ sending g to (g_{ij}). Introducing a height function on $G(\mathbb{A})$ by setting $\|g\| := \prod_{v\in S}\sup\{|g_{ij}|_v : \forall i,j\}$. It is well known that up to $O(1)$, height functions are unique. This implies that the following growth conditions do not depend on the height function we choose.

A function $f : G(\mathbb{A})\to\mathbb{C}$ is said to have moderate growth if there exist $c,r\in\mathbb{R}$ such that $|f(g)|\leq c\cdot\|g\|^r$ for all $g\in G(\mathbb{A})$. Similarly, for a standard parabolic subgroup $P = MU$, a function $f : U(\mathbb{A})M(F)\backslash G(\mathbb{A})\to\mathbb{C}$ is said to have moderate growth if there exist $c,r\in\mathbb{R}$, $\lambda\in\mathrm{Re}X_{M_0}$ such that for any $a\in A_{M(\mathbb{A})}, k\in\mathbb{K}, m\in M(\mathbb{A})^1\cap S^P(\omega;t_0)$,

$$|f(amk)|\leq c\cdot\|a\|^r\cdot m_{P_0}(m)^\lambda.$$

Also a function $f : G(\mathbb{A}) \to \mathbb{C}$ is said to be smooth if for any $g = g_f \cdot g_\infty \in G(\mathbb{A}_f) \times G(\mathbb{A}_\infty)$, there exist open neighborhoods V_* of g_* in $G(\mathbb{A})$ and a C^∞-function $f' : V_\infty \to \mathbb{C}$ such that $f(g'_f \cdot g'_\infty) = f'(g'_\infty)$ for all $g'_f \in V_f$ and $g'_\infty \in V_\infty$.

By contrast, a function $f : S(\omega; t_0) \to \mathbb{C}$ is said to be rapidly decreasing if there exists $r > 0$ and for all $\lambda \in \mathrm{Re}X_{M_0}$ there exists $c > 0$ such that for $a \in A_{M(\mathbb{A})}, g \in G(\mathbb{A})^1 \cap S(\omega; t_0)$, $|\phi(ag)| \leq c \cdot \|a\| \cdot m_{P_0}(g)^\lambda$. And a function $f : G(F)\backslash G(\mathbb{A}) \to \mathbb{C}$ is said to be rapidly decreasing if $f|_{S(\omega;t_0)}$ is so.

By definition, a function $\phi : U(\mathbb{A})M(F)\backslash G(\mathbb{A}) \to \mathbb{C}$ is called automorphic if

(i) ϕ has moderate growth;

(ii) ϕ is smooth;

(iii) ϕ is $\mathbb{K}$-finite, i.e, the $\mathbb{C}$-span of all $\phi(k_1 \cdot * \cdot k_2)$ parametrized by $(k_1, k_2) \in \mathbb{K} \times \mathbb{K}$ is finite dimensional; and

(iv) ϕ is $\mathfrak{z}$-finite, i.e, the $\mathbb{C}$-span of all $\delta(X)\phi$ parametrized by all $X \in \mathfrak{z}$ is finite dimensional. Here $\mathfrak{z}$ denotes the center of the universal enveloping algebra $\mathfrak{u} := \mathfrak{U}(\mathrm{Lie}G(\mathbb{A}_\infty))$ of the Lie algebra of $G(\mathbb{A}_\infty)$ and $\delta(X)$ denotes the derivative of ϕ along X.

For such a function ϕ, set $\phi_k : M(F)\backslash M(\mathbb{A}) \to \mathbb{C}$ by $m \mapsto m^{-\rho_P}\phi(mk)$ for all $k \in \mathbb{K}$. Then one checks that ϕ_k is an automorphic form in the usual sense. Set $A(U(\mathbb{A})M(F)\backslash G(\mathbb{A}))$ be the space of automorphic forms on $U(\mathbb{A})M(F)\backslash G(\mathbb{A})$.

For a measurable locally L^1-function $f : U(F)\backslash G(\mathbb{A}) \to \mathbb{C}$, define its constant term along with the standard parabolic subgroup $P = UM$ to be $f_P : U(\mathbb{A})\backslash G(\mathbb{A}) \to \mathbb{C}$ given by $g \mapsto \int_{U(F)\backslash G(\mathbb{A})} f(ng)dn$. Then an automorphic form $\phi \in A(U(\mathbb{A})M(F)\backslash G(\mathbb{A}))$ is called a cusp form if for any standard parabolic subgroup P' properly contained in P, $\phi_{P'} \equiv 0$. Denote by $A_0(U(\mathbb{A})M(F)\backslash G(\mathbb{A}))$ the space of cusp forms on $U(\mathbb{A})M(F)\backslash G(\mathbb{A})$. One checks easily that

(i) all cusp forms are rapidly decreasing; and hence

(ii) there is a natural pairing

$$\langle \cdot, \cdot \rangle : A_0(U(\mathbb{A})M(F)\backslash G(\mathbb{A})) \times A(U(\mathbb{A})M(F)\backslash G(\mathbb{A})) \to \mathbb{C}$$

defined by $\langle \psi, \phi \rangle := \int_{Z_{M(\mathbb{A})}U(\mathbb{A})M(F)\backslash G(\mathbb{A})} \psi(g)\bar{\phi}(g)\, dg$.

Moreover, for a (complex) character $\xi : Z_{M(\mathbb{A})} \to \mathbb{C}^*$ of $Z_{M(\mathbb{A})}$ set

$$\begin{aligned} A(U(\mathbb{A})M(F)\backslash G(\mathbb{A}))_\xi := \{&\phi \in A(U(\mathbb{A})M(F)\backslash G(\mathbb{A})) : \\ &\phi(zg) = z^{\rho_P} \cdot \xi(z) \cdot \phi(g), \forall z \in Z_{M(\mathbb{A})}, g \in G(\mathbb{A})\} \end{aligned}$$

and

$$\begin{aligned} &A_0(U(\mathbb{A})M(F)\backslash G(\mathbb{A}))_\xi \\ &:= A_0(U(\mathbb{A})M(F)\backslash G(\mathbb{A})) \cap A(U(\mathbb{A})M(F)\backslash G(\mathbb{A}))_\xi. \end{aligned}$$

Set now

$$A(U(\mathbb{A})M(F)\backslash G(\mathbb{A}))_Z := \sum_{\xi \in \mathrm{Hom}(Z_{M(\mathbb{A})}, \mathbb{C}^*)} A(U(\mathbb{A})M(F)\backslash G(\mathbb{A}))_\xi$$

and

$$A_0(U(\mathbb{A})M(F)\backslash G(\mathbb{A}))_Z := \sum_{\xi \in \mathrm{Hom}(Z_{M(\mathbb{A})}, \mathbb{C}^*)} A_0(U(\mathbb{A})M(F)\backslash G(\mathbb{A}))_\xi.$$

One checks that the natural morphism

$$\mathbb{C}[\mathrm{Re}\mathfrak{a}_M] \otimes A(U(\mathbb{A})M(F)\backslash G(\mathbb{A}))_Z \to A(U(\mathbb{A})M(F)\backslash G(\mathbb{A}))$$

defined by $(Q, \phi) \mapsto (g \mapsto Q(\log_M(m_P(g))) \cdot \phi(g)$ is an isomorphism, using the special structure of $A_{M(\mathbb{A})}$-finite functions and the Fourier analysis over the compact space $A_{M(\mathbb{A})}\backslash Z_{M(\mathbb{A})}$. Consequently, we also obtain a natural isomorphism

$$\mathbb{C}[\mathrm{Re}\mathfrak{a}_M] \otimes A_0(U(\mathbb{A})M(F)\backslash G(\mathbb{A}))_Z \to A_0(U(\mathbb{A})M(F)\backslash G(\mathbb{A}))_\xi.$$

(See 4.2 below for a more concrete and useful form.)

Set also $\Pi_0(M(\mathbb{A}))_\xi$ be isomorphism classes of irreducible representations of $M(\mathbb{A})$ occurring in the space $A_0(M(F)\backslash M(\mathbb{A}))_\xi$, and

$$\Pi_0(M(\mathbb{A}) := \cup_{\xi \in \mathrm{Hom}(Z_{M(\mathbb{A})}, \mathbb{C}^*)} \Pi_0(M(\mathbb{A}))_\xi.$$

(More precisely, we should use $M(\mathbb{A}_f) \times (M(\mathbb{A}) \cap \mathbb{K}, \mathrm{Lie}(M(\mathbb{A}_\infty)) \otimes_\mathbb{R} \mathbb{C}))$ instead of $M(\mathbb{A})$.) For any $\pi \in \Pi_0(M(\mathbb{A}))_\xi$ set $A_0(M(F)\backslash M(\mathbb{A}))_\pi$ to be the isotypic component of type π of $A_0(M(F)\backslash M(\mathbb{A}))_\xi$, i.e, the set of cusp forms of

$M(\mathbb{A})$ generating a semi-simple isotypic $M(\mathbb{A}_f)\times(M(\mathbb{A})\cap\mathbb{K}, \mathrm{Lie}(M(\mathbb{A}_\infty))\otimes_{\mathbb{R}} \mathbb{C}))$-module of type π. Set

$$\begin{aligned}&A_0(U(\mathbb{A})M(F)\backslash G(\mathbb{A}))_\pi \\ &:= \{\phi \in A_0(U(\mathbb{A})M(F)\backslash G(\mathbb{A})) : \phi_k \in A_0(M(F)\backslash M(\mathbb{A}))_\pi, \forall k \in \mathbb{K}\}.\end{aligned}$$

Clearly

$$A_0(U(\mathbb{A})M(F)\backslash G(\mathbb{A}))_\xi = \oplus_{\pi\in\Pi_0(M(\mathbb{A}))_\xi} A_0(U(\mathbb{A})M(F)\backslash G(\mathbb{A}))_\pi.$$

More generally, let $V \subset A(M(F)\backslash M(\mathbb{A}))$ be an irreducible $M(\mathbb{A}_f) \times (M(\mathbb{A}) \cap \mathbb{K}, \mathrm{Lie}(M(\mathbb{A}_\infty)) \otimes_{\mathbb{R}} \mathbb{C}))$-module with π_0 the induced representation of $M(\mathbb{A}_f) \times (M(\mathbb{A}) \cap \mathbb{K}, \mathrm{Lie}(M(\mathbb{A}_\infty)) \otimes_{\mathbb{R}} \mathbb{C}))$. Then we call π_0 an automorphic representation of $M(\mathbb{A})$. Denote $A(M(F)\backslash M(\mathbb{A}))_{\pi_0}$ the isotypic subquotient module of type π_0 of $A(M(F)\backslash M(\mathbb{A}))$. One checks that

$$\begin{aligned}V \otimes \mathrm{Hom}_{M(\mathbb{A}_f)\times(M(\mathbb{A})\cap\mathbb{K},\mathrm{Lie}(M(\mathbb{A}_\infty))\otimes_{\mathbb{R}}\mathbb{C}))}(V, A(M(F)\backslash M(\mathbb{A}))) \\ \simeq A(M(F)\backslash M(\mathbb{A}))_{\pi_0}.\end{aligned}$$

Set

$$\begin{aligned}&A(U(\mathbb{A})M(F)\backslash G(\mathbb{A}))_{\pi_0} \\ &:= \{\phi \in A(U(\mathbb{A})M(F)\backslash G(\mathbb{A})) : \phi_k \in A(M(F)\backslash M(\mathbb{A}))_{\pi_0}, \forall k \in \mathbb{K}\}.\end{aligned}$$

Moreover if $A(M(F)\backslash M(\mathbb{A}))_{\pi_0} \subset A_0(M(F)\backslash M(\mathbb{A}))$, we call π_0 a cuspidal representation.

Two automorphic representations π and π_0 of $M(\mathbb{A})$ are said to be equivalent if there exists $\lambda \in X_M^G$ such that $\pi \simeq \pi_0 \otimes \lambda$. This, in practice, means that $A(M(F)\backslash M(\mathbb{A}))_\pi = \lambda \cdot A(M(F)\backslash M(\mathbb{A}))_{\pi_0}$. That is, for any $\phi_\pi \in A(M(F)\backslash M(\mathbb{A}))_\pi$, there exists an element $\phi_{\pi_0} \in A(M(F)\backslash M(\mathbb{A}))_{\pi_0}$ such that $\phi_\pi(m) = m^\lambda \cdot \phi_{\pi_0}(m)$. Consequently,

$$A(U(\mathbb{A})M(F)\backslash G(\mathbb{A}))_\pi = (\lambda \circ m_P) \cdot A(U(\mathbb{A})M(F)\backslash G(\mathbb{A}))_{\pi_0}.$$

Denote by $\mathfrak{P} := [\pi_0]$ the equivalence class of π_0. Then $\mathfrak{P}$ is an X_M^G-principal homogeneous space, hence admits a natural complex structure. Usually we call $(M, \mathfrak{P})$ a cuspidal datum of G if π_0 is cuspidal. Also for $\pi \in \mathfrak{P}$ set $\mathrm{Re}\pi := \mathrm{Re}\chi_\pi = |\chi_\pi| \in \mathrm{Re}X_M$, where χ_π is the central character of π, and $\mathrm{Im}\pi := \pi \otimes (-\mathrm{Re}\pi)$.

Now fix an irreducible automorphic representation π of $M(\mathbb{A})$ and an automorphic form $\phi \in A(U(\mathbb{A})M(F)\backslash G(\mathbb{A}))_\pi$, define the associated Eisenstein series $E(\phi,\pi) : G(F)\backslash G(\mathbb{A}) \to \mathbb{C}$ by

$$E(\phi,\pi)(g) := \sum_{\delta\in P(F)\backslash G(F)} \phi(\delta g).$$

Then one checks that there is an open cone $C \subset \mathrm{Re}X_M^G$ such that if $\mathrm{Re}\pi \in C$, $E(\lambda\cdot\phi,\pi\otimes\lambda)(g)$ converges uniformly for g in a compact subset of $G(\mathbb{A})$ and λ in an open neighborhood of 0 in X_M^G. For example, if $\mathfrak{P} = [\pi]$ is cuspidal, we may even take C to be the cone $\{\lambda \in \mathrm{Re}X_M^G : \langle\lambda-\rho_P,\alpha^\vee\rangle > 0, \forall\alpha \in \Delta_P^G\}$. As a direct consequence, then $E(\phi,\pi) \in A(G(F)\backslash G(\mathbb{A}))$. That is, it is an automorphic form.

2.2 Non-Abelian L-Functions

2.2.1 Definition

Being automorphic forms, Eisenstein series are of moderate growth. Consequently, they are not integrable over $G(F)\backslash G(\mathbb{A})^1$. On the other hand, Eisenstein series are also smooth and hence integrable over compact subsets of $G(F)\backslash G(\mathbb{A})^1$. So it is very natural for us to search for compact domains which are intrinsically defined.

As such, let us now return to the group $G = GL_r$. Then form 1.3.8, we obtain compact moduli spaces

$$\mathcal{M}_{F,r}^{\leq p}[\Delta_F^{\frac{r}{2}}] := \{g \in GL_r(F)\backslash GL_r(\mathbb{A}) : \deg g = 0, \bar{p}^g \leq p\}$$

for a fixed convex polygon $p : [0,r] \to \mathbb{R}$. For example, $\mathcal{M}_{\mathbb{Q},r}^{\leq 0}[1] = \mathcal{M}_{\mathbb{Q},r}[1]$, (the adelic inverse image of) the moduli space of rank r semi-stable $\mathbb{Z}$-lattices of volume 1.

More generally, for a standard parabolic subgroup P of GL_r, we introduce moduli spaces

$$\mathcal{M}_{F,r}^{P;\leq p}[\Delta_F^{\frac{r}{2}}] := \{g \in P(F)\backslash GL_r(\mathbb{A}) : \deg g = 0, \bar{p}_P^g \leq p, \bar{p}_P^g \geq -p\}.$$

By 1.3.8, these moduli spaces $\mathcal{M}_{F,r}^{P;\leq p}[\Delta_F^{\frac{r}{2}}]$ are all compact.

As usual, take the minimal parabolic subgroup P_0 to be the one consisting of upper triangular matrices corresponding to the partition $(1,\ldots,1)$

with M_0 consisting of diagonal matrices. Then $P = P_I = U_I M_I$ corresponds to a certain partition $I = (r_1, \ldots, r_{|P|})$ of r with M_I the standard Levi and U_I the unipotent radical. (See e.g., 3.2.5.)

Now for a fixed irreducible automorphic representation π of $M_I(\mathbb{A})$, choose

$$\begin{aligned}\phi &\in A(U_I(\mathbb{A})M_I(F)\backslash G(\mathbb{A}))_\pi \cap L^2(U_I(\mathbb{A})M_I(F)\backslash G(\mathbb{A}))\\ &:= A^2(U_I(\mathbb{A})M_I(F)\backslash G(\mathbb{A}))_\pi,\end{aligned}$$

with $L^2(U_I(\mathbb{A})M_I(F)\backslash G(\mathbb{A}))$ the space of L^2 functions on the quotient space $Z_{G(\mathbb{A})}U_I(\mathbb{A})M_I(F)\backslash G(\mathbb{A})$. Denote the associated Eisenstein series by $E(\phi, \pi) \in A(G(F)\backslash G(\mathbb{A}))$.

Definition. *High rank L-function* $L_{F,r}^{\leq p}(\phi, \pi)$ for number field F associated to the L^2-automorphic form $\phi \in A^2(U_I(\mathbb{A})M_I(F)\backslash G(\mathbb{A}))_\pi$ is defined by the integration

$$L_{F,r}^{\leq p}(\phi, \pi) := \int_{\mathcal{M}_{F,r}^{\leq p}[\Delta_F^{\frac{r}{2}}]} E(\phi, \pi)(g)\, dg, \qquad \mathrm{Re}\,\pi \in C.$$

More generally, for a standard parabolic subgroup $P_J = U_J M_J \supset P_I$ (so that the partition J is a refinement of I), we obtain a relative Eisenstein series

$$E_I^J(\phi, \pi)(g) := \sum_{\delta \in P_I(F)\backslash P_J(F)} \phi(\delta g), \qquad \forall g \in P_J(F)\backslash G(\mathbb{A}).$$

There is an open cone C_I^J in $\mathrm{Re}X_{M_I}^{P_J}$ such that if $\mathrm{Re}\pi \in C_I^J$, then $E_I^J(\phi, \pi) \in A(P_J(F)\backslash G(\mathbb{A}))$, where $X_{M_I}^{P_J}$ is defined similarly as X_M^G with G replaced by P_J.

Definition. Define the *non-abelian L-function* by

$$L_{F,r}^{P_J;\leq p}(\phi, \pi) := \int_{\mathcal{M}_{F,r}^{P_J;\leq p}[\Delta_F^{\frac{r}{2}}]} E_I^J(\phi, \pi)(g)\, dg, \qquad \mathrm{Re}\,\pi \in C_I^J.$$

Remark. Here when defining non-abelian L-functions we assume that ϕ comes from a single irreducible automorphic representation. But this restriction is rather artificial and can be removed easily: such a restriction only serves the purpose of giving the constructions and results in a very neat form.

We end this subsection by pointing out that the discussion for rank r L-functions $L_{F,r}^{\leq p}$ holds for general non-abelian L-functions $L_{F,r}^{P_J;\leq p}$ as well. So from now on, we will leave the discussion on general $L_{F,r}^{P_J;\leq p}$ to the reader, while concentrate ourselves to the special case $L_{F,r}^{\leq p}$.

2.2.2 Meromorphic Extension and Functional Equations

With the same notation as above, set $\mathfrak{P} = [\pi]$. For $w \in W$ the Weyl group of $G = GL_r$, fix once and for all representative $w \in G(F)$ of w. Set $M' := wMw^{-1}$ and denote the associated parabolic subgroup by $P' = U'M'$. W acts naturally on automorphic representations, from which we obtain an equivalence classes $w\mathfrak{P}$ of automorphic representations of $M'(\mathbb{A})$. As usual, define the associated intertwining operator $M(w,\pi)$ by

$$(M(w,\pi)\phi)(g) := \int_{U'(F)\cap wU(F)w^{-1}\backslash U'(\mathbb{A})} \phi(w^{-1}n'g)dn', \ \forall g \in G(\mathbb{A}).$$

One checks that if $\langle \mathrm{Re}\pi, \alpha^\vee \rangle \gg 0, \forall \alpha \in \Delta_P^G$,
(i) for a fixed automorphic ϕ, $M(w,\pi)\phi$ depends only on the double coset $M'(F)wM(F)$. So $M(w,\pi)\phi$ is well-defined for $w \in W$;
(ii) the above integral converges absolutely and uniformly for g varying in a compact subset of $G(\mathbb{A})$;
(iii) $M(w,\pi)\phi \in A(U'(\mathbb{A})M'(F)\backslash G(\mathbb{A}))_{w\pi}$; and if ϕ is L^2, which from now on we always assume, so is $M(w,\pi)\phi$.

Basic Facts of Non-Abelian L-Functions.
(I) (Meromorphic Continuation) $L_{F,r}^{\leq p}(\phi,\pi)$ *for* $\mathrm{Re}\,\pi \in C$ *is well-defined and admits a unique meromorphic continuation to the whole space* $\mathfrak{P}$*;*
(II) (Functional Equation) *As meromorphic functions on* $\mathfrak{P}$*,*

$$L_{F,r}^{\leq p}(\phi,\pi) = L_{F,r}^{\leq p}(M(w,\pi)\phi, w\pi), \qquad \forall w \in W.$$

Proof. This is a direct consequence of the fundamental results of Langlands on Eisenstein series and spectrum decompositions and explains why only L^2-automorphic forms are used in the definition of our Ls. (See e.g, [Ar1], [La1], [MW] and/or [We7].) □

2.2.3 Holomorphicity and Singularities

Let $\pi \in \mathfrak{P}$ and $\alpha \in \Delta_M^G$. Define the function $h : \mathfrak{P} \to \mathbb{C}$ by $\pi \otimes \lambda \mapsto \langle \lambda, \alpha^\vee \rangle$, $\forall \lambda \in X_M^G \simeq \mathfrak{a}_M^G$. Here as usual, $\alpha^\vee$ denotes the coroot associated to α. Set $H := \{\pi' \in \mathfrak{P} : h(\pi') = 0\}$ and call it a root hyperplane. Clearly the function h is determined by H, hence we also denote h by h_H. Note also that root hyperplanes depend on the base point π we choose.

Let D be a set of root hyperplanes. Then
(i) the singularities of a meromorphic function f on $\mathfrak{P}$ is said to be carried out by D if for all $\pi \in \mathfrak{P}$, there exist $n_\pi : D \to \mathbb{Z}_{\geq 0}$ zero almost everywhere such that $\pi' \mapsto (\Pi_{H \in D} h_H(\pi')^{n_\pi(H)}) \cdot f(\pi')$ is holomorphic at π';
(ii) the singularities of f are said to be without multiplicity at π if $n_\pi \in \{0, 1\}$;
(iii) D is said to be locally finite, if for any compact subset $C \subset \mathfrak{P}$, $\{H \in D : H \cap C \neq \emptyset\}$ is finite.

Basic Facts of Non-Abelian L-Functions.
(III) (Holomorphicity) (i) *When* $\mathrm{Re}\,\pi \in C$, $L_{F,r}^{\leq p}(\phi, \pi)$ *is holomorphic;*
(ii) $L_{F,r}^{\leq p}(\phi, \pi)$ *is holomorphic at* π *where* $\mathrm{Re}\pi = 0$*;*
(IV) (Singularities) *Assume further that* ϕ *is a cusp form. Then*
(i) *There is a locally finite set of root hyperplanes D such that the singularities of* $L_{F,r}^{\leq p}(\phi, \pi)$ *are carried out by D;*
(ii) *The singularities of* $L_{F,r}^{\leq p}(\phi, \pi)$ *are without multiplicities at* π *if* $\langle \mathrm{Re}\pi, \alpha^\vee \rangle \geq 0, \forall \alpha \in \Delta_M^G$*;*
(iii) *There are only finitely many of singular hyperplanes of* $L_{F,r}^{\leq p}(\phi, \pi)$ *which intersect* $\{\pi \in \mathfrak{P} : \langle \mathrm{Re}\pi, \alpha^\vee \rangle \geq 0, \forall \alpha \in \Delta_M\}$.

Proof. As above, this is a direct consequence of the fundamental results of Langlands on Eisenstein series and spectrum decompositions. (See e.g, [Ar1], [La1], [MW], and/or [We7]). □

3 Geometric and Analytic Truncations

3.1 Geometric Truncation: Revised

3.1.1 Slopes, Canonical Filtrations

Following Lafforgue [Laf], we call an abelian category $\mathcal{A}$ together with two additive morphisms

$$\mathrm{rk} : \mathcal{A} \to \mathbb{N}, \qquad \deg : \mathcal{A} \to \mathbb{R}$$

a *category with slope structure*. In particular, for non-zero $A \in \mathcal{A}$,

(1) define the *slope* of A by $\mu(A) := \frac{\deg(A)}{\mathrm{rk} A}$;

(2) If $0 = A_0 \subset A_1 \subset \cdots \subset A_l = A$ is a filtration of A in $\mathcal{A}$ with $\mathrm{rk}(A_0) < \mathrm{rk}(A_1) < \cdots < \mathrm{rk}(A_l)$, define the *associated polygon* to be the function $[0, \mathrm{rk} A] \to \mathbb{R}$ such that
(i) its values at 0 and $\mathrm{rk}(A)$ are 0;
(ii) it is affine on the intervals $[\mathrm{rk}(A_{i-1}), \mathrm{rk}(A_i)]$ with slope $\mu(A_i/A_{i-1}) - \mu(A)$;

(3) If $\mathfrak{a}$ is a collection of subobjects of A in $\mathcal{A}$, then $\mathfrak{a}$ is said to be *nice* if
(i) $\mathfrak{a}$ is stable under intersection and finite summation;
(ii) $\mathfrak{a}$ is Noetherian, i.e., every increasing chain of elements in $\mathfrak{a}$ has a maximal element in $\mathfrak{a}$;
(iii) if $A_1 \in \mathfrak{a}$ then $A_1 \neq 0$ if and only if $\mathrm{rk}(A_1) \neq 0$; and
(iv) for $A_1, A_2 \in \mathfrak{a}$ with $\mathrm{rk}(A_1) = \mathrm{rk}(A_2)$. Then $A_1 \subset A_2$ is proper implies that $\deg(A_1) < \deg(A_2)$;

(4) For any nice $\mathfrak{a}$, set

$$\mu^+(A) := \sup\Big\{\mu(A_1) : A_1 \in \mathfrak{a}, \mathrm{rk}(A_1) \geq 1\Big\},$$
$$\mu^-(A) := \inf\Big\{\mu(A/A_1) : A_1 \in \mathfrak{a}, \mathrm{rk}(A_1) < \mathrm{rk}(A)\Big\}.$$

Then we say $(A, \mathfrak{a})$ is *semi-stable* if $\mu^+(A) = \mu(A) = \mu^-(A)$. Moreover if $\mathrm{rk}(A) = 0$, set also $\mu^+(A) = -\infty$ and $\mu^-(A) = +\infty$.

Proposition 1. ([Laf]) *Let $\mathcal{A}$ be a category with slope structure, A an object in $\mathcal{A}$ and $\mathfrak{a}$ a nice family of subobjects of A in $\mathcal{A}$. Then*

(1) (**Canonical Filtration**) *A admits a unique filtration* $0 = \overline{A}_0 \subset \overline{A}_1 \subset \cdots \subset \overline{A}_l = A$ *with elements in* $\mathfrak{a}$ *such that*
(i) $\overline{A}_i, 0 \leq i \leq k$ *are maximal in* $\mathfrak{a}$*;*
(ii) $\overline{A}_i/\overline{A}_{i-1}$ *are semi-stable; and*
(iii) $\mu(\overline{A}_1/\overline{A}_0) > \mu(\overline{A}_2/\overline{A}_1 > \cdots > \mu(\overline{A}_k/\overline{A}_{k-1})$;
(2) (**Boundness**) *All polygons of filtrations of A with elements in* $\mathfrak{a}$ *are bounded from above by* $\overline{p}$*, where* $\overline{p} := \overline{p}^A$ *is the associated polygon for the canonical filtration in (1);*
(3) *For any* $A_1 \in \mathfrak{a}, \mathrm{rk}(A_1) \geq 1$ *implies* $\mu(A_1) \leq \mu(A) + \frac{\overline{p}(\mathrm{rk}(A_1))}{\mathrm{rk}(A_1)}$;
(4) *The polygon* $\overline{p}$ *is convex with maximal slope* $\mu^+(A) - \mu(A)$ *and minimal slope* $\mu^-(A) - \mu(A)$*;*
(5) *If* $(A', \mathfrak{a}')$ *is another pair, and* $u : A \to A'$ *is a homomorphism such that* $\mathrm{Ker}(u) \in \mathfrak{a}$ *and* $\mathrm{Im}(u) \in \mathfrak{a}'$*. Then* $\mu^-(A) \geq \mu^+(A')$ *implies that* $u = 0$*.*

This results from a Harder-Narasimhan type filtration consideration. A detailed proof may be found at pp. 87-88 in [Laf].

As an example, we have the following
Proposition 2. *Let F be a number field. Then*
(1) *the abelian category of hermitian vector sheaves on* $\mathrm{Spec}\, O_F$ *together with the natural rank and the Arakelov degree is a category with slopes;*
(2) *For any hermitian vector sheaf* (E,ρ)*,* $\mathfrak{a}$ *consisting of pairs* (E_1,ρ_1) *with* E_1 *sub vector sheaves of E and* ρ_1 *the restrictions of* ρ*, forms a nice family.*

Proof. This is a simple reformulation of Prop. 1.3.6. Indeed, (1) is obvious, while (2) is a direct consequence of the following standard facts:
(i) For a fixed (E,ρ), $\big\{\deg(E_1,\rho_1) : (E_1,\rho_1) \in \mathfrak{a}\big\}$ is discrete subset of $\mathbb{R}$; and
(ii) for any two sublattices Λ_1, Λ_2 of Λ,

$$\mathrm{Vol}\big(\Lambda_1/(\Lambda_1 \cap \Lambda_2)\big) \geq \mathrm{Vol}\big((\Lambda_1 + \Lambda_2)/\Lambda_2\big).$$

□

Thus we get canonical filtrations of Harder-Narasimhan type for hermitian vector sheaves over $\mathrm{Spec}\, O_F$. Recall that hermitian vector sheaves over $\mathrm{Spec}\, O_F$ are O_F-lattices in $(\mathbb{R}^{r_1} \times \mathbb{C}^{r_2})^{r=\mathrm{rk}(E)}$: Say, corresponding O_F-lattices are induced from their H^0 via the natural embedding $F^r \hookrightarrow (\mathbb{R}^{r_1} \times \mathbb{C}^{r_2})^r$ where r_1 (resp. r_2) denotes the real (resp. complex) embeddings of F.

3.1.2 A Micro-Global Relation

In algebraic geometry, or better in Geometric Invariant Theory, a fundamental principle, the Micro-Global Principle, claims that if a point is not GIT stable then there exists a parabolic subgroup which destroys the corresponding stability.

Here even we do not have a proper definition of GIT stability for lattices, in terms of intersection stability, an analogue of the Micro-Global Principle holds.

Let $\Lambda = \Lambda^g$ be a rank r lattice associated to $g \in \mathrm{GL}_r(\mathbb{A})$ and P a parabolic subgroup. Denote the sublattices filtration associated to P by

$$0 = \Lambda_0 \subset \Lambda_1 \subset \Lambda_2 \subset \cdots \subset \Lambda_{|P|} = \Lambda.$$

Assume that P corresponds to the partition $I = (d_1, d_2, \ldots, d_{n=:|P|})$. Consequently, we have

$$\mathrm{rk}(\Lambda_i) = r_i := d_1 + d_2 + \cdots + d_i, \qquad \text{for } i = 1, 2, \ldots, |P|.$$

Let $p, q : [0, r] \to \mathbb{R}$ be two polygons such that $p(0) = q(0) = p(r) = q(r) = 0$. Then following Lafforgue, we say q is *bigger than p with respect to P* and denote it by $q >_P p$, if $q(r_i) - p(r_i) > 0$ for all $i = 1, \ldots, |P| - 1$. Introduce also the characteristic function $\mathbf{1}(\overline{p}^* \leq p)$ by

$$\mathbf{1}(\overline{p}^g \leq p) = \begin{cases} 1, & \text{if } \overline{p}^g \leq p; \\ 0, & \text{otherwise.} \end{cases}$$

Here as before, $\overline{p}^g$ denotes the canonical polygon for the lattice corresponding to g. Recall that for a parabolic subgroup P, p_P^g denotes the polygon induced by P for (the lattice corresponding to) the element $g \in G(\mathbb{A})$.

Fundamental Relation. *For a fixed convex polygon $p : [0, r] \to \mathbb{R}$ such that $p(0) = p(r) = 0$, we have*

$$\mathbf{1}(\overline{p}^g \leq p) = \sum_{P:\,\text{standard parabolic}} (-1)^{|P|-1} \sum_{\delta \in P(F)\backslash G(F)} \mathbf{1}(p_P^{\delta g} >_P p).$$

Remarks. (1) This result and its proof below are motivated by a similar result of Lafforgue for vector bundles over function fields.

(2) The right hand side may be naturally decomposite into two parts according to whether $P = G$ or not. In such away, the right hand side becomes

$$\mathbf{1}_G - \sum_{P:\,\text{proper standard parabolic}} (-1)^{|P|-1} \cdots .$$

This then exposes two aspects of our geometric truncation: First of all, if a lattice is not stable, then there will be parabolic subgroups which take the responsibility; Secondly, each parabolic subgroup has its fix role – Essentially, they should be counted only once each time. In other words, if more are substracted, then we need to add one fewer back to make sure the whole process is not overdone.

Proof. (Following Lafforgue) Note that all parabolic subgroups of G may be obtained from taking δ-conjugates of standard parabolic subgroups with $\delta \in P(F)\backslash G(F)$, and that $\overline{p}_P^{\delta g} = \overline{p}^g_{\delta P \delta^{-1}}$. Therefore, it suffices to prove the following

Fundamental Relation′.

$$\mathbf{1}(\overline{p}^g \leq p) = \sum_{P:\,\text{parabolic}} (-1)^{|P|-1} \mathbf{1}(p_P^g >_P p).$$

To establish this equality, we consider two cases depending on

whether (a) $\overline{p}^g \leq p$;

or (b) $\overline{p}^g$ is not bounded from above by p.

Assume that (a) holds. Then the LHS is simply 1. On the other hand, for a proper parabolic subgroup $Q \neq G$, by definition, $p_Q \leq p$. Hence $\mathbf{1}(p_Q^g >_Q p) = 0$. Consequently, the right hand side degenerates to the one consisting of a single term involving G only, which, by definition, it is simply 1 as well. We are done.

Next, consider (b), for which $\overline{p}^g$ is not bounded from above by p. This is a bit complicated. To start with, we set $\overline{\Lambda}_*^{g,P}$ to be the unique filtration of Λ^g such that the associated polygon $\overline{p}_P^g$ is **maximal** among p_Q^g where Q runs over all parabolic subgroups contained in P. That is

$$\overline{p}_P^g := \max\left\{p_Q^g : Q \subset P, \text{ parabolic subgroup}\right\}.$$

Denote the associated parabolic subgroup by $\overline{Q}^g_P$ and call it *P-canonical.* Clearly, *G*-canonical is simply canonical.

Moreover, since $\overline{p}^g \nleq p$, the LHS $=0$. So we should prove that the RHS $=0$ as well. For this, regroup the parabolic subgroups P appeared in the RHS according to the refined canonical parabolic subgroup $\overline{Q}^g_P$. Then clearly, it suffices to establish the following

Proposition. *With the same notation as above, if* $\overline{p}^g \nleq p$*, then for any fixed parabolic subgroup* Q*,*

$$\sum_{P:\overline{Q}^g_P=Q} (-1)^{|P|}\mathbf{1}(p^P_P >_P p) = 0.$$

Proof. We break this into the following 3 steps.

Step A. Fix a non-negative real number μ. For the lattice $\Lambda = \Lambda^g$ and a fixed parabolic subgroup P of G, set

(1) ${}^{\mu}\overline{Q}^g$ to be the unique parabolic subgroup of G such that if ${}^{\mu}\overline{\Lambda}^g_*$ is the induced filtration of Λ^g, then

$$\left\{{}^{\mu}\overline{\Lambda}^g_*\right\} = \left\{\overline{\Lambda}^g_j : \mu\left(\overline{\Lambda}^g_j/\overline{\Lambda}^g_{j-1}\right) > \mu\left(\overline{\Lambda}^g_{j+1}/\overline{\Lambda}^g_j\right) + \mu\right\},$$

where $\overline{\Lambda}^g_*$ is the canonical filtration associated with Λ^g;

(2) ${}^{\mu}\overline{Q}^g_P$ to be the unique parabolic subgroup of G such that if ${}^{\mu}\overline{\Lambda}^{g,P}_*$ is the induced filtration of Λ^g, then

$$\begin{aligned}\left\{{}^{\mu}\Lambda^{g,P}_*\right\} &= \left\{\Lambda^{g,P}_j\right\}\\ &\cup\left\{\overline{\Lambda}^{g,P}_j : \mu\left(\overline{\Lambda}^{g,P}_j/\overline{\Lambda}^{g,P}_{j-1}\right) > \mu\left(\overline{\Lambda}^{g,P}_{j+1}/\overline{\Lambda}^{g,P}_j\right) + \mu\right\},\end{aligned}$$

where $\Lambda^{g,P}_j$ is the filtration of Λ^g induced from the parabolic subgroup P, and $\overline{\Lambda}^{g,P}_*$ is the P-canonical filtration associated with Λ^g.

In particular, from this construction, we have:

(i) For $\overline{\Lambda}^{g,P}_*$, so-called P-canonical filtration, the Harder-Narasimhan conditions hold, i.e., $\overline{\Lambda}^{g,P}_j/\overline{\Lambda}^{g,P}_{j-1}$ are semi-stable, and

$$\mu\left(\overline{\Lambda}^{g,P}_j/\overline{\Lambda}^{g,P}_{j-1}\right) > \mu\left(\overline{\Lambda}^{g,P}_{j+1}/\overline{\Lambda}^{g,P}_j\right),$$

except for those j appeared as that associated to P itself;

(ii) While the canonical filtration $\overline{\Lambda}^g_*$ of Λ^g satisfying

$$\mu\Big(\overline{\Lambda}^g_j\Big/\overline{\Lambda}^g_{j-1}\Big) > \mu\Big(\overline{\Lambda}^g_{j+1}\Big/\overline{\Lambda}^g_j\Big), \quad \forall j = 1, \ldots, |\overline{Q}^g| - 1,$$

as a partial filtration, $\Big\{{}^{\mu}\overline{\Lambda}^g_*\Big\}$ is the one obtained from this canonical one by selecting only the part where much stronger conditions

$$\mu\Big(\overline{\Lambda}^g_j\Big/\overline{\Lambda}^g_{j-1}\Big) > \mu\Big(\overline{\Lambda}^g_{j+1}\Big/\overline{\Lambda}^g_j\Big) + \mu$$

are satisfied. Consequently,

$$\overline{Q}^g \subset {}^{\mu}\overline{Q}^g;$$

(iii) The filtration $\{{}^{\mu}\Lambda^{g,P}_*\}$ contains not only that part of the P-canonical filtration where stronger conditions

$$\mu\Big(\overline{\Lambda}^{g,P}_j\Big/\overline{\Lambda}^{g,P}_{j-1}\Big) > \mu\Big(\overline{\Lambda}^{g,P}_{j+1}\Big/\overline{\Lambda}^{g,P}_j\Big) + \mu$$

are satisfied, it contains also the full filtration of Λ^g associated to P. So,

$$\overline{Q}^g_P \subset {}^{\mu}\overline{Q}^g_P, \qquad \text{while} \qquad {}^{\mu}\overline{Q}^g_P \subset P.$$

Step B. For $g \in G(\mathbb{A})$, set

$$\begin{aligned}
I(Q_1) &:= \Big\{ \mathrm{rk}\Lambda^{g,Q_1}_i \ : \ i = 1, 2, \ldots, |Q_1| - 1 \Big\};\\
I(Q_2) &:= \Big\{ \mathrm{rk}\Lambda^{g,Q_2}_j \ : \ j = 1, 2, \ldots, |Q_2| - 1 \Big\};\\
J_0(Q_1) &:= \Big\{ \mathrm{rk}\Lambda^{g,Q_1}_k \ : \ \mu\Big(\overline{\Lambda}^{g,Q_1}_k\Big/\overline{\Lambda}^{g,Q_1}_{k-1}\Big) > \mu\Big(\overline{\Lambda}^{g,Q_1}_{k+1}\Big/\overline{\Lambda}^{g,Q_1}_k\Big)\Big\};\\
J_\mu(Q_1) &:= \Big\{ \mathrm{rk}\Lambda^{g,Q_1}_l \ : \ \mu\Big(\overline{\Lambda}^{g,Q_1}_l\Big/\overline{\Lambda}^{g,Q_1}_{l-1}\Big) > \mu\Big(\overline{\Lambda}^{g,Q_1}_{l+1}\Big/\overline{\Lambda}^{g,Q_1}_l\Big) + \mu \Big\};\\
K_p(Q_2) &:= \Big\{ \mathrm{rk}\Lambda^{g,Q_2}_m \ : \ \Big(p^g_{Q_2} - p\Big)\Big(\mathrm{rk}\Lambda^{g,Q_2}_m\Big) > 0 \Big\}.
\end{aligned}$$

Lemma. *With the same notation as above, the following correspondence*

$$\begin{aligned}
&\Big\{ P \in \mathcal{P} : \overline{Q}^g_P = Q_1, \ {}^{\mu}\overline{Q}^g_P = Q_2, \ p^g_P >_P p \Big\}\\
&\quad \longrightarrow \Big\{ \Sigma : \ \Sigma \subset I(Q_2), \ J_\mu(Q_1) \subset I(Q_2),\\
&\qquad \Big(I(Q_1) - J_0(Q_1)\Big) \cup \Big(I(Q_2) - J_\mu(Q_1)\Big) \subset \Sigma \subset K_p(Q_2) \Big\}
\end{aligned}$$

defined by

$$P \mapsto I(P) := \left\{ \mathrm{rk}\Lambda_i^{g,P} \,:\, i = 1, 2, \ldots, |P| - 1 \right\}$$

is a well-defined bijection.

Proof of the Lemma. (I) **Well-defined**. For this, let us look at the conditions for P in

$$\left\{ P \in \mathcal{P} : \overline{Q}_P^g = Q_1,\ {}^{\mu}\overline{Q}_P^g = Q_2,\ p_P^g >_P p \right\}.$$

(i) $\overline{Q}_P^g = Q_1$ gives the filtration

$$\{0\} = \Lambda_0^{g,Q_1} \subset \Lambda_1^{g,Q_1} \subset \Lambda_2^{g,Q_1} \subset \cdots \subset \Lambda_{|Q_1|-1}^{g,Q_1} \subset \Lambda_{|Q_1|}^{g,Q_1} = \Lambda^g$$

with $p_{Q_1}^g$ be the maximal among all $\left\{ p_Q^g : Q \subset P \right\}$. Hence,
(a) $Q_1 \subset P$; and
(b) Except possibly at indices corresponding to the partion coming from P,

$$\mu\left(\overline{\Lambda}_k^{g,Q_1} \big/ \overline{\Lambda}_{k-1}^{g,Q_1}\right) > \mu\left(\overline{\Lambda}_{k+1}^{g,Q_1} \big/ \overline{\Lambda}_k^{g,Q_1}\right),$$

by the maximal property, or more clearly, the Harder-Narasimhan property. Therefore, we have

$$I(Q_1) - J_0(Q_1) \subset I(P).$$

(ii) ${}^{\mu}\overline{Q}_P^g = Q_2$ gives the filtration

$$\{0\} = \Lambda_0^{g,Q_2} \subset \Lambda_1^{g,Q_2} \subset \Lambda_2^{g,Q_2} \subset \cdots \subset \Lambda_{|Q_2|-1}^{g,Q_2} \subset \Lambda_{|Q_2|}^{g,Q_2} = \Lambda^g,$$

which by definition coincides with

$$\left\{ \Lambda_j^{g,P} : j = 1, 2, \ldots, |P| - 1 \right\} \\ \cup \left\{ \Lambda_k^{g,Q_1} \,:\, \mu\left(\overline{\Lambda}_k^{g,Q_1} \big/ \overline{\Lambda}_{k-1}^{g,Q_1}\right) > \mu\left(\overline{\Lambda}_{k+1}^{g,Q_1} \big/ \overline{\Lambda}_k^{g,Q_1}\right) + \mu \right\}.$$

By taking the rank for each lattices, we get the relation $I(Q_2) = I(P) \cup J_\mu(Q_1)$. Therefore,
(a) $Q_2 \subset P$ or equivalently $I(P) \subset I(Q_2)$; and
(b) $I(Q_2) - J_\mu(Q_1) \subset I(P)$.
(iii) $p_P^g >_P p$ is, by definition, equivalent to the following group of inequalities

$$\left(p_P^g - p\right)\left(\mathrm{rk}\Lambda_j^{g,P}\right) > 0.$$

Consequently, $I(P) \subset K_p(Q_2)$ since $I(P) \subset I(Q_2)$.

Therefore, $P \mapsto I(P)$ is well-defined.

(II) **Injectivity**. This is clear since with a fixed partition I, there exists only one unique standard parabolic subgroup P_I such that $I(P_I) = \Delta - I$.

(III) **Surjectivity**. For a subset Σ of $\{1, 2, \ldots, r\}$ such that

$$\Sigma \subset I(Q_2),\ J_\mu(Q_1) \subset I(Q_2),$$
$$\Big(I(Q_1) - J_0(Q_1)\Big) \cup \Big(I(Q_2) - J_\mu(Q_1)\Big) \subset \Sigma \subset K_p(Q_2),$$

clearly, there exists a parabolic subgroup $P = P_\Sigma$. So we need to check whether $\overline{Q}_P^g = Q_1$, ${}^\mu\overline{Q}_P^g = Q_2$ and $p_P^g >_P p$.

It is clear that $p_P^g >_P p$ since $\Sigma \subset K_p(Q_2)$. On the other hand, $\overline{Q}_P^g = Q_1$, ${}^\mu\overline{Q}_P^g = Q_2$ are direct consequences of the definition of P-canonical filtration and the conditions that $\Big(I(Q_1) - J_0(Q_1)\Big) \subset \Sigma$ and $\Big(I(Q_2) - J_\mu(Q_1)\Big) \subset \Sigma$ as explained in the proof above of the well-defined statement. This completes the proof of the Lemma.

Step C. With the Lemma, we have

$$\begin{aligned} &\Big|\sum_{P:\overline{Q}_P^g = Q_1,\, {}^\mu\overline{Q}_P^g = Q_2} (-1)^{|P|}\mathbf{1}(p_P >_P p)\Big| \\ =&\Big|\sum_{\Sigma:\big(I(Q_1)-J_0(Q_1)\big)\cup\big(I(Q_2)-J_\mu(Q_1)\big)\subset\Sigma\subset K_p(Q_2)} (-1)^{\#(\sigma)}\Big| \\ =&\begin{cases} 0, & \text{if } \Big(I(Q_1) - J_0(Q_1)\Big) \cup \Big(I(Q_2) - J_\mu(Q_1)\Big) \neq K_p(Q_2); \\ 1, & \text{if } \Big(I(Q_1) - J_0(Q_1)\Big) \cup \Big(I(Q_2) - J_\mu(Q_1)\Big) = K_p(Q_2). \end{cases} \end{aligned}$$

Surely, we want to show that the value 1 cannot be taken for g such that $\overline{p}^g$ is not bounded by p. To this end, as the final push to establish our Fundamental Relation, set $\mu = 0$ and $Q_1 = Q_2 = Q$ in the above discussion so that

$$\begin{aligned} &\sum_{P:\overline{Q}_P^g = Q_1,\, {}^\mu\overline{Q}_P^g = Q_2} (-1)^{|P|}\mathbf{1}(p_P >_P p) \\ &\qquad = \sum_{P:\overline{Q}_P^g = Q = {}^0\overline{Q}_P^g} (-1)^{|P|}\mathbf{1}(p_P >_P p) \\ &\qquad = \sum_{P:\overline{Q}_P^g = Q} (-1)^{|P|}\mathbf{1}(p_P >_P p) \end{aligned}$$

is the summation in the Proposition. Then, by definition and the convexity property of p, $\big(I(Q_1) - J_0(Q_1)\big) \cup \big(I(Q_2) - J_\mu(Q_1)\big) = K_p(Q_2)$ implies that $I(Q_1) = J_0(Q_1)$, or better $K_p^g(Q_2) = \emptyset$. Consequently, this then shows that $\overline{p}^g \leq p$, a contradiction. This completes the proof of the Proposition and hence the Fundamental Relation. □

3.2 Arthur's Analytic Truncation

3.2.1 Parabolic Subgroups

Let F be a number field with $\mathbb{A} = \mathbb{A}_F$ the ring of adeles. Let G be a connected reductive group defined over F. Recall that a subgroup P of G is called *parabolic* if G/P is a complete algebraic variety. Fix a minimal F-parabolic subgroup P_0 of G with its unipotent radical $N_0 = N_{P_0}$ and fix an F-Levi subgroup $M_0 = M_{P_0}$ of P_0 so as to have a Levi decomposition $P_0 = M_0N_0$. An F-parabolic subgroup P is called *standard* if it contains P_0. For such parabolic subgroups P, there exists a unique Levi subgroup $M = M_P$ containing M_0 which we call the *standard Levi subgroup* of P. Let $N = N_P$ be the unipotent radical. Let us agree to use the term parabolic subgroups and Levi subgroups to denote standard F-parabolic subgroups and standard Levi subgroups respectively, unless otherwise is stated.

Let P be a parabolic subgroup of G. Write T_P for the maximal split torus in the center of M_P and T_P' for the maximal quotient split torus of M_P. Set $\tilde{\mathfrak{a}}_P := X_*(T_P) \otimes \mathbb{R}$ and denote its real dimension by $d(P)$, where $X_*(T)$ is the lattice of 1-parameter subgroups in the torus T. Then it is known that $\tilde{\mathfrak{a}}_P = X_*(T_P') \otimes \mathbb{R}$ as well. The two descriptions of $\tilde{\mathfrak{a}}_P$ show that if $Q \subset P$ is a parabolic subgroup, then there is a canonical injection $\tilde{\mathfrak{a}}_P \hookrightarrow \tilde{\mathfrak{a}}_Q$ and a natural surjection $\tilde{\mathfrak{a}}_Q \twoheadrightarrow \tilde{\mathfrak{a}}_P$. We thus obtain a canonical decomposition $\tilde{\mathfrak{a}}_Q = \tilde{\mathfrak{a}}_Q^P \oplus \tilde{\mathfrak{a}}_P$ for a certain subspace $\tilde{\mathfrak{a}}_Q^P$ of $\tilde{\mathfrak{a}}_Q$. In particular, $\tilde{\mathfrak{a}}_G$ is a summand of $\tilde{\mathfrak{a}} = \tilde{\mathfrak{a}}_P$ for all P. Set $\mathfrak{a}_P := \tilde{\mathfrak{a}}_P/\tilde{\mathfrak{a}}_G$ and $\mathfrak{a}_Q^P := \tilde{\mathfrak{a}}_Q^P/\tilde{\mathfrak{a}}_G$. Then we have

$$\mathfrak{a}_Q = \mathfrak{a}_Q^P \oplus \mathfrak{a}_P$$

and $\mathfrak{a}_P$ is canonically identified as a subspace of $\mathfrak{a}_Q$. Set $\mathfrak{a}_0 := \mathfrak{a}_{P_0}$ and $\mathfrak{a}_0^P = \mathfrak{a}_{P_0}^P$ then we also have $\mathfrak{a}_0 = \mathfrak{a}_0^P \oplus \mathfrak{a}_P$ for all P.

3.2.2 Logarithmic Map

For a real vector space V, write V^* its dual space over $\mathbb{R}$. Then dually we have the spaces $\mathfrak{a}_0^*, \mathfrak{a}_P^*, \left(\mathfrak{a}_0^P\right)^*$ and hence the decompositions

$$\mathfrak{a}_0^* = \left(\mathfrak{a}_0^Q\right)^* \oplus \left(\mathfrak{a}_Q^P\right)^* \oplus \mathfrak{a}_P^*.$$

So $\mathfrak{a}_P^* = X(M_P) \otimes \mathbb{R}$ with $X(M_P)$ the group $\mathrm{Hom}_F\big(M_P, GL(1)\big)$ i.e., collection of characters on M_P. It is known that $\mathfrak{a}_P^* = X(A_P) \otimes \mathbb{R}$ where A_P denotes the split component of the center of M_P. Clearly, if $Q \subset P$, then $M_Q \subset M_P$ while $A_P \subset A_Q$. Thus via restriction, the above two expressions of $\mathfrak{a}_P^*$ also naturally induce an injection $\mathfrak{a}_P^* \hookrightarrow \mathfrak{a}_Q^*$ and a surjection $\mathfrak{a}_Q^* \twoheadrightarrow \mathfrak{a}_P^*$, compactible with the decomposition $\mathfrak{a}_Q^* = \left(\mathfrak{a}_Q^P\right)^* \oplus \mathfrak{a}_P^*$.

Every $\chi = \sum s_i\chi_i$ in $\mathfrak{a}_{P,\mathbb{C}}^* := \mathfrak{a}_P^* \otimes \mathbb{C}$ determines a morphism $P(\mathbb{A}) \to \mathbb{C}^*$ by $p \mapsto p^\chi := \prod |\chi_i(p)|^{s_i}$. Consequently, we have a natural logarithmic map $H_P : P(\mathbb{A}) \to \mathfrak{a}_P$ defined by

$$\langle H_P(p), \chi\rangle = p^\chi, \qquad \forall \chi \in \mathfrak{a}_P^*.$$

The kernel of H_P is denoted by $P(\mathbb{A})^1$ and we set $M_P(\mathbb{A})^1 := P(\mathbb{A})^1 \cap M_P(\mathbb{A})$.

Let also A_+ be the set of $a \in A_P(\mathbb{A})$ such that
(1) $a_v = 1$ for all finite places v of F; and
(2) $\chi(a_\sigma)$ is a positive number independent of infinite places σ of F for all $\chi \in X(M_P)$.
Then $M(\mathbb{A}) = A_+ \cdot M(\mathbb{A})^1$.

3.2.3 Roots, Coroots, Weights and Coweights

We now introduce standard bases for above spaces and their duals. Let Δ_0 and $\widehat{\Delta}_0$ be the subsets of simple roots and simple weights in $\mathfrak{a}_0^*$ respectively. (Recall that elements of $\widehat{\Delta}_0$ are non-negative linear combinations of elements in Δ_0.) Write $\Delta_0^\vee$ (resp. $\widehat{\Delta}_0^\vee$) for the basis of $\mathfrak{a}_0$ dual to $\widehat{\Delta}_0$ (resp. Δ_0). Being the dual of the collection of simple weights (resp. of simple roots), $\Delta_0^\vee$ (resp. $\widehat{\Delta}_0^\vee$) is the set of coroots (resp. coweights).

For every P, let $\Delta_P \subset \mathfrak{a}_0^*$ be the set of non-trivial *restrictions* of elements of Δ_0 to $\mathfrak{a}_P$. Denote the dual basis of Δ_P by $\widehat{\Delta}_P^\vee$. For each $\alpha \in \Delta_P$, let

$\alpha^\vee$ be the projection of $\beta^\vee$ to $\mathfrak{a}_P$, where β is the root in Δ_0 whose restriction to $\mathfrak{a}_P$ is α. Set $\Delta_P^\vee := \{\alpha^\vee : \alpha \in \Delta_P\}$, and define the dual basis of $\Delta_P^\vee$ by $\widehat{\Delta}_P$.

More generally, if $Q \subset P$, write Δ_Q^P to denote the *subset* $\alpha \in \Delta_Q$ appearing in the action of T_Q in the unipotent radical of $Q \cap M_P$. (Indeed, $M_P \cap Q$ is a parabolic subgroup of M_P with nilpotent radical $N_Q^P := N_Q \cap M_P$. Thus Δ_Q^P is simply the set of roots of the parabolic subgroup $(M_P \cap Q, A_Q)$. And one checks that the map $P \mapsto \Delta_Q^P$ gives a natural bijection between parabolic subgroups P containing Q and subsets of Δ_Q.) Then $\mathfrak{a}_P$ is the subspace of $\mathfrak{a}_Q$ annihilated by Δ_Q^P. Denote by $(\widehat{\Delta^\vee})_Q^P$ the dual of Δ_Q^P. Let $(\Delta_Q^P)^\vee := \{\alpha^\vee : \alpha \in \Delta_Q^P\}$ and denote by $\widehat{\Delta}_Q^P$ the dual of $(\Delta_Q^P)^\vee$.

3.2.4 Positive Cone and Positive Chamber

Let $Q \subset P$ be two parabolic subgroups of G. We extend the linear functionals in Δ_Q^P and $\widehat{\Delta}_Q^P$ to elements of the dual space $\mathfrak{a}_0^*$ by means of the canonical projection from $\mathfrak{a}_0$ to $\mathfrak{a}_Q^P$ given by the decomposition $\mathfrak{a}_0 = \mathfrak{a}_0^Q \oplus \mathfrak{a}_Q^P \oplus \mathfrak{a}_P$. Let τ_Q^P be the characteristic function of the *positive chamber*

$$\begin{aligned}&\{H \in \mathfrak{a}_0 : \langle\alpha, H\rangle > 0\ \forall \alpha \in \Delta_Q^P\}\\ &= \mathfrak{a}_0^Q \oplus \{H \in \mathfrak{a}_Q^P : \langle\alpha, H\rangle > 0 \text{ for all } \alpha \in \Delta_Q^P\} \oplus \mathfrak{a}_P\end{aligned}$$

and let $\widehat{\tau}_Q^P$ be the characteristic function of the *positive cone*

$$\begin{aligned}&\{H \in \mathfrak{a}_0 : \langle\varpi, H\rangle > 0\ \forall\ \varpi \in \widehat{\Delta}_Q^P\}\\ &= \mathfrak{a}_0^Q \oplus \{H \in \mathfrak{a}_Q^P : \langle\varpi, H\rangle > 0 \text{ for all } \varpi \in \widehat{\Delta}_Q^P\} \oplus \mathfrak{a}_P.\end{aligned}$$

Note that elements in $\widehat{\Delta}_Q^P$ are non-negative linear combinations of elements in Δ_Q^P, so we have

$$\widehat{\tau}_Q^P \geq \tau_Q^P.$$

3.2.5 Example with SL_r

Set $G = SL_r(\mathbb{R})$. Let $P_0 = B$ be the Borel subgroup of G consisting of upper triangular matrices, and M_0 be the Levi component consisting of diagonal matrices. One checks easily that the unipotent radical N_0 consists of upper triangular matrices whose diagonal entries are all equal to 1 and that (the positive part A_{0+} of) the split component A_0 consists of diagonal matrices $\mathrm{diag}\big(a_1, a_2, \ldots, a_r\big)$ with $a_1, a_2, \ldots, a_r \in \mathbb{R}_{>0}$ and $\prod_{i=1}^r a_i = 1$. Hence $\mathfrak{a}_0 = \Big\{H := (H_1, H_2, \ldots, H_r) \in \mathbb{R}^r : \sum_{i=1}^r H_i = 0\Big\}$ and the natural logarithmic map $H_{P_0} : P_0 \to \mathfrak{a}_0$ is given by $g \mapsto H_{P_0}\big(a(g)\big) = \log a_1 + \log a_2 + \cdots + \log a_r$, where we have used the Iwasawa decomposition $g = n \cdot m \cdot a(g) \cdot k$ with $n \in N_0$, $m \in M_0^1$, $a(g) = \mathrm{diag}\big(a_1, a_2, \ldots, a_r\big) \in A_{0+}$ and $k \in K = SO(r)$.

Set $e_i \in \mathfrak{a}_0^*$ such that $e_i(H) = H_i$, then

$$\Delta_0 = \Big\{e_1 - e_2, e_2 - e_3, \ldots, e_{r-1} - e_r\Big\}.$$

We identify Δ_0 with the set $\Delta := \{1, 2, \ldots, r-1\}$ by the map $\alpha_i := e_i - e_{i+1} \mapsto i, i = 1, 2, \ldots, r-1$.

It is well known that there is a natural bijection between partitions of r and parabolic subgroups of G. More precisely, the partition $I := \big(d_1, d_2, \ldots, d_n\big)$ of r (so that $r = d_1 + d_2 + \cdots + d_n$ and $d_i \in \mathbb{Z}_{>0}$) corresponds to the parabolic subgroup P_I consisting of blocked upper triangular matrices whose diagonal entries are blocked matrices of sizes $d_1, d_2, \ldots, d_n$. Then clearly, the unipotent radical N_I consists of blocked upper triangular matrices whose blocked diagonal entries are identity matrices of sizes $d_1, d_2, \ldots, d_n$, the Levi component M_I consists of blocked diagonal matrices whose blocked diagonal entries are of sizes $d_1, d_2, \ldots, d_r$, and the corresponding A_{I+} consits of matrices of the form $\mathrm{diag}\big(a_1 I_{d_1}, a_2 I_{d_2}, \ldots, a_n I_{d_n}\big)$ where I_{d_i} denotes the identity matrix of size d_i, $a_i > 0$ for $i = 1, 2, \ldots, n$, and $\prod_{i=1}^n a_i^{d_i} = 1$. As such, $\mathfrak{a}_I$ is a subspace of $\mathfrak{a}_0$ defined by

$$\begin{aligned}\mathfrak{a}_I := \Big\{(H_1, H_2, \ldots, H_r) \in \mathfrak{a}_0 : H_1 &= H_2 = \cdots = H_{r_1}, \\ H_{r_1+1} &= H_{r_1+2} = \cdots = H_{r_2}, \\ \ldots, H_{r_{n-1}+1} &= H_{r_{n-1}+2} = \cdots = H_{r_n}\Big\}\end{aligned}$$

where we set $r_i = d_1 + d_2 + \cdots + d_i$ for $i = 1, 2, \ldots, n$. Moreover, with the help of Iwasawa decomposition, the corresponding logarithmic map may be simply seen to be the map $H_I : A_{I+} \to \mathfrak{a}_I$ defined by

$$\begin{aligned}&\mathrm{diag}\big(a_1 I_{d_1}, a_2 I_{d_2}, \ldots, a_n I_{d_n}\big) \\ &\quad\mapsto \Big((\log a_1)^{(d_1)}, (\log a_2)^{(d_2)}, \ldots, (\log a_n)^{(d_n)}\Big).\end{aligned}$$

Obviously, the natural bijection between parabolic subgroups P of G and subsets of Δ_0 or better of Δ is given explicitly as follows: The parabolic subgroup $P = P_I$ corresponds to the subset $I(P)$ of $\Delta = \{1, 2, \ldots, r-1\}$ such that $\Delta - I(P) = \{r_1, r_2, \ldots, r_n\}$. That is,

$$I(P) = \{1, 2, \ldots, r_1 - 1\} \bigcup \{r_1 + 1, r_1 + 2, \ldots, r_2 - 1\} \bigcup \cdots \bigcup \{r_{n-1} + 1, r_{n-1} + 2, \ldots, r_n - 1\}.$$

For simplicity, from now on, we will use I to indicate both the partition and its corresponding subset $I(P_I)$. For example, the subset $I = \emptyset$ corresponds to the partition $= (1, 1, \ldots, 1)$ of $r = 1 + 1 + \cdots + 1$ (since $\Delta - \emptyset = \{1, 2, \ldots, r-1\}$) whose corresponding parabolic subgroup is simply P_0 the minimal parabolic subgroup.

Consider now $\emptyset \subset I \subset J$. From $\Delta - J \subset \Delta - I \subset \Delta$, we get $P_0 = P_\emptyset \subset P_I \subset P_J$. Set $\Delta - J = \{s_1, s_2, \ldots, s_m\}$ corresponding to the partition $r = f_1 + f_2 + \cdots + f_m$. Then $f_1 = d_1 + d_2 + \cdots + d_{k_1}$, $f_2 = d_{k_1+1} + d_{k_1+2} + \cdots + d_{k_2}, \ldots, f_m = d_{k_{m-1}+1} + d_{k_{m-1}+2} + \cdots + d_{k_m}$ and $\mathfrak{a}_J$ is a subspace of $\mathfrak{a}_0$ defined by

$$\mathfrak{a}_J := \{(H_1^{(f_1)}, H_2^{(f_2)}, \ldots, H_m^{(f_m)}) \in \mathfrak{a}_0\}.$$

Clearly $\mathfrak{a}_J$ is in $\mathfrak{a}_I$ since $\mathfrak{a}_I$ is a subspace of $\mathfrak{a}_0$ defined by

$$\mathfrak{a}_I := \{\left(A_1^{(d_1)}, A_2^{(d_2)}, \ldots, A_n^{(d_n)}\right) \in \mathfrak{a}_0\}$$

and we have the map

$$\begin{array}{ccc} \mathfrak{a}_J & \to & \mathfrak{a}_I \\ \left(H_1^{(f_1)}, H_2^{(f_2)}, \ldots, H_m^{(f_m)}\right) & \mapsto & \left(H_1^{(d_1)}, H_1^{(d_2)}, \ldots, H_1^{(d_{k_1})}, \ldots, H_m^{(d_{k_m})}\right). \end{array}$$

Moreover, with the help of Iwasawa decomposition, the corresponding logarithmic map may be simply viewed as the map $H_J : A_{J+} \to \mathfrak{a}_J$ defined by

$$\operatorname{diag}\left(b_1 I_{f_1}, b_2 I_{f_2}, \ldots, b_m I_{f_m}\right) \mapsto \left((\log b_1)^{(f_1)}, (\log b_2)^{(f_2)}, \ldots, (\log b_m)^{(f_m)}\right).$$

To go further, for $i \in \Delta$, set

$$\varpi_i := \varpi_i^\Delta = \frac{r-i}{r} \sum_{k=1}^{i} k \cdot \alpha_k + \frac{i}{r} \sum_{k=i+1}^{r-1} (r-k) \cdot \alpha_k$$

be the corresponding fundamental weight. ϖ_i is the $\mathbb{Q}$-linear form on $\mathfrak{a}_\emptyset = \mathfrak{a}_0$ given by

$$\begin{aligned}\varpi_i(H) &= H_1 + H_2 + \cdots + H_i - \frac{i}{r}\Big(H_1 + H_2 + \cdots + H_r\Big)\\ &= H_1 + H_2 + \cdots + H_i\end{aligned}$$

for all $H = \Big(H_1, H_2, \ldots, H_r\Big) \in \mathfrak{a}_0$, so that $\widehat{\Delta}_I = \Big\{\varpi_i : i \in \Delta - I\Big\}$ since

$$\Delta_I = \Big\{\alpha_i\Big|_{\mathfrak{a}_I} : i \in \Delta - I\Big\}.$$

Note that $\mathfrak{a}_I^* = \langle\Delta_I\rangle_{\mathbb{R}}$ and elements of $\widehat{\Delta}_I$ are non-degenerate positive linear combination of elements in Δ_I, so both Δ_I and $\widehat{\Delta}_I$ generate $\mathfrak{a}_I^*$. Consequently, we have

$$\mathfrak{a}_I^* = \Big\langle\varpi_i : i \in \Delta - I\Big\rangle_{\mathbb{R}} \supset \Big\langle\varpi_j : j \in \Delta - J\Big\rangle_{\mathbb{R}} = \mathfrak{a}_J^*.$$

Moreover, by definition, the space $\mathfrak{a}_I^J$ is the subspace of $\mathfrak{a}_I$ annihilated by $\mathfrak{a}_J^*$. Therefore,

$$\begin{aligned}\mathfrak{a}_I^J &= \Big\{H \in \mathfrak{a}_I : \alpha(H) = 0\ \forall\alpha \in \mathfrak{a}_J^*\Big\}\\ &= \Big\{H \in \mathfrak{a}_I : \varpi_j(H) = 0\ \forall j \in \Delta - J\Big\}\\ &= \Big\{(H_1^{(d_1)}, H_2^{(d_2)}, \ldots, H_n^{(d_n)}) \in \mathfrak{a}_0 :\\ &\qquad d_1H_1 + d_2H_2 + \cdots + d_{k_1}H_{k_1} = 0,\\ &\qquad\quad d_{k_1+1}H_{k_1+1} + d_{k_1+2}H_{k_1+2} + \cdots + d_{k_2}H_{k_2} = 0, \ldots\Big\}.\end{aligned}$$

Note that clearly $\dim\mathfrak{a}_I^J + \dim\mathfrak{a}_J = \dim\mathfrak{a}_I$ and $\mathfrak{a}_I^J \cap \mathfrak{a}_J = \emptyset$, so we have the natural decomposition

$$\mathfrak{a}_I = \mathfrak{a}_I^J \oplus \mathfrak{a}_J.$$

Furthermore, note that

$$\begin{aligned}I &= \bigcup_{i=1}^{n}\Big\{r_{j-1} + 1, r_{j-1} + 2, \ldots, r_j - 1\Big\}\\ &= \Big\{1, 2, \ldots, r - 1\Big\} - \Big\{r_1, r_2, \ldots, r_{n-1}\Big\}.\end{aligned}$$

Set, for $k = 1, 2, \cdots, d_j - 1$,

$$\varpi^I_{r_{j-1}+k} = \frac{d_j - k}{d_j} \sum_{l=1}^{k} l \cdot \alpha_{r_{j-1}+l} + \frac{k}{d_j} \sum_{l=k+1}^{d_j-1} \left(d_j - l\right) \cdot \alpha_{r_{j-1}+l},$$

motivated by the construction of fundamental weights above. Then if, as usual, we define the coroot $\alpha_i^\vee \in \mathfrak{a}_0$ by

$$(\alpha_i^\vee)_k = \begin{cases} 1 & \text{if k=i,} \\ -1 & \text{if k=i+1,} \\ 0 & \text{otherwise,} \end{cases}$$

then $\left\{\alpha_i^\vee : i \in I\right\}$ is a basis of the subspace $\mathfrak{a}_\emptyset^I$ of $\mathfrak{a}_\emptyset$ and easily one checks that $\varpi_i(\alpha_j^\vee) = \delta_{ij}$ for all $i, j \in I$. Consequently, $\left\{\varpi_i^I\Big|_{\mathfrak{a}_\emptyset^I} : i \in I\right\}$ is its dual basis in $(\mathfrak{a}_\emptyset^I)^*$. In particular, we also get

$$\widehat{\Delta}_I^J = \left\{\varpi_k^J : k \in J - I = (\Delta - I) - (\Delta - J)\right\}.$$

We end this preperation with the following remarks about the notation. With respect to the decomposition

$$\left(\mathfrak{a}_\emptyset^J\right)^* = \left(\mathfrak{a}_\emptyset^I\right)^* \oplus \left(\mathfrak{a}_I^J\right)^*,$$

we have that the space $\left(\mathfrak{a}_\emptyset^J\right)^*$ is generated by

$$\Delta_\emptyset^J = \left\{\alpha_l : l \in J - \emptyset = J\right\} \subset \Delta_\emptyset$$

the root system, similarly $\left(\mathfrak{a}_\emptyset^I\right)^*$ is generated by

$$\Delta_\emptyset^I = \left\{\alpha_k : k \in I - \emptyset = I\right\} \subset \Delta_\emptyset$$

while $\left(\mathfrak{a}_I^J\right)^*$ is generated by Δ_I^J which is not a subset of $\Delta_\emptyset$ but we have $\Delta_I^J \subset \Delta_I$ consists of the restriction of α_k to $\mathfrak{a}_I^J$ with $k \in J - I = (\Delta - I) - (\Delta - J)$. That is

$$\Delta_I^J = \left\{\alpha_k\Big|_{\mathfrak{a}_I^J} : k \in J - I = (\Delta - I) - (\Delta - J)\right\}.$$

In particular, if $J = \Delta$ we have

$$\mathfrak{a}_\emptyset^\Delta = \mathfrak{a}_\emptyset,\ \mathfrak{a}_I^\Delta = \mathfrak{a}_I, \qquad \text{and} \qquad \Delta_\emptyset^\Delta = \Delta_\emptyset,\ \Delta_I^\Delta = \Delta_I.$$

3.2.6 Partial Truncation and First Estimations

Denote τ_P^G and $\widehat{\tau}_P^G$ simply by τ_P and $\widehat{\tau}_P$.

Basic Estimation. ([Ar2]) *Suppose that we are given a parabolic subgroup P, and a Euclidean norm $\|\cdot\|$ on $\mathfrak{a}_P$. Then there are constants c and N such that for all $x \in G(\mathbb{A})^1$ and $X \in \mathfrak{a}_P$,*

$$\sum_{\delta \in P(F)\backslash G(F)} \widehat{\tau}_P\Big(H(\delta x) - X\Big) \leq c\Big(\|x\|e^{\|X\|}\Big)^N.$$

Moreover, *the sum is finite.*

As a direct consequence, we have the following

Corollary. ([Ar2]) *Suppose that $T \in \mathfrak{a}_0$ and $N \geq 0$. Then there exist constants c' and N' such that for any function ϕ on $P(F)\backslash G(\mathbb{A})^1$, and $x, y \in G(\mathbb{A})^1$,*

$$\sum_{\delta \in P(F)\backslash G(F)} \Big|\phi(\delta x)\Big| \cdot \widehat{\tau}_P\Big(H(\delta x) - H(y) - X\Big)$$

is bounded from above by

$$c'\|x\|^{N'} \cdot \|y\|^{N'} \cdot \sup_{u \in G(\mathbb{A})^1} \Big(|\phi(u)| \cdot \|u\|^{-N}\Big).$$

3.2.7 Langlands' Combinatorial Lemma

In this subsection, following Arthur, we recall Langlands' combinatorial lemma.

If $P_1 \subset P_2$, following Arthur [Ar2], set

$$\sigma_1^2(H) := \sigma_{P_1}^{P_2} := \sum_{P_3 : P_2 \supset P_2} (-1)^{\dim(A_3/A_2)} \tau_1^3(H) \cdot \hat{\tau}_3(H),$$

for $H \in \mathfrak{a}_0$. Then we have

Lemma 1. If $P_1 \subset P_2$, *σ_1^2 is a characteristic function of the subset of* $H \in \mathfrak{a}_1$ such that
(i) $\alpha(H) > 0$ for all $\alpha \in \Delta_1^2$;
(ii) $\sigma(H) \leq 0$ for all $\sigma \in \Delta_1 \backslash \Delta_1^2$; and

(iii) $\varpi(H) > 0$ for all $\varpi \in \hat{\Delta}_2$.

As a special case, with $P_1 = P_2$, we get the following important consequence:

Langlands' Combinatorial Lemma. *If $Q \subset P$ are parabolic subgroups, then for all $H \in \mathfrak{a}_0$,*

$$\sum_{R:Q\subset R\subset P} (-1)^{\dim(A_R/A_P)}\tau_Q^R(H)\hat{\tau}_R^P(H) = \delta_{QP};$$
$$\sum_{R:Q\subset R\subset P} (-1)^{\dim(A_Q/A_R)}\hat{\tau}_Q^R(H)\tau_R^P(H) = \delta_{QP}.$$

We next give an application of this combinatorial lemma, which will be used in 3.2.8. Suppose that $Q \subset P$ are parabolic subgroups. Fix a vector $\Lambda \in \mathfrak{a}_0^*$. Let

$$\varepsilon_Q^P(\Lambda) := (-1)^{\#\{\alpha\in\Delta_Q^P:\Lambda(\alpha^\vee)\leq 0\}},$$

and let

$$\phi_Q^P(\Lambda, H), \qquad H \in \mathfrak{a}_0,$$

be the characteristic function of the set

$$\left\{H \in \mathfrak{a}_0 : \begin{matrix} \varpi(H) > 0, & \text{if } \Lambda(\alpha^\vee) \leq 0 \\ \varpi(H) \leq 0, & \text{if } \Lambda(\alpha^\vee) > 0 \end{matrix}, \forall \alpha \in \Delta_Q^P\right\}.$$

Lemma 2. With the same notation as above,

$$\sum_{R:Q\subset R\subset P} \varepsilon_Q^R(\Lambda)\cdot\phi_Q^R(\Lambda, H)\cdot\tau_R^P(H) = \begin{cases} 0, & \text{if } \Lambda(\alpha^\vee) \leq 0,\ \exists\alpha \in \Delta_Q^P \\ 1, & \text{otherwise} \end{cases}.$$

3.2.8 Langlands-Arthur's Partition: Reduction Theory

Our aim is to derive Langlands-Arthur's partition of $G(F)\backslash G(\mathbb{A})$ into disjoint subsets, one for each (standard) parabolic subgroup.

To start with, suppose that ω is a compact subset of $N_0(\mathbb{A})M_0(\mathbb{A})^1$ and that $T_0 \in -\mathfrak{a}_0^+$. For any parabolic subgroup P_1, introduce the associated Siegel set $\mathfrak{s}^{P_1}(T_0, \omega)$ as the collection of

$$pak, \qquad p \in \omega,\ a \in A_0(\mathbb{R})^0,\ k \in K,$$

where $\alpha\big(H_0(a)-T_0\big)$ is positive for each $\alpha\in\Delta_0^1$. Then from classical reduction theory, we conclude that *for sufficiently big* ω *and sufficiently small* T_0, $G(\mathbb{A})=P_1(F)\mathfrak{s}^{P_1}(T_0,\omega)$.

Suppose now that P_1 is given. Let $\mathfrak{s}^{P_1}(T_0,T,\omega)$ be the set of x in $\mathfrak{s}^{P_1}(T_0,\omega)$ such that $\varpi\big(H_0(x)-T\big)\leq 0$ for each $\varpi\in\hat{\Delta}_0^{P_1}$. Let $F^{P_1}(x,T):=F^1(x,T)$ be the characteristic function of the set of $x\in G(\mathbb{A})$ such that δx belongs to $\mathfrak{s}^{P_1}(T_0,T,\omega)$ for some $\delta\in P_1(F)$.

As such, $F^1(x,T)$ is left $A_1(\mathbb{R})^0N_1(\mathbb{A})M_1(F)$-invariant, and can be regarded as the characteristic function of the projection of the space $\mathfrak{s}^{P_1}(T_0,T,\omega)$ onto the space $A_1(\mathbb{R})^0N_1(\mathbb{A})M_1(F)\backslash G(\mathbb{A})$, a compact subset of $A_1(\mathbb{R})^0N_1(\mathbb{A})M_1(F)\backslash G(\mathbb{A})$.

For example, $F(x,T):=F^G(x,T)$ admits the following more direct description which will play a key role in our study of Arthur's periods:

If $P_1\subset P_2$ are (standard) parabolic subgroups, we write $A_1^\infty:=A_{P_1}^\infty$ for $A_{P_1}(\mathbb{A})^0$, the identity component of $A_{P_1}(\mathbb{R})$, and

$$A_{1,2}^\infty:=A_{P_1,P_2}^\infty:=A_{P_1}\cap M_{P_2}(\mathbb{A})^1.$$

Then the logarithmic map H_{P_1} maps $A_{1,2}^\infty$ isomorphically onto $\mathfrak{a}_1^2$, the orthogonal complement of $\mathfrak{a}_2$ in $\mathfrak{a}_1$. If T_0 and T are points in $\mathfrak{a}_0$, set $A_{1,2}^\infty(T_0,T)$ to be the set

$$\Big\{a\in A_{1,2}^\infty:\alpha\big(H_1(a)-T\big)>0,\ \alpha\in\Delta_1^2;\ \varpi\big(H_1(a)-T\big)<0,\ \varpi\in\hat{\Delta}_1^2\Big\},$$

where $\Delta_1^2:=\Delta_{P_1\cap M_2}$ and $\hat{\Delta}_1^2:=\hat{\Delta}_{P_1\cap M_2}$. In particular, for T_0 such that $-T_0$ is suitably regular, $F(x,T)$ *is the characteristic function of the compact subset of* $G(F)\backslash G(\mathbb{A})^1$ *obtained by projecting*

$$N_0(\mathbb{A})\cdot M_0(\mathbb{A})^1\cdot A_{P_0,G}^\infty(T_0,T)\cdot K$$

onto $G(F)\backslash G(\mathbb{A})^1$.

All in all, by Lemma 2 of 3.2.7, we arrive at the following

Arthur's Partition. ([Ar2]) *Fix* P *and let* T *be any suitably point in* $T_0+\mathfrak{a}_0^+$. *Then*

$$\sum_{P_1:P_0\subset P_1\subset P}\ \sum_{\delta\in P_1(F)\backslash G(F)}F^1(\delta x)\cdot\tau_1^P\big(H_0(\delta x)-T\big)=1\qquad\forall x\in G(\mathbb{A}).$$

3.2.9 Arthur's Analytic Truncation

Recall that an element $T \in \mathfrak{a}_0^+$ is called *sufficiently regular*, if for any $\alpha \in \Delta_0, \alpha(T) \gg 0$.

Definition. Fix a suitably regular point $T \in \mathfrak{a}_0^+$. If ϕ is a continuous function on $G(F)\backslash G(\mathbb{A})^1$, define *Arthur's analytic truncation* $\big(\Lambda^T\phi\big)(x)$ to be the function

$$\big(\Lambda^T\phi\big)(x) := \sum_P (-1)^{\dim(A/Z)} \sum_{\delta\in P(F)\backslash G(F)} \phi_P(\delta x)\cdot\hat{\tau}_P\big(H(\delta x)-T\big),$$

where

$$\phi_P(x) := \int_{N(F)\backslash N(\mathbb{A})} \phi(nx)\,dn$$

denotes the constant term of ϕ along P, and the sum is over all (standard) parabolic subgroups.

The main purpose for introducing analytic truncation is to give a natural way to construct integrable functions: even from the example of GL_2, we know that automorphic forms are generally not integrable over the total fundamental domain $G(F)\backslash G(\mathbb{A})^1$ mainly due to the fact that in Fourier expansions of such functions, constant terms are only of moderate growth (hence not integrable on A part). Thus in order to naturally obtain integrable functions, we should truncate the original function along cuspidal regions by removing constant terms. Simply put, Arthur's analytic truncation is a well-designed device in which constant terms are tackled in such a way that different levels of parabolic subgroups are suitably counted at the corresponding cuspidal region so that the whole truncation will not be overdone while there will be no parabolic subgroups left untackled.

Note that all parabolic subgroups of G can be obtained from standard parabolic subgroups by taking conjugations with elements from $P(F)\backslash G(F)$. So we have:

(a) $\big(\Lambda^T\phi\big)(x) = \sum_P (-1)^{\dim(A/Z)}\phi_P(x)\cdot\hat{\tau}_P\big(H(x)-T\big)$, *where the sum is over all, both standard and non-standard, parabolic subgroups*;

(b) *If* ϕ *is a cusp form, then* $\Lambda^T\phi = \phi$;

This is because by definition, all constant terms along proper $P : P \neq G$ are zero. Moreover, it is a direct consequence of the Basic Estimation in 3.2.6 for partial truncation, (see e.g. the Corollary there), we have

(c) *If ϕ is of moderate growth* in the sense that there exist some constants C, N such that $|\phi(x)| \leq c\|x\|^N$ for all $x \in G(\mathbb{A})$, *then so is* $\Lambda^T\phi$.

Fundamental properties of Arthur's analytic truncation may be summarized as follows:
Theorem. ([Ar2,3]) *For sufficiently regular T in $\mathfrak{a}_0$,*
(1) *Let $\phi : G(F)\backslash G(\mathbb{A}) \to \mathbb{C}$ be a locally L^1 function. Then*

$$\Lambda^T\Lambda^T\phi(g) = \Lambda^T\phi(g)$$

for almost all g. If ϕ is also locally bounded, then the above is true for all g;
(2) *Let ϕ_1, ϕ_2 be two locally L^1 functions on $G(F)\backslash G(\mathbb{A})$. Suppose that ϕ_1 is of moderate growth and ϕ_2 is rapidly decreasing. Then*

$$\int_{Z_{G(\mathbb{A})}G(F)\backslash G(\mathbb{A})} \overline{\Lambda^T\phi_1(g)} \cdot \phi_2(g)\, dg = \int_{Z_{G(\mathbb{A})}G(F)\backslash G(\mathbb{A})} \overline{\phi_1(g)} \cdot \Lambda^T\phi_2(g)\, dg;$$

(3) *Let K_f be an open compact subgroup of $G(\mathbb{A}_f)$, and r, r' are two positive real numbers. Then there exists a finite subset $\left\{X_i : i = 1, 2, \ldots, N\right\} \subset \mathcal{U}$, the universal enveloping algebra of $\mathfrak{g}_\infty$, such that the following is satisfied: Let ϕ be a smooth function on $G(F)\backslash G(\mathbb{A})$, right invariant under K_f and let $a \in A_{G(\mathbb{A})}$, $g \in G(\mathbb{A})^1 \cap S$. Then*

$$\left|\Lambda^T\phi(ag)\right| \leq \|g\|^{-r} \sum_{i=1}^{N} \sup\left\{|\delta(X_i)\phi(ag')|\,\|g'\|^{-r'} : g' \in G(\mathbb{A})^1\right\},$$

where S is a Siegel domain with respect to $G(F)\backslash G(\mathbb{A})$; and
(4) *The function $\Lambda^T\mathbf{1}$ is a characteristic function of a compact subset of $G(F)\backslash G(\mathbb{A})^1$.*

The key to the proof is that nilpotent groups have simple structure. Say, for (1-3), while a bit complicated, it is based on the discussion in 3.2.6-8. As for (4), which has been considered to be only secondary so far, please see 3.3.2 below.

3.3 Arthur's Periods

3.3.1 Arthur's Periods

Fix a sufficiently regular $T \in \mathfrak{a}_0$ and let ϕ be an automorphic form of G. Then by Theorem 3.2.9, $\Lambda^T\phi$ is rapidly decreasing, and hence integrable. In particular, the integration

$$A(\phi;T) := \int_{G(F)\backslash G(\mathbb{A})} \Lambda^T\phi(g)\,dg$$

makes sense. We claim that $A(\phi;T)$ can be written as an integration of the original automorphic form ϕ over a certain compact subset.

To start with, note that for Arthur's analytic truncation Λ^T, we have $\Lambda^T \circ \Lambda^T = \Lambda^T$. Hence,

$$\begin{aligned} A(\phi;T) &= \int_{Z_{G(\mathbb{A})}G(F)\backslash G(\mathbb{A})} \Lambda^T\phi\; d\mu(g) \\ &= \int_{Z_{G(\mathbb{A})}G(F)\backslash G(\mathbb{A})} \Lambda^T\big(\Lambda^T\phi\big)(g)\; d\mu(g). \end{aligned}$$

Moreover, by the self-adjoint property, for the constant function $\mathbf{1}$ on $G(\mathbb{A})$,

$$\begin{aligned} &\int_{Z_{G(\mathbb{A})}G(F)\backslash G(\mathbb{A})} \mathbf{1}(g)\cdot\Lambda^T\big(\Lambda^T\phi\big)(g)\; d\mu(g) \\ =&\int_{Z_{G(\mathbb{A})}G(F)\backslash G(\mathbb{A})} \big(\Lambda^T\mathbf{1}\big)(g)\cdot\big(\Lambda^T\phi\big)(g)\; d\mu(g) \\ =&\int_{Z_{G(\mathbb{A})}G(F)\backslash G(\mathbb{A})} \Lambda^T\big(\Lambda^T\mathbf{1}\big)(g)\cdot\phi(g)\; d\mu(g), \end{aligned}$$

since $\Lambda^T\phi$ and $\Lambda^T\mathbf{1}$ are rapidly decreasing. Therefore, using $\Lambda^T\circ\Lambda^T = \Lambda^T$ again, we arrive at

$$A(\phi;T) = \int_{Z_{G(\mathbb{A})}G(F)\backslash G(\mathbb{A})} \Lambda^T\mathbf{1}(g)\cdot\phi(g)\; d\mu(g). \qquad (*)$$

To go further, let us give a much more detailed study of Authur's analytic truncation for the constant function $\mathbf{1}$. Introduce the truncated subset $\Sigma(T) := \big(Z_{G(\mathbb{A})}G(F)\backslash G(\mathbb{A})\big)_T$ of the space $G(F)\backslash G(\mathbb{A})^1$ by $\Sigma(T) :=$

$$\big(Z_{G(\mathbb{A})}G(F)\backslash G(\mathbb{A})\big)_T := \big\{g \in Z_{G(\mathbb{A})}G(F)\backslash G(\mathbb{A}) : \Lambda^T\mathbf{1}(g) = 1\big\}.$$

We claim that $\Sigma(T)$ or the same $\big(Z_{G(\mathbb{A})}G(F)\backslash G(\mathbb{A})\big)_T$, is compact. In fact, much stronger result is correct. Namely, we have the following
Basic Fact. ([Ar4]) *For sufficiently regular $T \in \mathfrak{a}_0^+$, $\Lambda^T\mathbf{1}(x) = F(x,T)$. That is to say, $\Lambda^T\mathbf{1}$ is the characteristic function of the compact subset $\Sigma(T)$ of $G(F)\backslash G(\mathbb{A})^1$ obtained by projecting $N_0(\mathbb{A})\cdot M_0(\mathbb{A})^1\cdot A^\infty_{P_0,G}(T_0,T)\cdot K$ onto $G(F)\backslash G(\mathbb{A})^1$.*

Thus, by the equation $(*)$ and the Basic Fact above,

$$\begin{aligned}&\int_{Z_{G(\mathbb{A})}G(F)\backslash G(\mathbb{A})}\Lambda^T\phi(g)\,d\mu(g)\\ =&\int_{Z_{G(\mathbb{A})}G(F)\backslash G(\mathbb{A})}\Lambda^T\mathbf{1}(g)\cdot\phi(g)\,d\mu(g)\\ =&\int_{\Sigma(T)}\phi(g)\,d\mu(g).\end{aligned}$$

That is to say, we have obtained the following very beautiful relation:

Arthur's Periods. *For a sufficiently regular $T \in \mathfrak{a}_0$, and an automorphic form ϕ on $G(F)\backslash G(\mathbb{A})$,*

$$\int_{\Sigma(T)}\phi(g)\,d\mu(g) = \int_{G(F)\backslash G(\mathbb{A})^1}\Lambda^T\phi(g)\,d\mu(g).$$

The importance of this fundamental relation can hardly be over-estimated. Say, when taking ϕ to be Eisenstein series associated with an L^2-automorphic forms,

(i) (*Analytic Evaluation for RHS*) The right hand side can be evaluated using certain analytic methods. For example, in the case for Eisenstein series associated to cusp forms, as done in the next chapter, the RHS can be evaluated in terms of integrations of cusp forms twisted by intertwining operators, which normally are close related with L-functions;
(ii) (*Geometric Evaluation for LHS*) The left hand, by taking the residue (for suitable Eisenstein series), can be evaluated using geometric methods totally independently.
A good example is given in [KW], as to be explained in 4.6. See also 4.7-8, where this is used to obtain certain new yet fundamental relations among special values of zeta functions.

3.3.2 $\Lambda^T \mathbf{1}$ Is a Characteristic Function

Proof of the Basic Fact is based on Arthur's partition for $G(F)\backslash G(\mathbb{A})$ and an inversion formula, which itself is a direct consequence of Langlands' Combinatorial lemma.

Recall that for $T \in \mathfrak{a}_0$, level *P-Arthur's analytic truncation* of ϕ is defined by the formula

$$\Lambda^{T,P}\phi(g) := \sum_{R:R\subset P}(-1)^{d(R)-d(P)} \sum_{\delta\in R(F)\backslash P(F)} \phi_R(\delta g)\cdot \widehat{\tau}_R^P\Big(H(\delta g)-T\Big).$$

As such Λ^T stands simply for $\Lambda^{T,G}$. Thus, together with Langlands' combinatorial Lemma, we have the following:

(a) $\Lambda^T\phi(g) := \sum_R(-1)^{d(R)-d(G)} \sum_{\delta\in R(F)\backslash G(F)} \phi_R(\delta g)\cdot \widehat{\tau}_R\Big(H(\delta g)-T\Big)$; and

(b) (**Inversion Formula**) *For a $G(F)$-invariant function ϕ,*

$$\phi(g) = \sum_P \sum_{\delta\in P(F)\backslash G(F)} \Lambda^{T,P}\phi(\delta g)\cdot \tau_P\Big(H(\delta g)-T\Big).$$

(c) *Assume that $T \in C_P$ and ϕ is automorphic on $N(\mathbb{A})M(F)\backslash G(\mathbb{A})$. Then $\Lambda^{T,P}\phi(m)$ is rapidly decreasing on $M(F)\backslash M(\mathbb{A})^1$. In particular, the integration $\int_{M(F)\backslash M(\mathbb{A})^1} \Lambda_{T,P}\phi(m)\,dm$ is well-defined.*

Proof of the Basic Fact. ([Ar4]) By Arthur's partition for $G(F)\backslash G(\mathbb{A})$, we have

$$\sum_P \sum_{\delta\in P(F)\backslash G(F)} F^P\Big(\delta x, T\Big)\cdot \tau_P\Big(H_P(\delta x)-T\Big) = 1$$

where τ_P is the characteristic function of

$$\Big\{H \in \mathfrak{a}_0 : \alpha(H) > 0,\ \alpha \in \Delta_P\Big\}$$

and

$$F^P\Big(nmk, T\Big) = F^{M_P}\Big(m, T\Big), \qquad n \in N_P(\mathbb{A}), m \in M_P(\mathbb{A}), k \in K.$$

On the other hand, by the inversion formula, applying to the constant function $\mathbf{1}$, we have

$$\sum_P \sum_{\delta\in P(F)\backslash G(F)} \Big(\Lambda^{T,P}\mathbf{1}\Big)(\delta x)\cdot \tau_P\Big(H_P(\delta x)-T\Big) = 1,$$

where $\Lambda^{T,P}$ is the partial analytic truncation operator defined above. With this, the desired result is immediately obtained by induction.

As a direct consequence of this proof, we conclude also the following

Proposition 1. ([Ar4]) *For sufficiently regular T, the characteristic function $F^Q(g,T)$ coincides with the partial truncation along P for the constant function* $\mathbf{1}$. *That is to say,*

$$F^P(g,T) = \Lambda^{P,T}\mathbf{1} = \sum_{R:P_0\subset R\subset P} \varepsilon_R^P \sum_{\delta\in R(F)\backslash P(F)} \widehat{\tau}_R^P\big(H(\delta x) - T\big).$$

Note that the right hand side makes sense even when T is not sufficiently regular. This then leads to the definition of $F^P(g,T)$ for general T:

$$F^P(g,T) = \Lambda^{P,T}\mathbf{1} = \sum_{R:P_0\subset R\subset P} \varepsilon_R^P \sum_{\delta\in R(F)\backslash P(F)} \widehat{\tau}_R^P\big(H(\delta x) - T\big).$$

When preparing our Arthur period paper, we note that such a function is compactly supported. (See the lemma below.) As such, a natural question is whether $F^P(g,T)$ is a characteristic function for a natural compact subset. As it appears, the answer is well-known among those working with trace formula: When I addressed it to Arthur in 2005, he immediately gave me his paper [Ar4], which leads the discussion above for sufficiently regular T. For general T, we have the following

Proposition 2. ([Ar5], see also [Ho]) *As a function of $g \in G(\mathbb{A})^1$, $F^P(g,T)$ is bounded and compactly supported modulo $U(\mathbb{A})M(F)$, uniformly for T varying in a compcat subset.*

Proof. The proof is standard. Following Arthur, let

$$\Gamma_P^Q(H,X) := \sum_{R:P\subset R\subset Q} \varepsilon_R^Q \tau_P^R(H)\widehat{\tau}_R^Q(H-X).$$

Basic properties of $\Gamma_P^Q(H,X)$ are given in Lemma 2.1 of [Ar5], from which, in particular, we know that $\Gamma_P^Q(H,X)$ is compactly supported in $H \in \mathfrak{a}_P^Q$, uniformly for X varying in a compact subset. On the other hand, by the Langlands Combinatorial Lemma, easily,

$$F^P(g,T+X) = \sum_{R:P_0\subset R\subset P} \sum_{\delta\in R(F)\backslash P(F)} F^R(\delta g,T)\Gamma_R^P(H(\delta g) - T, X).$$

As such, the final result comes from Prop. 1: for sufficiently regular T, $F^P(g,T)$ is a characteristic function of a compact subset. □

3.4 Geometric and Analytic Truncations

3.4.1 A Micro Bridge

For simplicity, we in this subsection work only with the field of rationals $\mathbb{Q}$ and use mixed languages of adeles and lattices. Also, without loss of generality, we assume that $\mathbb{Z}$-lattices are of volume one. Accordingly, set $G = SL_r$.

For a rank r lattice Λ of volume one, denote its sublattices filtration associated to a parabolic subgroup P by

$$0 = \Lambda_0 \subset \Lambda_1 \subset \Lambda_2 \subset \cdots \subset \Lambda_{|P|} = \Lambda.$$

Assume that P corresponds to the partition $I = (d_1, d_2, \ldots, d_{|P|})$.

Recall that a polygon $p : [0, r] \to \mathbb{R}$ is called *normalized* if $p(0) = p(r) = 0$. For a (normalized) polygon $p : [0, r] \to \mathbb{R}$, define the associated (real) character $T = T(p)$ of M_0 by the condition that

$$\alpha_i(T) = \big[p(i) - p(i-1)\big] - \big[p(i+1) - p(i)\big]$$

for all $i = 1, 2, \ldots, r-1$.

Lemma. *With the same notation as above,*

$$T(p) = \big(p(1) - p(0), p(2) - p(1), \ldots, p(i) - p(i-1), \\ \ldots, p(r-1) - p(r-2), p(r) - p(r-1)\big).$$

Proof. Let

$$T = T(p) = \big(t_1(p), t_2(p), \ldots, t_r(p)\big) = \big(t_1, t_2, \ldots, t_r\big) \in \mathfrak{a}_0.$$

Note that$\alpha_i(T) = t_i - t_{i+1}$. Hence

$$\begin{aligned} t_i - t_{i+1} =& [p(i) - p(i-1)] - [p(i+1) - p(i)] \\ =& \Delta p(i) - \Delta p(i+1), \qquad i = 1, 2, \ldots, r-1 \end{aligned}$$

where $\Delta p(i) := p(i) - p(i-1)$. Consequently,

$$t_1 - t_{i+1} = \Delta p(1) - \Delta p(i+1), \qquad i = 1, 2, \ldots, r-1.$$

In particular,

$$t_1 - t_r = \Delta p(1) - \Delta p(r) = p(1) - p(r-1),$$

and

$$t_{i+1} = t_1 + \Delta p(i+1) - p(1), \qquad i = 1, 2, \ldots, r-1,$$

since $\Delta p(1) = p(1) - p(0) = p(1)$ and $\Delta(r) = p(r) - p(r-1) = -p(r-1)$. But

$$\begin{aligned} 0 = \sum_{i=1}^{r} t_i = &\sum_{i=0}^{r-1} \Big(t_1 - p(1) + \Delta p(i+1)\Big) \\ =&(r-1)\Big(t_1 - p(1)\Big) + \Big(p(0) - p(r)\Big) \\ =&(r-1)\Big(t_1 - p(1)\Big). \end{aligned}$$

Therefore $t_1 = p(1)$ and hence $t_{i+1} = \Delta p(i+1)$. □

Now take $g = g(\Lambda) \in G(\mathbb{A})$. Denote its lattice by Λ^g, and its induced filtration from P by

$$0 = \Lambda_0^{g,P} \subset \Lambda_1^{g,P} \subset \cdots \subset \Lambda_{|P|}^{g,P} = \Lambda^g.$$

Clearly, the polygon $p_P^g = p_P^{\Lambda^g} : [0, r] \to \mathbb{R}$ is characterized by
(1) $p_P^g(0) = p_P^g(r) = 0$;
(2) p_P^g is affine on $[r_i, r_{i+1}]$, $i = 1, 2, \ldots, |P| - 1$; and
(3) $p_P^g(r_i) = \deg(\Lambda_i^{g,P}) - r_i \cdot \frac{\deg(\Lambda^g)}{r}$, $i = 1, 2, \ldots, |P| - 1$.
Note that the volume of Λ is assumed to be one, therefore (3) is equivalent to
(3)′ $p_P^g(r_i) = \deg(\Lambda_i^{g,P})$, $i = 1, 2, \ldots, |P| - 1$.

The advantage of partially using adelic language is that the values of p_P^g may be written down precisely. Indeed, using Langlands decomposition $g = n \cdot m \cdot a(g) \cdot k$ with $n \in N_P(\mathbb{A}), m \in M_P(\mathbb{A})^1, a \in A_+$ and $k \in K := \prod_p SL(O_{\mathbb{Q}_p}) \times SO(r)$. Write

$$a = a(g) = \operatorname{diag}\Big(a_1 I_{d_1}, a_2 I_{d_2}, \ldots, a_{|P|} I_{d_{|P|}}\Big)$$

where $r = d_1 + d_2 + \cdots + d_{|P|}$ is the partition corresponding to P. Then it is a standard fact that

$$\deg\Big(\Lambda_i^{g,P}\Big) = -\log\Big(\prod_{j=1}^{i} a_j^{d_j}\Big) = -\sum_{j=1}^{i} d_j \log a_j, \qquad i = 1, \ldots, |P|.$$

Set now $\mathbf{1}(p_P^* >_P p)$ to be the characteristic function of the subset of g's such that $p_P^g >_P p$.

Micro Bridge. *For a fixed convex normalized polygon $p : [0, r] \to \mathbb{R}$, and $g \in SL_r(\mathbb{A})$, with respect to any parabolic subgroup P, we have*

$$\hat{\tau}_P\Big(- H_0(g) - T(p)\Big) = \mathbf{1}\Big(p_P^g >_P p\Big).$$

Proof. First let us consider the right hand side. By definition,

$$\mathbf{1}(p_P^g >_P p) = 1$$

if and only if

$$p_P^g(r_1) - p(r_1) > 0, \qquad p_P^g(r_2) - p(r_2) > 0, \\ \ldots, \qquad p_P^g(r_{|P|-1}) - p(r_{|P|-1}) > 0.$$

That is,

$$\deg\Big(\Lambda_1^{g,P}\Big) - p(r_1) > 0, \qquad \deg\Big(\Lambda_2^{g,P}\Big) - p(r_2) > 0, \\ \ldots, \qquad \deg\Big(\Lambda_{|P|-1}^{g,P}\Big) - p(r_{|P|-1}) > 0.$$

In this way, we have established the following

Sublemma 1. *With the same notation as above,*

$$\mathbf{1}(p_P^g >_P p) = 1$$

if and only if

$$\Big(- d_1 \log a_1\Big) - p(r_1) > 0,$$
$$\Big(- d_1 \log a_1 - d_2 \log a_2\Big) - p(r_2) > 0,$$
$$\cdots\cdots\cdots\cdots$$
$$\Big(- d_1 \log a_1 - d_2 \log a_2 - \cdots - d_{|P|-1} \log a_{|P|-1}\Big) - p(r_{|P|-1}) > 0.$$

Next let us consider the left hand side, that is, the function $\widehat{\tau}_P(-H_0(g)-T(p))$. When does it take the value one?

Recall that for $P = P_I, \widehat{\tau}_P$ is the characteristic function of the following subset of $\mathfrak{a}_0 = \mathfrak{a}_0^I \oplus \mathfrak{a}_I$ defined to be

$$\mathfrak{a}_0^I \oplus \Big\{H \in \mathfrak{a}_I : \varpi_i(H) > 0 \quad \forall i \in \Delta - I\Big\}.$$

That is to say, if $H \in \mathfrak{a}_0$ belongs to

$$\mathfrak{a}_0^I \oplus \Big\{H \in \mathfrak{a}_I : \varpi_i(H) > 0 \quad i = r_1, r_2, \ldots, r_{|P|-1}\Big\},$$

then $\widehat{\tau}_P(H) = 1$.

But $H \in \mathfrak{a}_I$ means $H = \Big(H_1^{(d_1)}, H_2^{(d_2)}, \ldots, H_{|P|}^{(d_{|P|})}\Big)$ and $\sum_{i=1}^{|P|} d_i H_i = 0$. Therefore, the conditions that

$$H \in \mathfrak{a}_I \quad \text{such that} \qquad \varpi_i(H) > 0 \quad \text{for all } i = r_1, r_2, \ldots, r_{|P|-1}$$

is equivalent to the conditions that

$$H = \Big(H_1^{(d_1)}, H_2^{(d_2)}, \ldots, H_{|P|}^{(d_{|P|})}\Big) \qquad \text{where} \qquad \sum_{i=1}^{|P|} d_i H_i = 0$$

and

$$d_1H_1 > 0, d_1H_1 + d_2H_2 > 0, \ldots, d_1H_1 + d_2H_2 + \cdots + d_{|P|-1}H_{|P|-1} > 0.$$

On the other hand, for $H = \Big(H_1, H_2, \ldots, H_r\Big) \in \mathfrak{a}_0$, the projections of H to $\mathfrak{a}_I$ (resp. to $\mathfrak{a}_0^I$) is given by

$$\Big(\Big(\frac{H_1 + H_2 + \cdots + H_{r_1}}{d_1}\Big)^{(d_1)}, \Big(\frac{H_{r_1+1} + H_{r_1+2} + \cdots + H_{r_2}}{d_2}\Big)^{(d_2)},$$
$$\ldots, \Big(\frac{H_{r_{|P|-1}+1} + H_{r_{|P|-1}+2} + \cdots + H_{r_{|P|}}}{d_{|P|}}\Big)^{(d_{|P|})}\Big) \in \mathfrak{a}_I$$

(resp.

$$\Big(H_1-\frac{H_1+H_2+\cdots+H_{r_1}}{d_1},H_2-\frac{H_1+H_2+\cdots+H_{r_1}}{d_1},\\ \ldots,H_{r_1}-\frac{H_1+H_2+\cdots+H_{r_1}}{d_1},\\ \ldots,H_{r_{|P|-1}+1}-\frac{H_{r_{|P|-1}+1}+H_{r_{|P|-1}+2}+\cdots+H_{r_{|P|}}}{d_{|P|}},\\ H_{r_{|P|-1}+2}-\frac{H_{r_{|P|-1}+1}+H_{r_{|P|-1}+2}+\cdots+H_{r_{|P|}}}{d_{|P|}},\\ \ldots,H_{r_{|P|}}-\frac{H_{r_{|P|-1}+1}+H_{r_{|P|-1}+2}+\cdots+H_{r_{|P|}}}{d_{|P|}}\Big)$$

$$=\Big(H_1,H_2,\ldots,H_r\Big)\\ -\Big(\Big(\frac{H_1+H_2+\cdots+H_{r_1}}{d_1}\Big)^{(d_1)},\Big(\frac{H_{r_1+1}+H_{r_1+2}+\cdots+H_{r_2}}{d_2}\Big)^{(d_2)},\\ \ldots,\Big(\frac{H_{r_{|P|-1}+1}+H_{r_{|P|-1}+2}+\cdots+H_{r_{|P|}}}{d_{|P|}}\Big)^{(d_{|P|})}\Big)\in\mathfrak{a}_0^I.)$$

Therefore,

$$\widehat{\tau}_P\Big(-H_0(g)-T(p)\Big)=1$$

if and only if the coordinates H_i of $-H_0(g)-T(p)=:H=\Big(H_i\Big)$ satisfy the conditions that

$$H_1+H_2+\cdots+H_{r_1}>0,\\ \Big(H_1+H_2+\cdots+H_{r_1}\Big)+\Big(H_{r_1+1}+H_{r_1+2}+\cdots+H_{r_2}\Big)>0,\\ \cdots\cdots\cdots\cdots\\ \Big(H_1+H_2+\cdots+H_{r_1}\Big)+\Big(H_{r_1+1}+H_{r_1+2}+\cdots+H_{r_2}\Big)\\ +\cdots+\Big(H_{r_{|P|-1}+1}+H_{r_{|P|-1}+2}+\cdots+H_{r_{|P|-1}}\Big)>0.$$

Recall now that, by definition,

$$-H_0(g)-T(p)=\Big(-t_1-\log b_1,-t_2-\log b_2,\ldots,-t_r-\log b_r\Big)$$

where

$$a(g)=\mathrm{diag}\Big(b_1,b_2,\ldots,b_r\Big)=\mathrm{diag}\Big(a_1^{(d_1)},a_2^{(d_2)},\ldots,a_{|P|}^{(d_{|P|})}\Big)$$

and

$$\begin{aligned} T(p) &= \big(t_1, t_2, \ldots, t_r\big) \\ &= \big(p(1), p(2) - p(1), \ldots, p(r-1) - p(r-2), -p(r-1)\big). \end{aligned}$$

As such, this latest group of inequalities, after replacing H_i's with the precise coordinates of $-H_0(g) - T(p)$, are then changed to the following group conditions:

$$\begin{aligned} &\big(-\log b_1 - p(1)\big) + \big(-\log b_2 - p(2) + p(1)\big) \\ &\qquad + \cdots + \big(-\log b_{r_1} - p(r_1) + p(r_1 - 1)\big) > 0 \\ &\Big(\big(-\log b_1 - p(1)\big) + \big(-\log b_2 - p(2) + p(1)\big) \\ &\qquad + \cdots + \big(-\log b_{r_1} - p(r_1) + p(r_1 - 1)\big)\Big) \\ &\quad + \Big(\big(-\log b_{r_1+1} - p(r_1 + 1) + p(r_1)\big) + \big(-\log b_{r_1+2} - p(r_1 + 2) + p(r_1 + 1)\big) \\ &\qquad + \cdots + \big(-\log b_{r_2} - p(r_2) + p(r_2 - 1)\big)\Big) > 0 \\ &\qquad \cdots\cdots\cdots\cdots \\ &\Big(\big(-\log b_1 - p(1)\big) + \big(-\log b_2 - p(2) + p(1)\big) \\ &\qquad + \cdots + \big(-\log b_{r_1} - p(r_1) + p(r_1 - 1)\big)\Big) \\ &\quad + \Big(\big(-\log b_{r_1+1} - p(r_1 + 1) + p(r_1)\big) + \big(-\log b_{r_1+2} - p(r_1 + 2) + p(r_1 + 1)\big) \\ &\qquad + \cdots + \big(-\log b_{r_2} - p(r_2) + p(r_2 - 1)\big)\Big) \\ &\qquad + \cdots\cdots\cdots\cdots \\ &\quad + \Big(\big(-\log b_{r_{|P|-2}+1} - p(r_{|P|-2} + 1) + p(r_{|P|-1})\big) \\ &\qquad + \big(-\log b_{r_{|P|-2}+2} - p(r_{|P|-2} + 2) + p(r_{|P|-2} + 1)\big) \\ &\qquad + \cdots + \big(-\log b_{r_{|P|-1}} - p(r_{|P|-1}) + p(r_{|P|-1} - 1)\big) > 0. \end{aligned}$$

This, after the obvious simplification with

$$B_1 := -\log b_1, B_2 = -\log b_2, \ldots, B_r = -\log b_r, \; r_0 := 0,$$

becomes the group of conditions

$$\begin{aligned}
&B_1 + B_2 + \cdots + B_{r_1} - p(r_1) + p(0) > 0,\\
&\Big(B_1 + B_2 + \cdots + B_{r_1}\Big) + \Big(B_{r_1+1} + B_{r_1+2} + \cdots + B_{r_2}\Big) - p(r_2) > 0,\\
&\cdots\cdots\cdots\cdots\\
&\Big(B_1 + B_2 + \cdots + B_{r_1}\Big) + \Big(B_{r_1+1} + B_{r_1+2} + \cdots + B_{r_2}\Big)\\
&+ \cdots + \Big(B_{r_{|P|-2}+1} + B_{r_{|P|-2}+2} + \cdots + B_{r_{|P|-1}}\Big) - p(r_{|P|-1}) > 0.
\end{aligned}$$

Thus note that

$$\Big(b_1, b_2, \ldots, b_r\Big) = a(g) = \Big(a_1^{(r_1-r_0)}, a_2^{(r_2-r_1)}, \ldots, a_{|P|}^{(r_{|P|}-r_{|P|-1})}\Big),$$

we arrive at the following

Sublemma 2. *With the same notation as above,*

$$\widehat{\tau}_P\Big(-H_0(g) - T(p)\Big) = 1$$

if and only if

$$\begin{aligned}
&-d_1 \log a_1 - p(r_1) > 0,\\
&\Big(-d_1 \log a_1 - d_2 \log a_2\Big) - p(r_2) > 0\\
&\cdots\cdots\cdots\cdots\\
&\Big(-d_1 \log a_1 - d_2 \log a_2 - \cdots - d_{|P|-1} \log a_{|P|-1}\Big) - p(r_{|P|-1}) > 0.
\end{aligned}$$

By comparing Sublemmas 1 and 2, we see that

$$\hat{\tau}_P\Big(-H_0(g) - T(p)\Big) = \mathbf{1}\Big(p_P^g >_P p\Big).$$

□

3.4.2 Global Bridge Between Geometric and Analytic Truncations

With the micro bridge above, now we are ready to expose a beautiful intrinsic relation between algebraic and analytic truncations.

Global Bridge. *For a fixed normalized convex polygon* $p : [0, r] \to \mathbb{R}$, *let* $T(p) :=$

$$\big(p(1), p(2) - p(1), \ldots, p(i) - p(i-1), \ldots, p(r-1) - p(r-2), -p(r-1)\big)$$

be the associated vector in $\mathfrak{a}_0$. *If* $T(p)$ *is sufficiently positive, then*

$$\mathbf{1}(\overline{p}^g \leq p) = \big(\Lambda^{T(p)}\mathbf{1}\big)(g).$$

Proof. This is a direct consequence of the Fundamental Relation in 3.1.2, the definition of Arthur's analytic truncation and the Micro Bridge in 3.4.1. □

4 Rankin-Selberg & Zagier Method

Global bridge between geometric and analytic truncations in 3.4.2 shows that at least for polygons p whose associated $T(p) \in \mathfrak{a}_0$ are sufficiently positive, the corresponding non-abelian L-functions may be viewed as a special kinds of Arthur's periods. Therefore, to understand our Ls, it is quite helpful to understand the structure of Arthur's periods.

However, to write down them precisely, Arthur's periods for automorphic forms seems to be too general. It is for this purpose that we then narrow our class by considering Eisenstein periods: By definition, *Eisenstein periods* are integrations of Eisenstein series over truncated compact domains. Simply put, the aim of this chapter is to show that a special kind of Eisenstein periods and hence also certain classes of Ls can be evaluated via an advanced version of Rankin-Selberg & Zagier method.

4.1 Regularized Integration over Cones

To begin with, we here give a review of Jacquet-Lapid-Rogawski's beautiful yet elementary treatment of integrations over cones.

Let V be a real r dimensional vector space. Let V^* be the space of complex linear forms on V. Denote by $S(V^*)$ the symmetric algebra of V^*, which may be regarded as the space of polynomial functions on V as well. By definition, an *exponential polynomial* function on V is a function of the form

$$f(x) = \sum_{i=1}^{r} e^{\langle \lambda_i, x\rangle} P_i(x)$$

where the λ_i are distinct elements of V^* and the $P_i(x)$ are non-zero elements of $S(V^*)$. One checks easily that

(1) *The λ_i are uniquely determined and called the exponents of f.*

By a *cone* in V we shall mean a closed subset of the form

$$C := \left\{x \in V : \langle \mu_i, x\rangle \geq 0\ \forall i\right\}$$

where $\{\mu_i\}$ is a basis of $\mathrm{Re}(V^*)$, the space of *real* linear forms on V. Let $\{e_j\}$ be the dual basis of V. We shall say that $\lambda \in V^*$ is *negative* (resp.

non-degenerate) with respect to C if $\mathrm{Re}\,\langle \lambda, e_j \rangle < 0$ (resp. $\langle \lambda, e_j \rangle \neq 0$) for each $j = 1, \ldots, r$.

(2) *The function* $f(s) = \sum_{i=1}^r e^{\langle \lambda_i, x\rangle} P_i(x)$ *is integrable over* C *if and only if* λ_i *are negative with respect to* C *for all* i.

To define regularized integrals over C, we study the integral

$$I_C\big(f;\lambda\big) := \int_C f(x) e^{\langle \lambda, x\rangle} dx.$$

It converges absolutely for λ in the open set

$$\Big\{\lambda \in V^* : \mathrm{Re}\langle \lambda_i - \lambda, e_j\rangle < 0\ \forall 1 \le i \le n, 1 \le j \le r\Big\},$$

and can be analytically continued as follows:
In the case when $\mathrm{Re}(\lambda) > 0$ and P is a polynomial of one variable,

$$\int_0^\infty e^{-\lambda x} P(x) dx = \sum_{m \ge 0} \frac{\big(D^m P\big)(0)}{\lambda^{m+1}}.$$

We then can use the right hand side to obtain the continuation.
To deal with more general case, fix an index k and let C_k be the intersection of C with the hyperplane $V_k := \Big\{x : \langle \mu_k, x\rangle = 0\Big\}$. Then, we can write $x = \langle \mu_k, x\rangle e_k + y$ with $y \in V_k$. Thus, for a suitable choice of Haar measures,

$$I_C\big(f;\lambda\big) = \sum_{i=1}^n \sum_{m \ge 0} \frac{1}{\Big(\langle \lambda - \lambda_i, e_k\rangle\Big)^{m+1}} \int_{C_k} e^{-\langle \lambda - \lambda_i, y\rangle} \big(D_{e_k}^m\big) P_i(y)\, dy.$$

This formula then gives the analytic continuation of $I_C\big(f;\lambda\big)$ to the tube domain defined by $\mathrm{Re}\langle \lambda_i - \lambda, e_k\rangle > 0$ for $j \neq k$, $1 \le i \le n$ with singularities on hyperplanes $H_{k,i} := \Big\{\lambda : \langle \lambda, e_k\rangle = \langle \lambda_i, e_k\rangle\Big\}, 1 \le i \le n$. Moreover, since this is true for all $k = 1, 2, \ldots, r$, the function $I_C\big(f;\lambda\big)$ has analytic continuation to V^* by Hartogs' lemma with (only) hyperplane singularities along $H_{i,k}$, $1 \le k, \le r, 1 \le i \le r$. In particular, from such an induction, one checks

(3) *Suppose that* f *is absolutely integrable over* C. *Then* $I_C\big(f;\lambda\big)$ *is holomorphic at 0 and* $I_C\big(f;0\big) = \int_C f(x)\,dx$; and

(4) *The function* $I_C\big(f;\lambda\big)$ *is holomorphic at 0 if and only if for all i,* λ_i *are non-degenerate with respect to C, i.e.,*$\langle\lambda_i, e_k\rangle \neq 0$ *for all pairs* (i,k).

Denote the characteristic function of a set Y by τ^Y. For $\lambda \in V^*$ such that $\lambda_i - \lambda$ are negative with respect to C for all $i = 1, 2, \ldots, n$, set

$$\begin{aligned}\widehat{F}\big(\lambda; C, T\big) &:= \int_V f(x)\cdot\tau^C(x-T)\cdot e^{-\langle\lambda,x\rangle}dx\\ &= \sum_{i=1}^n \int_V P_i(x)\tau^C(x-T)e^{-\langle\lambda-\lambda_i,x\rangle}dx\\ &= \sum_{i=1}^n e^{-\langle\lambda_i-\lambda,T\rangle}I_C\big(P_i(*+T)e^{\langle\lambda_i,*\rangle};\lambda\big).\end{aligned}$$

The integrals are absolutely convergent and $\widehat{F}\big(\lambda; C, T\big)$ extends to a meromorphic function on V^*. With this, following [JLR], we call the function $f(x)\cdot\tau^C(x-T)$ *#-integrable* if $\widehat{F}\big(\lambda; C, T\big)$ is holomorphic at $\lambda = 0$. In this case, set

$$\int_V^{\#} f(x)\cdot\tau^C(x-T)dx := \widehat{F}\big(0; C, T\big).$$

By (4), the #-integral exists if and only if each exponent λ_j is non-degenerate with respect to C.

To go further, suppose that $V = W_1 \oplus W_2$ is a decomposition of V as a direct sum, and let C_j be a cone in W_j. Write $T = T_1 + T_2$ and $x = x_1 + x_2$ relative to this decomposition. If the exponents λ_j of f are non-degenerate with respect to $C = C_1 + C_2$, then the function

$$w_2 \mapsto \int_{W_1}^{\#} f(w_1+w_2)\tau_1^{C_1}(w_1-T_1)\,dw_1$$

is defined and is an exponential polynomial. And it follows by analytic continuation that

$$(5)\ \int_V^{\#} f(x)\tau^C(x-T)\,dx$$

$$= \int_{W_2}^{\#}\left(\int_{W_1}^{\#} f(w_1+w_2)\tau_1^{C_1}(w_1-T_1)\,dw_1\right)\cdot\tau_2^{C_2}(w_2-T_2)\,dw_2;$$

and as a function of T, it is an exponential polynomial with the same exponents as f.

More generally, let $g(x)$ be a compactly supported function on W_1 and consider functions of the form $g(w_1 - t_1)\tau_C(w_2 - T_2)$, which we call functions of *type (C)*. It is clear that the integral

$$\widehat{F}(\lambda) := \int_V f(x) \cdot g(w_1 - t_1)\tau_C(w_2 - T_2) \cdot e^{-\langle\lambda,x\rangle} dx$$

converges absolutely for an open set of λ whose restriction to W_2 is negative with respect to C_2. Furthermore, $\widehat{F}(\lambda)$ has a meromorphic continuation to V^*. Following JLR, the function $f(x)g(w_1 - t_1)\tau_C(w_2 - T_2)$ is called #-*integrable* if $\widehat{F}(\lambda)$ is holomorphic at $\lambda = 0$. If so we define

$$\int_V^{\#} f(x) \cdot g(w_1 - t_1)\tau_C(w_2 - T_2)\, dx := \widehat{F}(0).$$

Note that as a function of w_2, $\int_{W_1} f(w_1+w_2)g(w_1-T_1)\, dw_1$ is an exponential polynomial on W_2, hence

$$\begin{aligned}&\int_V^{\#} f(x)\, g(w_1 - t_1)\tau_C(w_2 - T_2)\, dx \\ &= \int_{W_2}^{\#} \left(\int_{W_1} f(w_1 + w_2)g(w_1 - T_1)\, dw_1 \right) \tau_2^C(w_2 - T_2)\, dw_2.\end{aligned}$$

We end this discussion by the following explicit formulas:

(6) $\displaystyle\int_V^{\#} e^{\langle\lambda,x\rangle}\tau^C(x - T)\, dx = (-1)^r \mathrm{Vol}\big(e_1, e_2, \ldots, e_r\big) \cdot \frac{e^{\langle\lambda,T\rangle}}{\prod_{j=1}^r \langle\lambda, e_j\rangle}$, *where* $C := \left\{ \sum_{j=1}^r a_j e_j \,:\, a_j \geq 0 \right\}$ *and* $\mathrm{Vol}\big(e_1, e_2, \ldots, e_r\big)$ *is the volume of the associated parallelepiped* $\left\{ \sum_{j=1}^r a_j e_j : a_j \in [0, 1] \right\}$.

In particular, if $V = \mathbb{R}$, we have $\displaystyle\int_T^{\infty} e^{\lambda t} dt = -\frac{e^{\lambda T}}{\lambda}$ and hence for any cone $C \subset \mathbb{R}$,

$$\int_V^{\#} e^{\langle\lambda,x\rangle}\big(1 - \tau^C(x - T)\big)\, dx = - \int_V^{\#} e^{\langle\lambda,x\rangle}\tau^C(x - T)\, dx$$

since $1 - \tau^C$ is the characteristic function of the cone $-C$.

4.2 JLR Periods of Automorphic Forms

Now let us follow [JLR] to apply the above discussion on integration over cones to periods of automorphic forms.

Let $\tau_k(X)$ be a function of type (C) on $\mathfrak{a}_P$ that depends continuously on $k \in K$, i.e., we assume that there is a decomposition $\mathfrak{a}_P = W_1 \oplus W_2$ such that $\widehat{\tau}_k$ has the form

$$g_k(w_1 - T_1)\tau_{C_{2k}}(w_2 - T_2)$$

where the compactly supported function g_k varies continuously in the L^1-norm and linear inequalities defining the cone C_{2k} vary continuously.

Also with the application to automorphic forms in mind, let f be a function on the space $M(F)N(\mathbb{A})\backslash G(\mathbb{A})$ of the form

$$f(namk) = \sum_{j=1}^{s} \phi_j(m,k) \cdot \alpha_j\Big(H_P(a), k\Big) e^{\langle \lambda_j + \rho_P, H_P(a)\rangle}$$

for $n \in N(\mathbb{A}), a \in A_P, m \in M(\mathbb{A})^1$ and $k \in K$, where, for all j,
(a) $\phi_j(m,k)$ is absolutely integrable on $M(F)\backslash M(\mathbb{A})^1 \times K$; and
(b) $\lambda_j \in \mathfrak{a}_P^*$ and $\alpha_j(X,k)$ is a continuous family of polynomials on $\mathfrak{a}_P$ such that, for all $k \in K$, $\alpha_J(X,k)e^{\langle \lambda_j, X\rangle}\tau_k(X)$ is #-integrable.
Remark. This is really nothing but the micro structure for automorphic forms stated more formally at 2.1. The point is also related to the reduction theory: with $M(F)\backslash M(\mathbb{A})^1$ of finite volume, ϕ_j behaves very well. So the problem when say taking integration is about the A-part, which is affine – the exponential polynomial function can be controlled nicely provided the exponents are negative.

As such, following [JLR], define the *#-integral*

$$\int_{P(F)\backslash G(\mathbb{A})}^{\#} f(g) \cdot \tau_k\Big(H_P(g)\Big)\, dg :=$$
$$\sum_{j=1}^{s} \int_K \Big(\int_{M(F)\backslash M(\mathbb{A})^1} \phi_j(mk)\, dm\Big) \cdot \Big(\int_{\mathfrak{a}_P}^{\#} \alpha_j(X,k)\tau_k(X)\, dX\Big)\, dk.$$

Remark. The reader should notice that in this definition there is no integration involving $\int_{N(F)\backslash N(\mathbb{A})} dn$: In practice, f is supposed to be the constant term along P, so is $N(\mathbb{A})$-invariant.

Clearly, the set $\mathcal{E}_P(f)$ of distinct exponents $\{\lambda_j\}$ of f is uniquely determined by f, but the functions ϕ_j and α_j are not. So we need to show that the #-integral is independent of the choices of these functions. This goes as follows: The function

$$X \mapsto \int_{M(F)\backslash M(\mathbb{A})^1} f(e^X mk)\, dm$$

is an exponential polynomial on $\mathfrak{a}_P$ and the above #-integral is equal to

$$\int_K \int_{\mathfrak{a}_P}^{\#} \left(\int_{M(F)\backslash M(\mathbb{A})^1} f(e^X mk)\, dm \right) e^{-\langle \rho_P, X\rangle} \tau_k(X)\, dX\, dk.$$

This shows in particular that the original #-integral is independent of the decomposition of f as used in the definition. Moreover, if each of the exponents λ_j is negative with respect to C_{2k} for all $k \in K$, then the ordinary integral

$$\int_{P(F)\backslash G(\mathbb{A})^1} f(g) \cdot \tau_k(H_P(g))\, dg$$

is absolutely convergent and equals to the #-integral by Basic Property (3) of integration over cones listed in 4.1. That is to say, we have the following

Lemma. (1) *The above #-integral* $\int_{P(F)\backslash G(\mathbb{A})}^{\#} f(g) \cdot \tau_k\big(H_P(g)\big)\, dg$ *is well-defined;* and
(2) *If each of the exponents* λ_j *of* f *is negative with respect to* C_{2k} *for all* $k \in K$*, then the ordinary integral*

$$\int_{P(F)\backslash G(\mathbb{A})^1} f(g) \cdot \tau_k\big(H_P(g)\big)\, dg$$

is absolutely convergent and equals to the #-integral.

Fix a sufficiently regular element $T \in \mathfrak{a}_0^+$. Then the above construction applies to Arthur's analytic truncation $\Lambda^{T,P}\Psi(g)$ and the **characteristic function** $\tau_P\big(H_P(g) - T\big)$ where $\Psi \in \mathcal{A}_P(G) := A(U(\mathbb{A})M(F)\backslash G(\mathbb{A}))$, the space of level $P = MU$ automorphic forms. Indeed, with

$$\Psi(namk) = \sum_{j=1}^{s} Q_j\big(H_P(a)\big) \cdot \psi_j(amk)$$

for $n \in N(\mathbb{A}), a \in A_P, m \in M(\mathbb{A})^1$ and $k \in K$, where Q_j are polynomials and $\psi_j \in \mathcal{A}_P(G)$ satisfies

$$\psi_j(ag) = e^{\langle \lambda_j + \rho_P, H_P(a) \rangle} \psi_j(g)$$

for some $\lambda_j \in \mathfrak{a}_P^*$ and all $a \in A_P$, $\int_{M(F)\backslash M(\mathbb{A})^1 \times K} \Lambda^{T,P} \psi_j(mk) dm\, dk$ is well-defined, since the function $m \mapsto \Lambda^{T,P}\psi_j(mk)$ is rapidly decreasing on the space $M(F)\backslash M(\mathbb{A})^1 \times K$. Now τ_P is the characteristic function of the cone spanned by the coweights $\widehat{\Delta}_P^\vee$, so

$$\int_{P(F)\backslash G(\mathbb{A})^1}^{\#} \Lambda^{T,P}\Psi(g) \cdot \tau_P\Big(H_P(g) - T\Big)\, dg$$

exists if and only if

$$\langle \lambda_j, \varpi^\vee \rangle \neq 0 \qquad \forall \varpi \in \widehat{\Delta}_P^\vee \quad \text{and} \quad \lambda_j \in \mathcal{E}_P(\Psi).$$

The same is true for $\Lambda^{T,P}\Psi(g) \cdot \tau_P(H_P(gx) - T)$ for $x \in G(\mathbb{A})$. (For $x \in G(\mathbb{A})$, let $K(g) \in K$ be an element such that $gK(g)^{-1} \in P_0(\mathbb{A})$, then $H_P(gx) = H_P(g) + H_P(K(g)x)$ and hence $\tau_P\Big(X - T - H_P(K(g)x)\Big)$ is the characteristic function of a cone depending continuously on g.) Set then

$$I_G^{P,T}(\Psi) := \int_{P(F)\backslash G(\mathbb{A})^1}^{\#} \Lambda^{T,P}\Psi(g) \cdot \tau_P\Big(H_P(g) - T\Big)\, dg.$$

For any automorphic form $\phi \in \mathcal{A}(G)$, write $\mathcal{E}_P(\phi)$ for the set of exponents $\mathcal{E}_P(\phi_P)$. Set $\mathcal{A}(G)^* :=$

$$\Big\{\phi \in \mathcal{A}(G) : \langle \lambda, \varpi^\vee \rangle \neq 0 \qquad \forall \varpi^\vee \in \widehat{\Delta}_P^\vee,\ \lambda \in \mathcal{E}_P(\phi),\ P \neq G\Big\}.$$

If $\phi \in \mathcal{A}(G)^*$, then $I_{P,T}^G(\phi_P)$ exists for all P; and define JLR's *regularized period* by

$$\int_{G(F)\backslash G(\mathbb{A})}^{*} \phi(g)\, dg = \sum_P I_G^{P,T}(\phi_P) := I_G^T(\phi).$$

In the follows, let us expose some of the basic properties of JLR's regularized periods. For this, we make the following preparation. As above, for $x \in G(\mathbb{A}_f)$, let $\rho(x)$ denote right translation by x: $\rho(x)\phi(g) = \phi(gx)$.

The space $\mathcal{A}(G)$ is stable under right translation by $G(\mathbb{A}_f)$. Furthermore, $\rho(x)\phi$ has the same set of exponents as ϕ. Consequently, the space $\mathcal{A}(G)_*$ is invariant under right translation by $G(\mathbb{A}_f)$. Indeed, for $k \in K$, write the Iwasawa decomposition of kx as $kx = n'a'm'K(kx)$ and write ϕ_P in the form

$$\phi_P\Big(namk\Big) = \sum_j Q_j\Big(H_P(a)\Big) \cdot e^{\langle \lambda_j+\rho_P, H(a)\rangle} \phi_j(mk).$$

Since $amkx = n^* aa'mm'K(kx)$ for some $n^* \in N(\mathbb{A})$, we have

$$\begin{aligned} &\phi_P\Big(namk \cdot x\Big) \\ &= \sum_j Q_j\Big(H_P(a) + H_P(a')\Big) \cdot e^{\langle \lambda_j+\rho_P, H(a)+H(a')\rangle} \phi_j\Big(mm'K(kx)\Big). \end{aligned}$$

This shows that $\mathcal{E}_P\Big(\rho(x)\phi\Big) = \mathcal{E}_P(\phi)$.

Theorem. ([JLR]) (1) *I_G^T defines a $G(\mathbb{A}_f)^1$-invariant linear functional on $\mathcal{A}(G)_*$;*
(2) *$I_G^T(\phi) := \int^*_{G(F)\backslash G(\mathbb{A})^1} \phi(g)\,dg$ is independent of the choice of T*; and
(3) *If $\phi \in \mathcal{A}(G)$ is integrable over $G(F)\backslash G(\mathbb{A})^1$, then $\phi \in \mathcal{A}(G)_*$ and*

$$\int^*_{G(F)\backslash G(\mathbb{A})^1} \phi(g)\,dg = \int_{G(F)\backslash G(\mathbb{A})^1} \phi(g)\,dg.$$

4.3 Arthur's Periods and JLR's Periods

For $f \in \mathcal{A}_P(G)$ with

$$f\Big(namk\Big) = \sum_{j=1}^{l} \phi_j(mk) \cdot \alpha_j\Big(H(a)\Big) e^{\langle \lambda_j+\rho_P, H(a)\rangle}$$

where $n \in N(\mathbb{A})$, $a \in A_P$, $m \in M(\mathbb{A})^1$, $k \in K$, and for all j, $\alpha_j(X)$ is a polynomial, and $\phi_j(g)$ is an automorphic form in $\mathcal{A}_P(G)$ such that $\phi_j(ag) =$

$\phi_j(g)$ for $a \in A_P$, following [JLR], define

$$\int_{P(F)\backslash G(\mathbb{A})^1}^{*} f(g)\cdot\tau_P\Big(H(g)-T\Big)\,dg$$
$$:=\sum_{j=1}^{l}\int_K\int_{M(F)\backslash M(\mathbb{A})^1}^{*}\phi_j(mk)\,dm\,dk$$
$$\times\left(\int_{\mathfrak{a}_P}^{\#}\alpha_j(X)e^{\langle\lambda_j,X\rangle}\tau_P(X-T)\,dX\right).$$

This is well-defined provided that the following two conditions are satisfied:
(i*) $\langle\mu,\varpi^\vee\rangle\neq 0$ $\quad\forall Q\subset P,\ \varpi^\vee\in(\widehat{\Delta^\vee})_Q^P,\ \mu\in\mathcal{E}_Q(\phi)$; and
(ii*) $\langle\lambda,\alpha^\vee\rangle\neq 0$ $\quad\forall\alpha\in\Delta_P,\ \lambda\in\mathcal{E}_P(\phi)$.

Let $\mathcal{A}(G)^{**}$ be the space of $\phi\in\mathcal{A}(G)$ such that (i*) and hence also (ii*) are satisfied for any P. As a direct consequence of the theorem in 5.2, we have the following

Theorem. ([JLR]) *For $\phi\in\mathcal{A}(G)^{**}$,*

$$\int_{G(F)\backslash G(\mathbb{A})^1}\Lambda^T\phi(g)\,dg$$
$$=\sum_P(-1)^{d(P)-d(G)}\int_{P(F)\backslash G(\mathbb{A})^1}^{*}\phi_P(g)\cdot\widehat{\tau}_P\Big(H(g)-T\Big)\,dg.$$

4.4 Bernstein Principle

With above, to understand our L-functions at least for $T\gg 0$, i.e., for $T\in\mathfrak{a}_0^+$ sufficiently positive, we need to study Eisenstein periods. For this, the following result plays a key role.

Bernstein's Principle. ([JLR]) *Let $P=MU$ be a proper parabolic subgroup and let σ be an irreducible cuspidal representation in the space $L^2\Big(M(F)\backslash M(\mathbb{A})^1\Big)$. Let $E(g,\phi,\lambda)$ be an Eisenstein series associated to a cusp form $\phi\in\mathcal{A}_P(G)_\sigma := A(U(\mathbb{A})M(F)\backslash G(\mathbb{A}))$. Then*

$$I_G^T\Big(E(g,\phi,\lambda)\Big)=0$$

for all λ such that $E(g,\phi,\lambda)$ and $I_G^T(E(g,\phi,\lambda))$ are defined.

We believe that this principle is the main reason behind special yet very important relations among multiple and ordinary zeta functions in literature.

4.5 Eisenstein Periods for Cusp Forms

As a direct consequence of Bernstein principle, with the explicit formula for constant terms of Eisenstein series associated to cusp forms, we have the following closed formula for Eisenstein periods associated with cusp forms:

Theorem. ([JLR]) *Fix a sufficiently positive $T \in \mathfrak{a}_0^+$. Let $P = MU$ be a parabolic subgroup and σ an irreducible* **cuspidal** *representation in $L^2(M(F)\backslash M(\mathbb{A})^1)$. Let $E(g,\phi,\lambda)$ be an Eisenstein series associated to $\phi \in \mathcal{A}_P(G)_\sigma$. Then the integration $\int_{G(F)\backslash G(\mathbb{A})} \Lambda^T E(g,\phi,\lambda)\,dg$ is equal to*
(1) *0 if $P \neq P_0$ is not minimal;* and
(2) $\mathrm{Vol}\Big(\Big\{\sum_{\alpha\in\Delta_0} a_\alpha\alpha^\vee : a_\alpha \in [0,1)\Big\}\Big)\cdot \sum_{w\in W} \frac{e^{\langle w\lambda-\rho,T\rangle}}{\prod_{\alpha\in\Delta_0}\langle w\lambda-\rho,\alpha^\vee\rangle}$

$$\times \int_{M_0(F)\backslash M_0(\mathbb{A})^1\times K} \Big(M(w,\lambda)\phi\Big)(mk)\,dm\,dk,$$

if $P = P_0 = M_0U_0$ is minimal.

Consequently, from the very definition, we have the following
Theorem′. *Fix a convex polygon p such that $T(p) \in \mathfrak{a}_0$ is sufficiently positive. Then for a P-level cusp form ϕ, $L_{F,r}^{\leq p}(\phi,\lambda)$ is equal to*
(1) 0 *if P is not minimal;*
(2) $\mathrm{Vol}\Big(\Big\{\sum_{\alpha\in\Delta_0} a_\alpha\alpha^\vee : a_\alpha \in [0,1)\Big\}\Big)\cdot \sum_{w\in W} \frac{e^{\langle w\lambda-\rho,T\rangle}}{\prod_{\alpha\in\Delta_0}\langle w\lambda-\rho,\alpha^\vee\rangle}$

$$\times \int_{M_0(F)\backslash M_0(\mathbb{A})^1\times K} \Big(M(w,\lambda)\phi\Big)(mk)\,dm\,dk,$$

if $P = P_0 = M_0N_0$ is minimal.

We remind the reader that this result is merely the beginning of a quantitive study about our L-functions. Say,
(1) When the polygon p satisfies $T(p) \gg 0$, we still need to understand $L_{F,r}^{P',\leq p}(\phi,\lambda)$ where $P' \supset P$ is a parabolic subgroup of G;

(2) We have to weak the restriction that $T(p) \gg 0$; and

(3) We need also to study our non-abelian L-functions for principal lattices, i.e., that for general reductive groups.

In fact all this proves to be very interesting and quite fruitful: In (1), Artin L-functions naturally appears; to (2), the above discussion is valid as long as $T(p) \in \mathfrak{a}_0^+ \cup \{0\}$; and for (3), no essentially difficulties appear despite its generality.

4.6 Volumes of Truncated Fundamental Domains

As an application, we here recall a recent result of H. Kim and the author on the volumes of Arthur's truncated domains [KW].

4.6.1 The Results

Let then G be a split, semi-simple group of rank r. Recall that for sufficiently regular $T \in \mathfrak{a}_0^+$, Arthur's analytic truncation $\Lambda^T \mathbf{1}$ coincides with the characteristic function of a compact subset $\Sigma(T) := \left(G(F)\backslash G(\mathbb{A})^1\right)_T \subset G(F)\backslash G(\mathbb{A})^1$. (See e.g., 3.3.1.) Our result calculates the volume of $\Sigma(T)$. In particular, letting $T \to \infty$, we recover Langlands' famous formula on the volumes of fundamental domains $G(F)\backslash G(\mathbb{A})^1$.

To state the result, for any $w \in W$, let $J_w := \{\alpha \in \Delta_0 : w\alpha \in \Delta_0\}$ and P_w be the standard parabolic subgroup corresponding to J_w. Let $r_w := \{\alpha \in \Delta_0 : w\alpha < 0\}$, and, for $i > 1$, set

$$\begin{aligned} n_i :=& \#\{\alpha > 0 : \langle \rho, \alpha^\vee \rangle = i\} - \#\{\alpha > 0, : \langle \rho, \alpha^\vee \rangle = i-1\}, \\ n_{i,w} :=& \#\{\alpha > 0, w\alpha < 0 : \langle \rho, \alpha^\vee \rangle = i\} \\ & - \#\{\alpha > 0, w\alpha < 0 : \langle \rho, \alpha^\vee \rangle = i-1\}. \end{aligned}$$

As usual, denote by Δ_F the discriminant of F, r_2 the number of complex embeddings of F, and $\xi_F(1)$ the residue of the completed Dedekind zeta function $\xi_F(s)$ at $s = 1$.

Theorem. ([KW]) *For a sufficiently positive $T \in \mathfrak{a}_0^+$, the volume of the* *truncated fundamental domain* $\left(G(F)\backslash G(\mathbb{A})^1\right)_T$ *is given by*

$$\mathrm{Vol}\Big(\Big\{\sum_{\alpha\in\Delta_0} a_\alpha\alpha^\vee : a_\alpha\in[0,1)\Big\}\Big)\cdot(2\pi)^{r_2 r}\Delta_F^{-\frac{r}{2}}\prod_{i>1}\xi_F(i)^{-n_i}$$
$$\times\sum_{w\in W_0}(-1)^{\mathrm{rk}P_w}\frac{\xi_F(1)^{r_w}\prod_{i>1}\xi_F(i)^{n_{i,w}}}{\prod_{\alpha\in\Delta_0-wJ_w}(1-\langle w\rho,\alpha^\vee\rangle)}\cdot e^{\langle w\rho-\rho,T\rangle}.$$

Note that $\Sigma(\infty)=G(F)\backslash G(\mathbb{A})^1$, and in the above formula, as $T\to\infty$, only the term corresponding to $w=1$ survives. In this way, we then recover Langlands' formula on volumes of fundamental domains;

Corollary. ([La2]) *If G is split and semi-stable, then the volume of a fundamental domain $\mathfrak{F}$ for $G(F)\backslash G(\mathbb{A})^1$ is given by*

$$\mathrm{Vol}(\mathfrak{F})=\mathrm{Vol}\Big(\Big\{\sum_{\alpha\in\Delta_0} a_\alpha\alpha^\vee : a_\alpha\in[0,1)\Big\}\Big)$$
$$\times(2\pi)^{r_2 r}\Delta_F^{-\frac{r}{2}}\cdot\prod_{i>1}\xi_F(i)^{-n_i}.$$

4.6.2 The Proof

Apply Theorem 4.5 to the special case where G is a split, semi-simple group and $\varphi=1$. Moreover, note that the constant term of $E(g,1,\lambda)$ along the Borel subgroup is given by

$$E_B(g,1,\lambda)=\sum_{w\in W}M(w,\lambda)e^{\langle w\lambda+\rho,H(g)\rangle},$$

where, by Gindikin-Karpelevich formula [La4],

$$M(w,\lambda)=\prod_{\alpha>0,w\alpha<0}\frac{\xi_F(\langle\lambda,\alpha^\vee\rangle)}{\xi_F(\langle\lambda,\alpha^\vee\rangle+1)}.$$

Consequently, we have

$$\int_{G(F)\backslash G(\mathbb{A})^1}\wedge^T E(g,1,\lambda)\,dg$$
$$=v\sum_{w\in\Omega}\frac{e^{\langle w\lambda-\rho,T\rangle}}{\prod_{\alpha\in\Delta_0}\langle w\lambda-\rho,\alpha^\vee\rangle}M(w,\lambda)\int_{M(F)\backslash M(\mathbb{A})^1}dm. \qquad (*)$$

since $e^{\langle w\lambda+\rho, H(m)\rangle} = 1$ for $m \in M(\mathbb{A})^1$. (Here $v := \mathrm{Vol}\Big(\Big\{ \sum_{\alpha\in\Delta_0} a_\alpha \alpha^\vee : a_\alpha \in [0,1)\Big\}\Big)$ appears because the measure on $\mathfrak{a}_0$ is defined using co-weights $\big\{\varpi_\alpha^\vee : \alpha \in \Delta_0\big\}$. Indeed, v is equal to the order of the fundamental group of Φ [Hu, page 68] for simply laced groups. For example, $v = n$ for SL_n.)

With this, we then look at the residue at $\lambda = \rho$. By Langlands [La2], we know that the residue of $E(g,1,\lambda)$ at $\lambda = \rho$ is constant. Hence this residue can be obtained by the residue of the above constant term along the Borel subgroup. To write it down explicitly, note that $\langle \rho, \alpha^\vee\rangle = 1$ for $\alpha \in \Delta_0$ and $\langle \rho, \alpha^\vee\rangle > 1$ if $\alpha \notin \Delta_0$, $\alpha > 0$. Hence

$$\mathrm{Res}_{\lambda=\rho} E(g,1,\lambda) = \lim_{\lambda\to\rho} \prod_{\alpha\in\Delta_0} \Big(\langle \lambda, \alpha^\vee\rangle - 1\Big) \cdot M(w_0, \lambda)$$

where w_0 is the longest Weyl group element. Since

$$M(w_0,\lambda) = \prod_{\alpha>0} \frac{\xi_F(\langle\lambda,\alpha^\vee\rangle)}{\xi_F(\langle\lambda,\alpha^\vee\rangle+1)},$$

we have

$$\mathrm{Res}_{\lambda=\rho} E(g,1,\lambda) = \xi_F(1)^r \cdot \prod_{i>1} \xi_F(i)^{n_i},$$

where r denotes the rank of G.

Now by 3.3.1,

$$\int_{G(F)\backslash G(\mathbb{A})^1} \wedge^T E(g,1,\lambda)\, dg = \int_{\Sigma(T)} E(g,1,\lambda)\, dg.$$

Hence

$$\begin{aligned} \mathrm{Res}_{\lambda=\rho}\Big(\int_{G(F)\backslash G(\mathbb{A})^1} \wedge^T E(g,1,\lambda)\, dg\Big) &= \int_{\Sigma(T)} \mathrm{Res}_{\lambda=\rho} E(g,1,\lambda)\, dg \\ &= \mathrm{Vol}(\Sigma(T)) \cdot \mathrm{Res}_{\lambda=\rho} E(g,1,\lambda) = \xi_F(1)^r \prod_{i>1} \xi_F(i)^{n_i} \cdot \mathrm{Vol}(\Sigma(T)). \end{aligned} \qquad (*)$$

On the other hand, by Theorem 4.5, the residue at $\lambda = \rho$ of the left hand side of $(*)$ is

$$\mathrm{Vol}\Big(T(F)\backslash T(\mathbb{A})^1\Big) \cdot \lim_{\lambda\to\rho} \Big(v \sum_{w\in W} e^{\langle w\lambda-\rho, T\rangle} M(w,\lambda) \prod_{\alpha\in\Delta_0} \frac{\langle\lambda,\alpha^\vee\rangle - 1}{\langle w\lambda,\alpha^\vee\rangle - 1}\Big).$$

Here following Tate [T], $\mathrm{Vol}\big(T(F)\backslash T(\mathbb{A})^1\big) = \big(\mathrm{Vol}(F^*\backslash \mathbf{I}_F^1)\big)^r = \zeta_F(1)^r$ with $\zeta_F(1) := \mathrm{Res}_{s=1}\zeta_F(s)$. (See e.g., [T].)

Note also that $\langle w\lambda, \alpha^\vee\rangle = \langle \lambda, (w^{-1}\alpha)^\vee\rangle$, and for a given $w \in W$, the sets $\{\alpha \in \Delta_0 : w\alpha < 0\}$ and $\{\alpha \in \Delta_0 : w\alpha \in \Delta_0\}$ are disjoint. Thus, if we let W_0 be the set of all $w \in W$ such that

$$\{\alpha \in \Delta_0 : w\alpha < 0\} \cup \{\alpha \in \Delta_0 : w\alpha \in \Delta_0\} = \Delta_0,$$

and $r_w := \#\{\alpha \in \Delta_0 : w\alpha < 0\}$, then we have

$$\lim_{\lambda\to\rho}\Big(\sum_{w\in W} e^{\langle w\lambda-\rho,T\rangle} M(w,\lambda)\prod_{\alpha\in\Delta_0}\frac{\langle\lambda,\alpha^\vee\rangle-1}{\langle w\lambda,\alpha^\vee\rangle-1}\Big)$$
$$= \sum_{w\in W_0} e^{\langle w\rho-\rho,T\rangle}\frac{\xi_F(1)^{r_w}\prod_{i>1}\xi_F(i)^{n_{i,w}}}{\prod_{\alpha\in\Delta_0, w^{-1}\alpha\notin\Delta_0}(\langle w\rho,\alpha^\vee\rangle-1)}.$$

Therefore,

$$\mathrm{Vol}\big(\Sigma(T)\big) = v\cdot(2\pi)^{r_2 r}\Delta_F^{-\frac{r}{2}}\prod_{i>1}\xi_F(i)^{-n_i}$$
$$\times\sum_{w\in W_0}\xi_F(1)^{r_w}\frac{\prod_{i>1}\xi_F(i)^{n_{i,w}}}{\prod_{\alpha\in\Delta_0, w^{-1}\alpha\notin\Delta_0}(\langle w\rho,\alpha^\vee\rangle-1)}\cdot e^{\langle w\rho-\rho,T\rangle}.$$

To go further, let us recall that:
there is a one-to-one and onto correspondence between the elements of W_0 and the standard parabolic subgroups of G.
Indeed, given $w \in W_0$, let $J_w = \{\alpha \in \Delta : w\alpha \in \Delta\}$ and denote by P_w the standard parabolic subgroup corresponding to J_w. Then $P := P_w$ is the standard parabolic subgroup of G associated to w. Conversely, any standard parabolic subgroup P of G is associated to a unique subset $J \subset \Delta$. Set then $D_J := \{w \in W/W_J : wJ > 0\}$. It is known that the set D_J gives distinguished coset representatives of W/W_J. Let w_J be the element of the maximal length in D_J. Then $w_J \in W_0$ and in fact $W_0 = \{w_J : J \subset \Delta\}$.

We claim that *if $w \in W_0$, $\langle w\rho, \alpha^\vee\rangle < 0$ for all $\alpha \in \Delta - wJ_w$*: if $\alpha \in \Delta - wJ_w$, then $w_l(\alpha) = -\beta$ for some $\beta \in \Delta$. But by [Ca], if w_l is the element of the maximal length in W, and $w_{l,J}$ is the element of the maximal length in W_J, then $w_J = w_l w_{l,J}$. So $w^{-1}(\alpha) = -w_{l,J_w}(\beta)$ since $w^{-1} = w_{l,J_w}w_l$. If now $\beta \in J_w$, then $-w_{l,J_w}(\beta) \in J_w$ and $w^{-1}(\alpha) \in J_w$. But $w^{-1}(\alpha) \notin J_w$, since

$\alpha \notin wJ_w$. Contradiction. Consequently, $\beta \notin J_w$ and $w_{l,J_w} > 0$. Therefore $w^{-1}(\alpha) < 0$. As such,

$$\prod_{\alpha\in\Delta_0, w^{-1}\alpha\notin\Delta_0} \Big(\langle w\rho, \alpha^\vee\rangle - 1\Big) = (-1)^{\mathrm{rk}P_w} \prod_{\alpha\in\Delta - wJ_w} \Big(1 - \langle w\rho, \alpha^\vee\rangle\Big).$$

This then completes the proof of the Theorem.

The result may also be written as follows, oriented by standard parabolic subgroups;

Theorem′. ([KW]) *For a sufficiently regular T,*

$$\mathrm{Vol}(\Sigma(T)) = \mathrm{Vol}\Big(\Big\{\sum_{\alpha\in\Delta_0} a_\alpha\alpha^\vee : a_\alpha \in [0,1)\Big\}\Big) \cdot (2\pi)^{r_2 r} d_F^{-\frac{r}{2}}$$
$$\times \sum_{J\subset\Delta_0} (-1)^{\mathrm{rk}\,P_J} \frac{\xi_F(1)^{\mathrm{rk}P_J} \prod_{i>1} \xi_F(i)^{-n_i+n_{i,w_J}}}{\prod_{\alpha\in\Delta_0 - w_J J}(1 - \langle w_J\rho, \alpha^\vee\rangle)} \cdot e^{\langle w_J\rho - \rho, T\rangle}.$$

Remark. The formula, as it stands, suggests that the volumes of cuspidal regions corresponding to parabolic subgroups can be written in terms of special zeta values as well.

Proof. By the Theorem, this is a direct consequence of the fact that for $w \in W_0$, $r_w + \sum n_{i,w} = 0$, and $-\sum n_i = r$, the rank of G. Indeed, $-\sum n_i$ is the number of positive roots such that $\langle\rho, \alpha^\vee\rangle = 1$. It is exactly r. Similarly, $-\sum n_{i,w}$ is the number of positive roots such that $w\alpha < 0$, and $\langle\rho, \alpha^\vee\rangle = 1$. It is exactly $r_w = \mathrm{rank}P_w$. So $r_w + \sum(-n_i + n_{i,w}) = r$ for each $w \in W_0$. □

4.7 When Parameters Are Small

By definition,

$$\Lambda^T\phi(g) := \sum_P (-1)^{d(P)-d(G)} \sum_{\delta\in P(F)\backslash G(F)} \phi_P(\delta g) \cdot \widehat{\tau}_P\Big(H(\delta g) - T\Big).$$

And, as indicated in 3.3.1, to have the formula

$$\int_{\Sigma(T)} \phi(g) d\mu(g) = \int_{G(F)\backslash G(\mathbb{A})^1} \Lambda^T\phi(g) d\mu(g),$$

we need the following basic properties of the truncation operator Λ^T.

Theorem. (Arthur) *For T sufficiently large,*
(1) $\wedge^T \circ \wedge^T = \wedge^T$;
(2) $\wedge^T$ *is self-adjoint;*
(3) $\wedge^T \phi$ *is rapidly decreasing, if ϕ is automorphic on $G(F)\backslash G(\mathbb{A})$;*
(4) $\wedge^T \mathbf{1}$ *is a characteristic function of a compact subset $\Sigma(T)$ of $G(F)\backslash G(\mathbb{A})^1$.*

However for geometric truncation using stability, parameters T may not be very regular. So we need to check whether properties (1) ~ (4) still hold for such Ts.

Let us look at (1) first. For T sufficiently large, this is proved by Author based on the following

Lemma. ([Ar2]) *Fix $P_1 = M_1 U_1$. Then for $\phi \in C(G(F)\backslash G(\mathbb{A})^1)$,*

$$\int_{U_1(F)\backslash U_1(\mathbb{A})} \Lambda^T \phi(u_1 g)\, du_1 = 0$$

unless $\varpi(H(g) - T) < 0$ for each $\varpi \in \widehat{\Delta}_{P_1}$.

By examining the proof of Arthur, we see that for SL_n, if T is taken to be in the positive chamber or $T = 0$, the whole proof still works. Indeed, only the part at p. 91 of Arthur's proof [Ar2] should be checked. But then this is quite obvious. (For general reductive groups, the extra element 0 should be replaced by a certain naturally defined element. For details see [Ar5].)

Next, let us turn to (2) and (3). Clearly, formal calculations in Arthur's proof work for any T. So the point here is really about the convergence of all integrals involved.

Put this in a more general term, convergence for the integrals appeared are really the central part of Arthur's truncation theory. Sure, one more key ingredient needed here is the combinatorial technique used. Luckily, this elementary yet highly non-trivial combinatorial part can be formally applied even when T is not that large. All in all, the up-shot is the so-called Arthur-Langlands partition for the fundamental domain listed at 3.2.8. It is here that we use the condition T being sufficiently regular to make sure that the function $F^P(g,T)$ is a characteristic function of a certain compact subset. While it is a bit involved, however, by checking the proof on convergence very carefully, we can conclude that what one really needs in proving (2) and (3) for smaller T is that, as a function of g, $F^P(g,T)$ is bounded and compactly supported modulo $U(\mathbb{A})M(F)$ uniformly for T

varying in a compact subset. But this is then exactly what Prop. 2 of 3.3.2 provides.

Finally, Let us turn our attention to (4). This is a very hard one: In our treatment, this is proved using the Global Bridge between geometric truncation and analytic truncation established in 3.4.2 and a micro-global principle called Fundamental Relation established in 3.1.2.

As such, we have proved the following

Theorem. *Let $P = MN$ be a parabolic subgroup of $G = SL_n$ and let (M, σ) be a cuspidal datum. Let $E(\varphi, \lambda)(g)$ be the Eisenstein series associated to a P-level cusp form $\varphi \in \mathcal{A}_P^0(G)_\sigma$. Let T be in the positive chamber or $T = 0$. Then*

$$\int_{G(F)\backslash G(\mathbb{A})^1} \wedge^T E(g, \varphi, \lambda)\, dg$$

(i) *is equal to zero if P is not a minimal parabolic;*
(ii) *is equal to*

$$\mathrm{Vol}\Big(\Big\{\sum_{\alpha\in\Delta_0} a_\alpha\alpha^\vee : 0 \le a_\alpha < 1\Big\}\Big)$$
$$\times \sum_{w\in W} \frac{e^{\langle w\lambda-\rho, T\rangle}}{\prod_{\alpha\in\Delta_0}\langle w\lambda-\rho, \alpha^\vee\rangle} \int_{M(F)\backslash M(\mathbb{A})^1\times K} M(w,\lambda)\varphi(mk)\, dmdk,$$

if P is minimal.

Consequently, we also have the following

Theorem. *For any T in the positive chamber $\mathfrak{a}_0^+$ or $T = 0$,*

$$\mathrm{Vol}\big(\Sigma(T)\big) = \mathrm{Vol}\Big(\Big\{\sum_{\alpha\in\Delta_0} a_\alpha\alpha^\vee : a_\alpha \in [0,1)\Big\}\Big)\cdot (2\pi)^{r_2 r} d_F^{-\frac{r}{2}}$$
$$\times \sum_{J\subset\Delta_0} (-1)^{\mathrm{rk}\,P_J} \frac{\xi_F(1)^{\mathrm{rk}P_J} \prod_{i>1}\xi_F(i)^{-n_i+n_{i,w_J}}}{\prod_{\alpha\in\Delta_0-w_J J}(1-\langle w_J\rho, \alpha^\vee\rangle)} \cdot e^{\langle w_J\rho-\rho, T\rangle}.$$

4.8 Fundamental Relations Between Special Zeta Values

By Corollary 4.7, with a detailed calculation, we then obtain the following result on the volumes of moduli space $\mathcal{M}_{F,r}(|\Delta_F|^{\frac{r}{2}})$, which reveals certain fundamental relations among special values of classical Dedekind and our zeta functions.

Theorem. *Let* $\xi_{F,r}(1) := \mathrm{Res}_{s=1}\xi_{F,r}(s) = \mathrm{Vol}\Big(\mathcal{M}_{F,r}(|\Delta_F|^{\frac{r}{2}})\Big)$.

Level 1. $\xi_{F,1}(1) = \xi_F(1);$

Level 2. $\xi_{F,2}(1) = \xi_F(2)\xi_F(1) - \xi_F(1)\xi_F(1);$

Level 3.

$$\xi_{F,3}(1) = \xi_F(3)\xi_F(2)\xi_F(1) - \frac{1}{3}\xi_F(2)\xi_F(1)^2 - \frac{1}{3}\xi_F(2)\xi_F(1)^2 + \frac{1}{4}\xi_F(1)^3;$$

Level 4.

$$\begin{aligned}\xi_{F,4}(1) &= \xi_F(4)\xi_F(3)\xi_F(2)\xi_F(1)\\ &- \frac{1}{4}\xi_F(3)\xi_F(2)\xi_F(1)^2 - \frac{1}{4}\xi_F(2)^2\xi_F(1)^2 - \frac{1}{4}\xi_F(3)\xi_F(2)\xi_F(1)^2\\ &+ \frac{1}{6}\xi_F(2)^2\xi_F(1)^2 + \frac{1}{9}\xi_F(2)\xi_F(1)^2 + \frac{1}{6}\xi_F(2)\xi_F(1)^3\\ &- \frac{1}{8}\xi_F(1)^4;\end{aligned}$$

Level r. *The* r*-level relation can be described as follows.*
(1) $\xi_{F,r}(1)$ *is a degree* r *homogenous polynomial in* $\xi_F(i)$*s, for* $i = 1, 2, \ldots, r$*;*
(2) *there is a one-to-one and onto correspondence between the terms of this polynomial and partitions of* r*, or better, standard parabolic subgroups. More precisely, in the case* $r = r_1 + r_2 + \cdots + r_N$*, the corresponding term consists of three parts, that is,*

$$\text{sign} \cdot \text{coefficient} \cdot \text{product of zeta values;}$$

(i) *the sign is given by* $(-1)^{N-1}$*;*
(ii) *The coefficient is* $\dfrac{1}{(r_1 + r_2)(r_2 + r_3)\cdots(r_{N-1} + r_N)}$; and
(iii) *the product of zeta values is given by: If* $r_i \geq 2$*, the product* $\xi_F(r_i)\xi_F(r_i - 1)\cdots\xi_F(2)$ *does appear. Moreover, these subfactors together with certain powers of* $\xi_F(1)$ *then gives all the product of zeta values (under the constrain that the final product is of degree* r*).*

It is well known that following a result of Siegel, in order to read the special value $\xi_F(n)$, we have to go to rank n lattices. The above relations may be viewed as a refined version of this basic result. Roughly put, our result says that

(i) among all rank n-lattices, only stable lattices are essential;
(ii) other rank n lattices can be built up from stable lattices of lower ranks;
(iii) corresponding modular points of lattices in (ii) are sitting in cuspidal regions; and
(iv) volumes of cupidal regions can be expressed in terms of $\xi_F(m)$'s, $m < n$.

We end this discussion by pointing out that our result above is closely related with the result of Harder and Narasimhan ([HN]) on counting stable bundles over curves on finite fields: Their method, an arithmetic geometry one, is algebraic while ours, a geometric arithmetic one, is analytic.

5 High Rank Zetas and Eisenstein Series

5.1 High Rank Zetas for Number Fields

Let K be an algebraic number field (of finite degree n) with Δ_K the absolute value of its discriminant. Denote by O_K the ring of integers. For a fixed positive integer $r \in \mathbb{N}$, denote by $\mathcal{M}_{F,r}$ the moduli space of semi-stable O_K-lattices of rank r with $d\mu$ the associated (Tamagawa type) measure (induced from that on GL). Recall that for each $\Lambda \in \mathcal{M}_{K,r}$, the associated *0-th geo-arithmetical cohomology* $h^0(K,\Lambda)$ is

$$h^0(K,\Lambda) := \log\Big(\sum_{x\in\Lambda}\exp\Big(-\pi\sum_{\sigma:\mathbb{R}}\|x_\sigma\|_{\rho_\sigma}^2 - 2\pi\sum_{\sigma:\mathbb{C}}\|x_\sigma\|_{\rho_\sigma}^2\Big)\Big)$$

where $x = (x_\sigma)_{\sigma\in S_\infty}$ and $(\rho_\sigma)_{\sigma\in S_\infty}$ denote the σ-component of the metric $\rho = \rho_\Lambda$ determined by the lattice Λ with S_∞ a collection Archimedean places of K; and that the *rank r zeta function $\xi_{K,r}(s)$ of K* is the integration

$$\xi_{K,r}(s) := \int_{\Lambda\in\mathcal{M}_{K,r}} \left(e^{h^0(K,\Lambda)-1}\right)\cdot\left(e^{-s}\right)^{-\log \mathrm{Vol}(\Lambda)}\, d\mu(\Lambda),\ \mathrm{Re}\,(s) > 1,$$

or the same,

$$\xi_{K,r}(s) := \left(\Delta_K^{\frac{r}{2}}\right)^s\cdot\int_{\Lambda\in\mathcal{M}_{K,r}} \left(e^{h^0(K,\Lambda)-1}\right)\cdot\left(e^{-s}\right)^{\deg(\Lambda)}\, d\mu(\Lambda), \mathrm{Re}\,(s) > 1.$$

Theorem. (1) (**Meromorphic Continuation**) *$\xi_{K,r}(s)$ is well-defined when* $\mathrm{Re}\,(s) > 1$ *and admits a meromorphic continuation, denoted also by $\xi_{K,r}(s)$, to the whole complex s-plane;*
(2) (**Functional Equation**) $\xi_{K,r}(1-s) = \xi_{K,r}(s)$;
(3) (**Singularities & Residues**) *$\xi_{K,r}(s)$ has two singularities, all simple poles, at $s = 0, 1$, with the same residues* $\mathrm{Vol}\big(\mathcal{M}_{K,r}([\Delta_K^{\frac{r}{2}}])\big)$.

Here, as before, $\mathcal{M}_{K,r}([\Delta_K^{\frac{r}{2}}])$ denotes the moduli space of rank r semi-stable O_K-lattices whose volumes are fixed to be $\Delta_K^{\frac{r}{2}}$. The proof of this theorem is fairly standard now since we have a good arithmetic cohomology which satisfies the Serre duality and the Riemann-Roch formula. For details, see e.g., 1.4.2.

5.2 Space of O_K-Lattices

Recall that an O_K-lattice Λ consists of a underlying projective O_K-module P and a metric structure on the space $V = \Lambda \otimes_{\mathbb{Z}} \mathbb{R} = \prod_{\sigma\in S_\infty} V_\sigma$. Moreover, in assuming that the O_K-rank of P is r, we can identify P with one of the $P_i := P_{r;\mathfrak{a}_i} := O_K^{(r-1)} \oplus \mathfrak{a}_i$, where $\mathfrak{a}_i, i = 1, \dots, h$, are chosen integral O_K-ideals so that $\{[\mathfrak{a}_1], [\mathfrak{a}_2], \dots, [\mathfrak{a}_h]\} = CL(K)$ the class group of K.

With this said, following Minkowski, we obtain a natural embedding for P:

$$P := O_K^{(r-1)} \oplus \mathfrak{a} \hookrightarrow K^{(r)} \hookrightarrow \left(\mathbb{R}^{r_1} \times \mathbb{C}^{r_2}\right)^r \cong \left(\mathbb{R}^r\right)^{r_1} \times \left(\mathbb{C}^r\right)^{r_2},$$

which is simply the space $V = \Lambda \otimes_{\mathbb{Z}} \mathbb{R}$ above. As a direct consequence, our lattice Λ then is determined by a metric structure on $V = \prod_{\sigma\in S_\infty} V_\sigma$, or the same, on $(\mathbb{R}^r)^{r_1} \times (\mathbb{C}^r)^{r_2}$. It is well known that all these metrices are parametrized by the space

$$\left(GL(r,\mathbb{R})\big/O(r)\right)^{r_1} \times \left(GL(r,\mathbb{C})\big/U(r)\right)^{r_2}.$$

Next let us shift our discussion from GL to SL. For this, we start with a local discussion on O_K-lattice structures: For complex places τ, by fixing a branch of the n-th root, we get

$$GL(r,\mathbb{C})\big/U(r) \cong \left(SL(r,\mathbb{C})\big/SU(r)\right) \times \mathbb{R}_+^*;$$

while for real places σ, we have

$$GL(r,\mathbb{R})\big/O(r) \cong \left(SL(r,\mathbb{R})\big/SO(r)\right) \times \mathbb{R}_+^*.$$

Consequently, metric structures on $V = \prod_{\sigma\in S_\infty} V_\sigma \simeq (\mathbb{R}^r)^{r_1} \times (\mathbb{C}^r)^{r_2}$ are parametrized by

$$\left(\left(SL(r,\mathbb{R})\big/SO(r)\right)^{r_1} \times \left(SL(r,\mathbb{C})\big/SU(r)\right)^{r_2}\right) \times (\mathbb{R}_+^*)^{r_1+r_2}.$$

Furthermore, when we work with O_K-lattice strucures $\mathbf{\Lambda} = \mathbf{\Lambda}(P)$ on P, from above parametrized space of metric structures on $V = \prod_{\sigma\in S_\infty} V_\sigma$, we need to also factor out $GL(P)$, the automorphism group $\mathrm{Aut}_{O_K}(O_K^{(r-1)} \oplus \mathfrak{a})$ of $O_K^{(r-1)} \oplus \mathfrak{a}$ as O_K-modules, to get

$$GL(P)\Big\backslash \left(\left(\left(SL(r,\mathbb{R})\big/SO(r)\right)^{r_1} \times \left(SL(r,\mathbb{C})\big/SU(r)\right)^{r_2}\right) \times (\mathbb{R}_+^*)^{r_1+r_2}\right).$$

Hence we need
(a) to study the structure of the group $\mathrm{Aut}_{O_K}(O_K^{(r-1)} \oplus \mathfrak{a})$ in terms of SL and units; and
(b) to see how this group acts on the space of metric structures

$$\Big(\big(SL(r,\mathbb{R})/SO(r)\big)^{r_1} \times \big(SL(r,\mathbb{C})/SU(r)\big)^{r_2}\Big) \times (\mathbb{R}_+^*)^{r_1+r_2}.$$

View $\mathrm{Aut}_{O_K}(O_K^{(r-1)} \oplus \mathfrak{a})$ as a subgroup of $GL(r, K)$. Then, $\mathrm{Aut}_{O_K}(O_K^{(r-1)} \oplus \mathfrak{a})$

$$= \left\{ A \in GL(r, K) \cap \begin{pmatrix} & & & \mathfrak{a} \\ & O_K & & \vdots \\ & & & \mathfrak{a} \\ \mathfrak{a}^{-1} & \dots & \mathfrak{a}^{-1} & O_K \end{pmatrix} : \det A \in U_K \right\}.$$

Introduce groups $GL^+(r,\mathbb{R}) := \{g \in GL(r,\mathbb{R}) : \det g > 0\}$ and $O^+(r) := \{A \in O(r,\mathbb{R}) : \det g > 0\}$, and also the subgroup $\mathrm{Aut}^+_{O_K}(O_K^{(r-1)} \oplus \mathfrak{a})$ of $\mathrm{Aut}_{O_K}(O_K^{(r-1)} \oplus \mathfrak{a})$ consisting of these elements whose local determinants at real places are all positive. Then, by checking directly, we obtain a natural identification of quotient spaces between

$$\mathrm{Aut}_{O_K}(O_K^{(r-1)} \oplus \mathfrak{a})\backslash\Big((GL(r,\mathbb{R})/O(r))^{r_1} \times (GL(r,\mathbb{C})/U(r))^{r_2}\Big)$$

and

$$\mathrm{Aut}^+_{O_K}(O_K^{(r-1)} \oplus \mathfrak{a})\backslash\Big((GL^+(r,\mathbb{R})/O^+(r))^{r_1} \times (GL(r,\mathbb{C})/U(r))^{r_2}\Big).$$

Recall that for a unit $\varepsilon \in U_K$,
(a) $\mathrm{diag}(\varepsilon, \dots, \varepsilon) \in \mathrm{Aut}_{O_K}(O_K^{(r-1)} \oplus \mathfrak{a})$; and
(b) $\det \mathrm{diag}(\varepsilon, \dots, \varepsilon) = \varepsilon^r \in U_F^r := \{\varepsilon^r : \varepsilon \in U_K\}$.
Similar to the subgroup GL^+, introduce a subgroup U_K^+ of U_K by setting

$$U_K^+ := \{\varepsilon \in U_K : \varepsilon_\sigma > 0, \forall \sigma \text{ real}\}$$

so as to get a well-controlled subgroup $U_K^{r,+} := U_K^+ \cap U_K^r$: by Dirichlet's Unit Theorem, the quotient group $U_K^+/(U_K^+ \cap U_K^r)$ is finite.

As such, next we use $U_K^+ \cap U_K^r$ to decomposite the automorphism group $\mathrm{Aut}_{O_K}(O_K^{(r-1)} \oplus \mathfrak{a})$. Thus, choose elements

$$u_1, \dots, u_{\mu(r,K)} \in U_K^+$$

such that $\left\{[u_1], \ldots, [u_{\mu(r,K)}]\right\}$ gives a complete representatives of the finite quotient group $U_K^+/\left(U_K^+ \cap U_K^r\right)$, where $\mu(r, K)$ denotes the cardinality of the group $U_K^+/(U_K^+ \cap U_K^r)$. Set also

$$SL(O_K^{(r-1)} \oplus \mathfrak{a}) := SL(r, K) \cap GL(O_K^{(r-1)} \oplus \mathfrak{a}).$$

Then we have the following

Lemma. ([We6,8]) *There exist* $A_1, \ldots, A_{\mu(r,K)}$ *in* $GL^+(O_K^{(r-1)} \oplus \mathfrak{a})$ *such that*
(i) $\det A_i = u_i$ for $i = 1, \ldots, \mu(r, K)$; and
(ii) $\left\{A_1, \ldots, A_{\mu(r,K)}\right\}$ *is a completed representative of the quotient resulting from* $\mathrm{Aut}^+_{O_K}\left(O_K^{(r-1)} \oplus \mathfrak{a}\right)$ *modulo* $SL(O_K^{(r-1)} \oplus \mathfrak{a}) \times \left(U_K^{r,+} \cdot \mathrm{diag}(1, \ldots, 1)\right)$.

That is to say, for automorphism groups,
(a) $\mathrm{Aut}^+_{O_K}(O_K^{(r-1)} \oplus \mathfrak{a})$ is naturally identified with the disjoint union

$$\cup_{i=1}^{\mu(r,K)} A_i \cdot \left(SL(O_K^{(r-1)} \oplus \mathfrak{a}) \times \left(U_K^{r,+} \cdot \mathrm{diag}(1, \ldots, 1)\right)\right);$$

and, consequently,
(b) The O_K-lattice structures $\mathbf{\Lambda}(P)$ on the projective O_K-module $P = O_K^{(r-1)} \oplus \mathfrak{a}$ are parametrized by the disjoint union

$$\cup_{i=1}^{\mu(r,K)} A_i \backslash \left(\left(SL(O_K^{(r-1)} \oplus \mathfrak{a}) \backslash \left(\left(SL(r, \mathbb{R})/SO(r)\right)^{r_1} \times \left(SL(r, \mathbb{C})/SU(r)\right)^{r_2}\right)\right) \times \left(\left|U_K^r \cap U_K^+\right| \backslash \left(\mathbb{R}_+^*\right)^{r_1+r_2}\right)\right).$$

Therefore, to understand the space of O_K-lattice structures, beyond the spaces $SL(r, \mathbb{R})/SO(r)$ and $SL(r, \mathbb{C})/SU(r)$, we further need to study
(i) the quotient space $\left|U_K^r \cap U_K^+\right| \backslash \left(\mathbb{R}_+^*\right)^{r_1+r_2}$, which is more or less standard (see e.g. [Ne] or [We6,8]); and more importantly,
(ii) the quotient space

$$SL(O_K^{(r-1)} \oplus \mathfrak{a}) \backslash \left(\left(SL(r, \mathbb{R})/SO(r)\right)^{r_1} \times \left(SL(r, \mathbb{C})/SU(r)\right)^{r_2}\right).$$

5.3 High Rank Zetas and Epstein Zetas

Choose integral O_K-ideals $\mathfrak{a}_1 = O_K, \mathfrak{a}_2, \dots, \mathfrak{a}_h$ such that the ideal class group $CL(K)$ is given by $\{[\mathfrak{a}_1], \dots, [\mathfrak{a}_h]\}$, and that any rank r projective O_K-module P is isomorphic to $P_{\mathfrak{a}_i}$ for a certain $i, 1 \leq i \leq h$. Here, $P_{\mathfrak{a}} := P_{r,\mathfrak{a}} := O_K^{(r-1)} \oplus \mathfrak{a}$ for a fractional O_K-ideal $\mathfrak{a}$. Introduce the partial high rank zeta function $\xi_{K,r;\mathfrak{a}}(s)$ associated to $\mathfrak{a}$ by setting

$$\xi_{K,r;\mathfrak{a}}(s) := \int_{\mathcal{M}_{K,r}(\mathfrak{a})} \left(e^{h^0(K,\Lambda)} - 1\right) \cdot \left(e^{-s}\right)^{-\log \mathrm{Vol}(\Lambda)} d\mu(\Lambda), \mathrm{Re}\,(s) > 1$$

where $\mathcal{M}_{K,r}(\mathfrak{a})$ denotes the part of the moduli space of semi-stable O_K-lattices whose corresponding modular points lie in

$$\left(SL(O_K^{(r-1)} \oplus \mathfrak{a})\Big\backslash\left(\left(SL(r,\mathbb{R})/SO(r)\right)^{r_1} \times \left(SL(r,\mathbb{C})/SU(r)\right)^{r_2}\right)\right)$$
$$\times \left(|U_K^r \cap U_K^+|\backslash(\mathbb{R}_+^*)^{r_1+r_2}\right).$$

Accordingly, for $\mathrm{Re}\,(s) > 1$, define the *completed Epstein zeta function* $\hat{E}_{K,r;\mathfrak{a}}(s)$ by

$$\hat{E}_{K,r;\mathfrak{a}}(s) := \left(\pi^{-\frac{rs}{2}}\Gamma\left(\frac{rs}{2}\right)\right)^{r_1} \cdot \left((2\pi)^{-rs}\Gamma(rs)\right)^{r_2}$$
$$\times \left[\left(N(\mathfrak{a}) \cdot \Delta_K^{\frac{r}{2}}\right)^s \cdot \sum_{x \in (\Lambda\backslash\{0\})/U_{r,F}^+} \frac{1}{\|x\|_\Lambda^{rs}}\right].$$

With this, from the discussion above, by applying the Mellin transform and the formula

$$\int_0^\infty e^{-At^B} t^s \frac{dt}{t} = \frac{1}{B} \cdot A^{-\frac{s}{B}} \cdot \Gamma\left(\frac{s}{B}\right)$$

(whenever both sides make sense,) we obtain the following

Proposition. ([We6,8]) (1) (**Decomposition**) *The rank r zeta funtion of K admits a natural decomposition*

$$\xi_{K,r}(s) = \mu(r,K) \cdot \sum_{i=1}^h \xi_{K,r;\mathfrak{a}_i}(s);$$

(2) (**High Rank Zeta = Integration of Epstein Zeta**) *The partial rank r zeta function* $\xi_{K,r;\mathfrak{a}}(s)$ *of K associated to* $\mathfrak{a}$ *is given by an integration of a completed Epstein type zeta function:*

$$\xi_{K,r;\mathfrak{a}}(s) = \Big(\frac{r}{2}\Big)^{r_1+r_2} \cdot \int_{\mathcal{M}_{F,r;\mathfrak{a}}[N(\mathfrak{a})\cdot\Delta_K^{\frac{r}{2}}]} \hat{E}_{K,r;\mathfrak{a}}(s)\, d\mu, \qquad \mathrm{Re}\,(s) > 1.$$

6 Stability and Distance to Cusps

Chapters 6, 7, 8 are dealing with rank two zeta functions.

6.1 Upper Half Plane

On the upper half plane $\mathcal{H} := \{z = x + iy \in \mathbb{C}, x \in \mathbb{R}, y \in \mathbb{R}_+^*\}$, the natural hyperbolic metric is given by the line element $ds^2 := \frac{dx^2+dy^2}{y^2}$ with volume form $d\mu := \frac{dx\wedge dy}{y^2}$. Consequently, the associated hyperbolic Laplace-Beltrami operator can be written as

$$\Delta := y^2\Big(\frac{\partial^2}{\partial x^2} + \frac{\partial^2}{\partial y^2}\Big).$$

The group $SL(2,\mathbb{R})$ naturally acts on $\mathcal{H}$ via:

$$M\,z := \frac{az+b}{cz+d}, \qquad \forall M = \begin{pmatrix} a & b \\ c & d \end{pmatrix} \in SL(2,\mathbb{R}), \;\; z \in \mathcal{H}.$$

The stablizer of $i = (0,1) \in \mathcal{H}$ is equal to $SO(2) := \{A \in O(2) : \det A = 1\}$. Since this action is transitive, we can identify the quotient $SL(2,\mathbb{R})/SO(2)$ with $\mathcal{H}$ with the quotient map induced from $SL(2,\mathbb{R}) \to \mathcal{H},\; g \mapsto g\cdot i$.

$\mathcal{H}$ admits the real line $\mathbb{R}$ as its boundary. Consequently, to compactify it, we add on it the real projective line $\mathbb{P}^1(\mathbb{R})$ with $\infty = \begin{bmatrix} 1 \\ 0 \end{bmatrix}$. Naturally, the above action of $SL(2,\mathbb{R})$ also extends to $\mathbb{P}^1(\mathbb{R})$ via

$$\begin{pmatrix} a & b \\ c & d \end{pmatrix}\begin{bmatrix} x \\ y \end{bmatrix} = \begin{bmatrix} ax+by \\ cx+dy \end{bmatrix}.$$

6.2 Upper Half Space

The 3-dimensional hyperbolic space may be written as

$$\begin{aligned}\mathbb{H} :=& \mathbb{C}\times]0,\infty[= \Big\{(z,r) : z = x+iy \in \mathbb{C}, r \in \mathbb{R}_+^*\Big\}\\ =&\Big\{(x,y,r) : x,y \in \mathbb{R}, r \in \mathbb{R}_+^*\Big\}.\end{aligned}$$

We will think of $\mathbb{H}$ as a subset of Hamilton's quaternions with $1,\ i,\ j,\ k$ the standard $\mathbb{R}$-basis. Write points P in $\mathbb{H}$ as

$$P = (z, r) = (x, y, r) = z + rj \qquad \text{where } z = x + iy,\ j = (0, 0, 1).$$

We equip $\mathbb{H}$ with the hyperbolic metric coming from the line element $ds^2 := \frac{dx^2+dy^2+dr^2}{r^2}$ with hyperbolic volume form $d\mu := \frac{dx\wedge dy\wedge dr}{r^3}$. Consequently, the hyperbolic Laplace-Beltrami operator associated to the hyperbolic metric ds^2 is simply

$$\Delta := r^2\Big(\frac{\partial^2}{\partial x^2} + \frac{\partial^2}{\partial y^2} + \frac{\partial^2}{\partial r^2}\Big) - r\frac{\partial}{\partial r}.$$

The natural action of $SL(2, \mathbb{C})$ on $\mathbb{H}$ and on its boundary $\mathbb{P}^1(\mathbb{C})$ may be described as follows: We represent an element of $\mathbb{P}^1(\mathbb{C})$ by $\begin{bmatrix} x \\ y \end{bmatrix}$ where $x, y \in \mathbb{C}$ with $(x, y) \neq (0, 0)$. Then the action of the matrix $M = \begin{pmatrix} a & b \\ c & d \end{pmatrix} \in SL(2, \mathbb{C})$ on $\mathbb{P}^1(\mathbb{C})$ is defined to be

$$\begin{bmatrix} x \\ y \end{bmatrix} \mapsto \begin{pmatrix} a & b \\ c & d \end{pmatrix}\begin{bmatrix} x \\ y \end{bmatrix} := \begin{bmatrix} ax + by \\ cx + dy \end{bmatrix}.$$

Moreover, if we represent points $P \in \mathbb{H}$ as quaternions whose fourth component equals zero, then the action of M on $\mathbb{H}$ is defined to be

$$P \mapsto M\,P := (aP + b)(cP + d)^{-1},$$

where the inverse on the right is taken in the skew field of quaternions.

Furthermore, with this action, the stabilizer of $j = (0, 0, 1) \in \mathbb{H}$ in $SL(2, \mathbb{C})$ is equal to $SU(2) := \{A \in U(2) : \det A = 1\}$. Since the action of $SL(2, \mathbb{C})$ on $\mathbb{H}$ is transitive, we obtain also a natural identification $\mathbb{H} \simeq SL(2, \mathbb{C})/SU(2)$ via the quotient map induced from $SL(2, \mathbb{C}) \to \mathbb{H},\ \ g \mapsto g \cdot j$.

6.3 Upper Half Space Model

Identifying $\mathcal{H}$ with $SL(2, \mathbb{R})/SO(2)$ and $\mathbb{H}$ with $SL(2, \mathbb{C})/SU(2)$, using the discussion in 5.2, we conclude that

$$\mathcal{M}_{K,2;\mathfrak{a}}[N(\mathfrak{a}) \cdot \Delta_K] \simeq \Big(SL(O_K \oplus \mathfrak{a})\Big\backslash\Big(\mathcal{H}^{r_1} \times \mathbb{H}^{r_2}\Big)\Big)_{ss},$$

where we use ss to denote the subset consisting of points corresponding to rank two semi-stable O_K-lattices in

$$SL(O_K \oplus \mathfrak{a}) \Big\backslash \Big(\big(SL(2,\mathbb{R})/SO(2) \big)^{r_1} \times \big(SL(2,\mathbb{C})/SU(2) \big)^{r_2} \Big).$$

More precisely, if the metric on $O_K \oplus \mathfrak{a}$ is given by $g = (g_\sigma)_{\sigma \in S_\infty}$ with $g_\sigma \in SL(2, K_\sigma)$, then the corresponding points on the right hand side is $g(\mathrm{ImJ})$ with $\mathrm{ImJ} := (i^{(r_1)}, j^{(r_2)})$, i.e., the point given by $(g_\sigma \tau_\sigma)_{\sigma \in S_\infty}$ where $\tau_\sigma = i_\sigma := (0,1)$ if σ is real and $\tau_\sigma = j_\sigma := (0,0,1)$ if σ is complex.

6.4 Cusps

As such, the working site becomes the space $SL(O_K \oplus \mathfrak{a}) \Big\backslash \big(\mathcal{H}^{r_1} \times \mathbb{H}^{r_2} \big)$. Here the action of $SL(O_K \oplus \mathfrak{a})$ is via the action of $SL(2, K)$ on $\mathcal{H}^{r_1} \times \mathbb{H}^{r_2}$. More precisely, K^2 admits natural embeddings $K^2 \hookrightarrow \big(\mathbb{R}^{r_1} \times \mathbb{C}^{r_2} \big)^2 \simeq (\mathbb{R}^2)^{r_1} \times (\mathbb{C}^2)^{r_2}$ so that $O_K \oplus \mathfrak{a}$ naturally embeds into $(\mathbb{R}^2)^{r_1} \times (\mathbb{C}^2)^{r_2}$ as a rank two O_K-lattice. As such, $SL(O_K \oplus \mathfrak{a})$ acts on the image of $O_K \oplus \mathfrak{a}$ in $(\mathbb{R}^2)^{r_1} \times (\mathbb{C}^2)^{r_2}$ as automorphisms. Our task here is to understand the cusps of this action of $SL(O_K \oplus \mathfrak{a})$ on $\mathcal{H}^{r_1} \times \mathbb{H}^{r_2}$. For this, we go as follows.

First, the space $\mathcal{H}^{r_1} \times \mathbb{H}^{r_2}$ admits a natural boundary $\mathbb{R}^{r_1} \times \mathbb{C}^{r_2}$, in which the field K is imbedded via Archmidean places in S_∞: $K \hookrightarrow \mathbb{R}^{r_1} \times \mathbb{C}^{r_2}$. Consequently, $\mathbb{P}^1(K) \hookrightarrow \mathbb{P}^1(\mathbb{R})^{r_1} \times \mathbb{P}^1(\mathbb{C})^{r_2}$ with $\begin{bmatrix} 1 \\ 0 \end{bmatrix} := \infty \mapsto (\infty^{(r_1)}, \infty^{(r_2)})$. As usual, via fractional linear transformations, $SL(2,\mathbb{R})$ acts on $\mathbb{P}^1(\mathbb{R})$, and $SL(2,\mathbb{C})$ acts on $\mathbb{P}^1(\mathbb{C})$, hence so does $SL(2, K)$ on

$$\mathbb{P}^1(K) \hookrightarrow \mathbb{P}^1(\mathbb{R})^{r_1} \times \mathbb{P}^1(\mathbb{C})^{r_2}.$$

Being a discrete subgroup of $SL(2,\mathbb{R})^{r_1} \times SL(2,\mathbb{C})^{r_2}$, for the action of $SL(O_K \oplus \mathfrak{a})$ on $\mathbb{P}^1(K)$, we call the corresponding orbits (of $SL(O_K \oplus \mathfrak{a})$ on $\mathbb{P}^1(K)$) the *cusps*. Very often we also call representatives cusps.

With this, we have the following fundamental result rooted back to Maaβ.

Cusp and Ideal Class Correspondence. (See e.g., [We6,8]) *There is a natural bijection between the ideal class group $CL(K)$ of K and the cusps*

C_Γ of $\Gamma = SL(O_K \oplus \mathfrak{a})$ acting on $\mathcal{H}^{r_1} \times \mathbb{H}^{r_2}$ given by

$$C_\Gamma \to CL(K), \qquad \begin{bmatrix}\alpha\\ \beta\end{bmatrix} \mapsto \big[O_K\,\alpha + \mathfrak{a}\beta\big].$$

Easily, one checks that the inverse map Π^{-1} is given as follows: For $\mathfrak{b}$, choose $\alpha_{\mathfrak{b}}, \beta_{\mathfrak{b}} \in K$ such that $O_K \cdot \alpha_{\mathfrak{b}} + \mathfrak{a} \cdot \beta_{\mathfrak{b}} = \mathfrak{b}$; then $\Pi^{-1}([\mathfrak{b}])$ is simply the class of the point $\begin{bmatrix}\alpha_{\mathfrak{b}}\\ \beta_{\mathfrak{b}}\end{bmatrix}$ in $SL(2, O_K \oplus a)\backslash\mathbb{P}^1(K)$. Moreover, there always exists $M_{\begin{bmatrix}\alpha\\ \beta\end{bmatrix}} := \begin{pmatrix}\alpha & \alpha^*\\ \beta & \beta^*\end{pmatrix} \in SL(2, K)$ such that $M_{\begin{bmatrix}\alpha\\ \beta\end{bmatrix}} \cdot \infty = \begin{bmatrix}\alpha\\ \beta\end{bmatrix}$.

6.5 Stablizer Groups of Cusps

Recall that under the Cusp-Ideal Class Correspondence, there are exactly h inequivalent cusps η_i, $i = 1, 2, \ldots, h$. Moreover, if we write the cusp $\eta := \eta_i = \begin{bmatrix}\alpha_i\\ \beta_i\end{bmatrix}$ for suitable α_i, $\beta_i \in K$, then the associated ideal class is exactly the one for the fractional ideal $O_K\alpha_i + \mathfrak{a}\beta_i =: \mathfrak{b}_i$. Denote the stablizer group of η in $SL(O_K \oplus \mathfrak{a})$ by Γ_η.

Lemma. ([We6,8]) *The associated 'lattice' for the cusp η is given by $\mathfrak{a}\mathfrak{b}^{-2}$. Moreover,*

$$A^{-1}\Gamma_\eta A = \left\{\begin{pmatrix}u & z\\ 0 & u^{-1}\end{pmatrix} : u \in U_K, z \in \mathfrak{a}\mathfrak{b}^{-2}\right\}.$$

Set $\Gamma'_\eta := \left\{A\begin{pmatrix}1 & z\\ 0 & 1\end{pmatrix}A^{-1} : z \in \mathfrak{a}\mathfrak{b}^{-2}\right\}$, Then

$$\Gamma_\eta = \Gamma'_\eta \times \left\{A\begin{pmatrix}u & 0\\ 0 & u^{-1}\end{pmatrix}A^{-1} : u \in U_K\right\}.$$

Note that also componentwisely, $\begin{pmatrix}u & 0\\ 0 & u^{-1}\end{pmatrix}z = \frac{uz}{u^{-1}} = u^2 z$. So, in practice, what we really get is the following decomposition

$$\Gamma_\eta = \Gamma'_\eta \times U_K^2$$

with

$$U_K^2 \simeq \left\{A \cdot \begin{pmatrix} u & 0 \\ 0 & u^{-1} \end{pmatrix} \cdot A^{-1} : u \in U_K\right\} \simeq \left\{A \begin{pmatrix} 1 & 0 \\ 0 & u^2 \end{pmatrix} A^{-1} : u \in U_K\right\}.$$

With this, we are ready to construct a fundamental domain for the action of $\Gamma_\eta \subset SL(O_K \oplus \mathfrak{a})$ on $\mathcal{H}^{r_1} \times \mathbb{H}^{r_2}$. This is based on a construction of a fundamental domain for the action of Γ_∞ on $\mathcal{H}^{r_1} \times \mathbb{H}^{r_2}$. More precisely, with an element $A = \begin{pmatrix} \alpha & \alpha^* \\ \beta & \beta^* \end{pmatrix} \in SL(2, K)$ (always exists!), we have

i) $A \cdot \infty = \begin{bmatrix} \alpha \\ \beta \end{bmatrix}$; and

ii) The isotropy group of η in $A^{-1} SL(O_K \oplus \mathfrak{a})A$ is generated by translations $\tau \mapsto \tau + z$ with $z \in \mathfrak{a}\mathfrak{b}^{-2}$ and by dilations $\tau \mapsto u\tau$ where u runs through the group U_K^2.

(Here, we use A, α, β, $\mathfrak{b}$ as running symbols for A_i, α_i, β_i, $\mathfrak{b}_i := O_K\alpha_i + \mathfrak{a}\beta_i$.)

Consider then the map $\mathcal{H}^{r_1} \times \mathbb{H}^{r_2} \overset{\mathrm{ImJ}}{\to} \mathbb{R}_{>0}^{r_1+r_2}$ defined by

$$(z_1, \ldots, z_{r_1}; P_1, \ldots, P_{r_2}) \mapsto (\mathrm{Im}\,(z_1), \ldots, \mathrm{Im}\,(z_{r_1}); J(P_1), \ldots, J(P_{r_2}))$$

where if $z = x + iy \in \mathcal{H}$ resp. $P = z + vj \in \mathbb{H}$, we set $\mathrm{Im}\,(z) = y$ resp. $J(P) = r$. It induces a map

$$\left(A^{-1} \cdot \Gamma_\eta \cdot A\right)\Big\backslash\left(\mathcal{H}^{r_1} \times \mathbb{H}^{r_2}\right) \to U_K^2\Big\backslash\mathbb{R}_{>0}^{r_1+r_2},$$

which exhibits $\left(A^{-1} \cdot \Gamma_\eta \cdot A\right)\Big\backslash\left(\mathcal{H}^{r_1} \times \mathbb{H}^{r_2}\right)$ as a torus bundle over $U_K^2\Big\backslash\mathbb{R}_{>0}^{r_1+r_2}$ with fiber the $n = r_1 + 2r_2$ dimensional torus $\left(\mathbb{R}^{r_1} \times \mathbb{C}^{r_2}\right)\big/\mathfrak{a}\mathfrak{b}^{-2}$. Having factored out the action of the translations, we only have to construct a fundamental domain for the action of U_K^2 on $\mathbb{R}_{>0}^{r_1+r_2}$. For this, we look first at the action of U_K^2 on the norm-one hypersurface $\mathbf{S} := \left\{y \in \mathbb{R}_{>0}^{r_1+r_2} : N(y) = 1\right\}$. By taking logarithms, it is transformed bijectively into a trace-zero hyperplane which is isomorphic to the space $\mathbb{R}^{r_1+r_2-1}$

$$\begin{aligned} \mathbf{S} \overset{\log}{\to} \mathbb{R}^{r_1+r_2-1} &:= \left\{(a_1, \ldots, a_{r_1+r_2}) \in \mathbb{R}^{r_1+r_2} : \sum a_i = 0\right\}, \\ y \mapsto &\quad \left(\log y_1, \ldots, \log y_{r_1+r_2}\right), \end{aligned}$$

where the action of U_K^2 on $\mathbf{S}$ is carried out over an action on $\mathbb{R}^{r_1+r_2-1}$ by the translations $a_i \mapsto a_i + \log \varepsilon^{(i)}$. By Dirichlet's Unit Theorem, the logarithm transforms U_K^2 into a lattice in $\mathbb{R}^{r_1+r_2-1}$. Accordingly, the exponential map transforms a fundamental domain, e.g., a fundamental parallelepiped, for this action back into a fundamental domain $\mathbf{S}_{U_K^2}$ for the action of U_K^2 on $\mathbf{S}$. The cone over $\mathbf{S}_{U_K^2}$, that is, $\mathbb{R}_{>0} \cdot \mathbf{S}_{U_K^2} \subset \mathbb{R}_{>0}^{r_1+r_2}$, is then a fundamental domain for the action of U_K^2 on $\mathbb{R}_{>0}^{r_1+r_2}$. Denote by $\mathcal{T}$ a fundamental domain for the action of the translations by elements of $\mathfrak{a}\mathfrak{b}^{-2}$ on $\mathbb{R}^{r_1} \times \mathbb{C}^{r_2}$, and set

$$\begin{aligned}&\mathrm{ReZ}\Big(z_1,\dots,z_{r_1};P_1,\dots,P_{r_2}\Big)\\ :=&\Big(\mathrm{Re}\,(z_1),\dots,\mathrm{Re}\,(z_{r_1});Z(P_1),\dots,Z(P_{r_2})\Big)\end{aligned}$$

with $\mathrm{Re}\,(z) := x$ resp. $Z(P) := z$ if $z = x + iy \in \mathcal{H}$ resp. $P = z + vj \in \mathbb{H}$, then what we have just said proves the following

Theorem. ([We6,8]) *A fundamental domain for the action of $A^{-1}\Gamma_\eta A$ on $\mathcal{H}^{r_1} \times \mathbb{H}^{r_2}$ is given by*

$$\mathbf{E} := \Big\{\tau \in \mathcal{H}^{r_1} \times \mathbb{H}^{r_2} : \mathrm{ReZ}\,(\tau) \in \mathcal{T},\ \mathrm{ImJ}\,(\tau) \in \mathbb{R}_{>0} \cdot \mathbf{S}_{U_K^2}\Big\}.$$

For later use, we also set $\mathcal{F}_\eta := A_\eta^{-1} \cdot \mathbf{E}$.

6.6 Fundamental Domain

Guided by Siegel's discussion on totally real fields [Sie] and the discussion above, we are now ready to construct fundamental domains for $SL(\mathcal{O}_K \oplus \mathfrak{a})\Big\backslash\Big(\mathcal{H}^{r_1} \times \mathbb{H}^{r_2}\Big)$.

As the first step, we generalize Siegel's 'distance to cusps'. For this, recall that for a cusp $\eta = \begin{bmatrix}\alpha\\ \beta\end{bmatrix} \in \mathbb{P}^1(K)$, by the Cusp-Ideal Class Correspondence, we obtain a natural ideal class associated to the fractional ideal $\mathfrak{b} := \mathcal{O}_K \cdot \alpha + \mathfrak{a} \cdot \beta$. Moreover, by assuming that α, β are all contained in $\mathcal{O}_K$, as we may, we know that the corresponding stablizer group Γ_η is given by

$$A^{-1} \cdot \Gamma_\eta \cdot A = \Big\{\gamma = \begin{pmatrix} u & z\\ 0 & u^{-1}\end{pmatrix} \in \Gamma : u \in U_K, z \in \mathfrak{a}\mathfrak{b}^{-2}\Big\},$$

where $A \in SL(2,K)$ satisfying $A\infty = \eta$ which may be further chosen in the form $A = \begin{pmatrix} \alpha & \alpha^* \\ \beta & \beta^* \end{pmatrix} \in SL(2,K)$ so that $O_K\beta^* + \mathfrak{a}^{-1}\alpha^* = \mathfrak{b}^{-1}$.

Now for $\tau = (z_1,\ldots,z_{r_1}; P_1,\ldots,P_{r_2}) \in \mathcal{H}^{r_1} \times \mathbb{H}^{r_2}$, set

$$\begin{aligned} N(\tau) &:= N\big(\mathrm{ImJ}(\tau)\big) \\ &= \prod_{i=1}^{r_1} \mathrm{Im}\,(z_i) \cdot \prod_{j=1}^{r_2} J(P_j)^2 = \big(y_1 \cdots y_{r_1}\big) \cdot \big(v_1 \cdots v_{r_2}\big)^2. \end{aligned}$$

Then for all $\gamma = \begin{pmatrix} a & b \\ c & d \end{pmatrix} \in SL(2,K)$,

$$N\big(\mathrm{ImJ}(\gamma \cdot \tau)\big) = \frac{N(\mathrm{ImJ}(\tau))}{\|N(c\tau + d)\|^2}. \tag{$*$}$$

(Note that here only the second row of γ appears.) Moreover, define the *reciprocal distance* $\mu(\eta,\tau)$ *from the point* $\tau \in \mathcal{H}^{r_1} \times \mathbb{H}^{r_2}$ *to the cusp* $\eta = \begin{bmatrix} \alpha \\ \beta \end{bmatrix}$ *in* $\mathbb{P}^1(K)$ by

$$\begin{aligned} \mu(\eta,\tau) &:= N\big(\mathfrak{a}^{-1} \cdot (O_K\alpha + \mathfrak{a}\beta)^2\big) \\ &\quad \cdot \frac{\mathrm{Im}\,(z_1) \cdots \mathrm{Im}\,(z_{r_1}) \cdot J(P_1)^2 \cdots J(P_{r_2})^2}{\prod_{i=1}^{r_1} |(-\beta^{(i)} z_i + \alpha^{(i)})|^2 \prod_{j=1}^{r_2} \|(-\beta^{(j)} P_j + \alpha^{(j)})\|^2} \\ &= \frac{1}{N(\mathfrak{a}\mathfrak{b}^{-2})} \cdot \frac{N(\mathrm{ImJ}(\tau))}{\|N(-\beta\tau + \alpha)\|^2}. \end{aligned}$$

Lemma 1. ([We6,8]) (i) μ *is well-defined;*
(ii) μ *is invariant under the action of* $SL(O_K \oplus \mathfrak{a})$. *That is to say,*

$$\mu(\gamma\eta, \gamma\tau) = \mu(\eta,\tau), \qquad \forall \gamma \in SL(O_K \oplus \mathfrak{a}).$$

(iii) *There exists a positive constant* C *depending only on* K *and* $\mathfrak{a}$ *such that if* $\mu(\eta,\tau) > C$ *and* $\mu(\eta',\tau) > C$ *for* $\tau \in \mathcal{H}^{r_1} \times \mathbb{H}^{r_2}$ *and* $\eta, \eta' \in \mathbb{P}^1(K)$, *then* $\eta = \eta'$.
(iv) *There exists a positive real number* $T := T(K)$ *depending only on* K *such that for* $\tau \in \mathcal{H}^{r_1} \times \mathbb{H}^{r_2}$, *there exists a cusp* η *such that* $\mu(\eta,\tau) > T$.

Now for the cusp $\eta = \begin{bmatrix}\alpha\\ \beta\end{bmatrix} \in \mathbb{P}^1(K)$, define the 'sphere of influence' of η by

$$F_\eta := \left\{\tau \in \mathcal{H}^{r_1} \times \mathbb{H}^{r_2} : \mu(\eta,\tau) \geq \mu(\eta',\tau), \forall \eta' \in \mathbb{P}^1(K)\right\}.$$

Lemma 2. ([We6,8]) *The action of $SL(O_K \oplus \mathfrak{a})$ in the interior F_η^0 of F_η reduces to that of the isotropy group Γ_η of η, i.e., if τ and $\gamma\tau$ both belong to F_η^0, then $\gamma\tau = \tau$.*

Consequently, we arrive at the following way to decompose the orbit space $SL(O_K \oplus \mathfrak{a})\backslash\left(\mathcal{H}^{r_1} \times \mathbb{H}^{r_2}\right)$ into h pieces glued in some way along pants of their boundary.

Theorem. (We]) *Let $i_\eta : \Gamma_\eta\backslash F_\eta \hookrightarrow SL(O_K \oplus \mathfrak{a})\backslash\left(\mathcal{H}^{r_1} \times \mathbb{H}^{r_2}\right)$ denote the natural map. Then*

$$SL(O_K \oplus \mathfrak{a})\backslash\left(\mathcal{H}^{r_1} \times \mathbb{H}^{r_2}\right) = \cup_\eta i_\eta\left(\Gamma_\eta\backslash F_\eta\right),$$

where the union is taken over a set of h cusps representing the ideal classes of K. Each piece corresponds to an ideal class of K.

Note that the action of Γ_η on $\mathcal{H}^{r_1} \times \mathbb{H}^{r_2}$ is free. Consequently, all fixed points of $SL(O_K \oplus \mathfrak{a})$ on $\mathcal{H}^{r_1} \times \mathbb{H}^{r_2}$ lie on the boundaries of F_η.

We can give a more precise description of the fundamental domain, based on our understanding of that for stabilizer groups of cusps. To state it, denote by $\eta_1, \ldots, \eta_h$ inequivalent cusps for the action of $SL(O_K \oplus \mathfrak{a})$ on $\mathcal{H}^{r_1} \times \mathbb{H}^{r_2}$. Choose $A_{\eta_i} \in SL(2,K)$ such that $A_{\eta_i}\infty = \eta_i,\ i = 1,2,\ldots,h$. Write $\mathbf{S}$ for the norm-one hypersurface $\mathbf{S} := \left\{y \in \mathbb{R}_{>0}^{r_1+r_2} : N(y) = 1\right\}$, and $\mathbf{S}_{U_K^2}$ for the action of U_K^2 on $\mathbf{S}$. Denote by $\mathcal{T}$ a fundamental domain for the action of the translations by elements of $\mathfrak{a}\mathfrak{b}^{-2}$ on $\mathbb{R}^{r_1} \times \mathbb{C}^{r_2}$, and

$$\mathbf{E} := \left\{\tau \in \mathcal{H}^{r_1} \times \mathbb{H}^{r_2} : \mathrm{ReZ}(\tau) \in \mathcal{T},\ \mathrm{ImJ}(\tau) \in \mathbb{R}_{>0} \cdot \mathbf{S}_{U_K^2}\right\}$$

a fundamental domain for the action of $A_\eta^{-1}\Gamma_\eta A_\eta$ on $\mathcal{H}^{r_1} \times \mathbb{H}^{r_2}$. The intersections of $\mathbf{E}$ with $i_\eta(F_\eta)$ are connected. Consequently, we have the following

Theorem′. ([We6,8]) (1) *$D_\eta := A_\eta^{-1}\mathbf{E} \cap F_\eta$ is a fundamental domain for the action of Γ_η on F_η;*

(2) *There exist* $\alpha_1, \ldots, \alpha_h \in SL(O_K \oplus \mathfrak{a})$ *such that* $\cup_{i=1}^h \alpha(D_{\eta_i})$ *is connected and hence a fundamental domain for* $SL(O_K \oplus \mathfrak{a})$.

That is to say, a fundamental domain may be given as $S_Y \cup \mathcal{F}_1(Y_1) \cup \cdots \cup \mathcal{F}_h(Y_h)$ with S_Y bounded, $\mathcal{F}_i(Y_i) = A_i \cdot \widetilde{\mathcal{F}_i}(Y_i)$ and

$$\widetilde{\mathcal{F}_i}(Y_i) := \left\{\tau \in \mathcal{H}^{r_1} \times \mathbb{H}^{r_2} : \mathrm{ReZ}(\tau) \in \Sigma,\ \mathrm{ImJ}(\tau) \in \mathbb{R}_{>T} \cdot \mathbf{S}_{U_K^2}\right\}.$$

Moreover, all $\mathcal{F}_i(Y_i)$'s are disjoint from each other when Y_i are sufficiently large.

6.7 Stability and Distances to Cusps

Define now the *distance of* τ *to the cusp* η by

$$d(\eta, \tau_\Lambda) := \frac{1}{\mu(\eta, \tau_\Lambda)} \geq 1.$$

Then we are ready to state the following fundamental result, which exposes a beautiful intrinsic relation between stability and the distance to cusps.

Theorem. ([We6,8]) *The lattice* Λ *is semi-stable if and only if the distances of corresponding point* $\tau_\Lambda \in \mathcal{H}^{r_1} \times \mathbb{H}^{r_2}$ *to all cusps are all bigger or equal to 1.*

Our proof uses an algebraic result of Tsukasa Hayashi [Ha].

One can never overestimate the importance of this relation. Being stable, lattices should be away from cusps. More generally, while the stability condition is defined in terms of sublattices, the relation above transforms these volumes inequalities in terms of distances to cusps. In a more theoretical term for higher rank lattices, the essence of this fact is that, sublattices and cusps, as two different aspects of parabolic subgroups, are naturally corresponding to each other: the stability conditions for various sublattices are naturally related with generalized distances to all types of cusps.

6.8 Moduli Space in Rank Two

For a rank two O_K-lattice Λ, denote by $\tau_\Lambda \in \mathcal{H}^{r_1} \times \mathbb{H}^{r_2}$ the corresponding point. Then, from above, Λ is semi-stable if and only if for all cusps η, $d(\eta, \tau_\Lambda) := \frac{1}{\mu(\eta,\tau_\Lambda)}$ are bigger than or equal to 1. This then leads to the consideration of the following truncation of the fundamental domain $\mathcal{D}$ of $SL(O_K \oplus \mathfrak{a})\Big\backslash\big(\mathcal{H}^{r_1} \times \mathbb{H}^{r_2}\big)$: For $T \geq 1$, denote by

$$\mathcal{D}_T := \Big\{\tau \in \mathcal{D} : d(\eta, \tau_\Lambda) \geq T^{-1},\ \forall \text{cusp } \eta\Big\}.$$

The space $\mathcal{D}_T$ may be precisely described in terms of $\mathcal{D}$ and certain neighborhood of cusps. To explain this, we first establish the following **Lemma.** *For a cusp η, denote by*

$$X_\eta(T) := \Big\{\tau \in \mathcal{H}^{r_1} \times \mathbb{H}^{r_2} : d(\eta, \tau) < T^{-1}\Big\}.$$

Then for $T \geq 1$,

$$X_{\eta_1}(T) \cap X_{\eta_2}(T) \neq \emptyset \qquad \Leftrightarrow \eta_1 = \eta_2.$$

With this, we are ready to state the following

Theorem. ([We6,8]) *There is a natural identification between*
(a) *the moduli space of rank two semi-stable O_K-lattices of volume $N(\mathfrak{a})\Delta_K$ with underlying projective module $O_K \oplus \mathfrak{a}$;* and
(b) *the truncated compact domain $\mathcal{D}_1$ consisting of points in the fundamental domain $\mathcal{D}$ whose distances to all cusps are bigger than 1.*

In other words, the truncated compact domain $\mathcal{D}_1$ is obtained from the fundamental domain $\mathcal{D}$ of $SL(O_K \oplus \mathfrak{a})\Big\backslash\big(\mathcal{H}^{r_1} \times \mathbb{H}^{r_2}\big)$ by delecting the disjoint open neighborhoods $\dot{\cup} \cup_{i=1}^h \mathcal{F}_i(1)$ associated to inequivalent cusps $\eta_1, \eta_2, \ldots, \eta_h$, where $\mathcal{F}_i(T)$ denotes the neighborhood of η_i consisting of $\tau \in \mathcal{D}$ whose distance to η_i is strictly less than T^{-1}.

For later use, we set also

$$\mathcal{D}_T := \mathcal{D} \backslash \cup \cup_{i=1}^h \mathcal{F}_i(T), \qquad T \geq 1.$$

7 Explicit Formulas for Rank Two Zetas

7.1 Epstein Zetas and Eisenstein Series

For a fixed integer $r \geq 1$ and a fractional ideal $\mathfrak{a}$ of a number field K, define *the Epstein type zeta function* $\widehat{E}_{r,\mathfrak{a};\Lambda}(s)$ associated to an O_K-lattice Λ with underlying projective module $P_\mathfrak{a} = O_K^{(r-1)} \oplus \mathfrak{a}$ to be

$$\begin{aligned}\widehat{E}_{r,\mathfrak{a};\Lambda}(s) :=&\Big(\pi^{-\frac{rs}{2}}\Gamma(\frac{rs}{2})\Big)^{r_1}\Big(2\pi^{-rs}\Gamma(rs)\Big)^{r_2}\\ &\times\Big(N(\mathfrak{a})\Delta_K^{\frac{r}{2}}\Big)^s\cdot \sum_{\mathbf{x}\in O_K^{(r-1)}\oplus\mathfrak{a}/U_{K,r}^+,\mathbf{x}\neq(0,\dots,0)} \frac{1}{\|\mathbf{x}\|_\Lambda^{rs}}\end{aligned}$$

where $U_{K,r}^+ := \Big\{\varepsilon^r : \varepsilon \in U_K,\ \varepsilon^r \in U_K^+\Big\} = U_K^+ \cap U_K^r$. For example, note that in the case $r = 2$, $U_{K,2}^+ = U_K^2$, we have

$$\begin{aligned}\widehat{E}_{2,\mathfrak{a};\Lambda}(s) :=&\Big(\pi^{-s}\Gamma(s)\Big)^{r_1}\Big(2\pi^{-2s}\Gamma(2s)\Big)^{r_2}\\ &\times\Big(N(\mathfrak{a})\Delta_K\Big)^s\cdot \sum_{\mathbf{x}\in O_K\oplus\mathfrak{a}/U_K^2,\mathbf{x}\neq(0,0)} \frac{1}{\|\mathbf{x}\|_\Lambda^{2s}}.\end{aligned}$$

Set also, for $\mathrm{Re}\,(s) > 1$,

$$\begin{aligned}\widehat{E}_{2,\mathfrak{a}}(\tau, s) :=& \Big(\pi^{-s}\Gamma(s)\Big)^{r_1}\Big(2\pi^{-2s}\Gamma(2s)\Big)^{r_2}\\ &\times\Big(N(\mathfrak{a})\Delta_K\Big)^s\cdot \sum_{(x,y)\in O_K\oplus\mathfrak{a}/U_K,(x,y)\neq(0,0)} \left(\frac{N(\mathrm{ImJ}(\tau_\Lambda))}{\big\|x\cdot\tau+y\big\|^2}\right)^s.\end{aligned}$$

Lemma. ([We6,8]) *For a rank two O_K-lattice $\Lambda = (O_K \oplus a, \rho_\Lambda)$, denote by τ_Λ the corresponding point in the moduli space $SL(O_K \oplus a)\Big\backslash\Big(\mathcal{H}^{r_1} \times \mathbb{H}^{r_2}\Big)$. Then $\widehat{E}_{2,\mathfrak{a};\Lambda}(s) = \widehat{E}_{2,\mathfrak{a}}(\tau_\Lambda, s)$.*

7.2 Fourier Expansions

For simplicity, introduce the standard Eisenstein series by setting

$$E_{2,\mathfrak{a}}(\tau, s) := \sum_{(x,y)\in O_K\oplus\mathfrak{a}/U_K,(x,y)\neq(0,0)} \left(\frac{N(\mathrm{ImJ}(\tau))}{\big\|x\cdot\tau+y\big\|^2}\right)^s, \qquad \mathrm{Re}\,(s) > 1.$$

Then the completed one becomes

$$\widehat{E}_{2,\mathfrak{a}}(\tau,s)=\Big(\pi^{-s}\Gamma(s)\Big)^{r_1}\Big(2\pi^{-2s}\Gamma(2s)\Big)^{r_2}\cdot\Big(N(\mathfrak{a})\Delta_K\Big)^s\cdot E_{2,\mathfrak{a}}(\tau,s). \qquad (*)$$

As before, for the cusp $\eta=\begin{bmatrix}\alpha\\ \beta\end{bmatrix}$, choose a (normalized) matrix $A=\begin{pmatrix}\alpha & \alpha^*\\ \beta & \beta^*\end{pmatrix}\in SL(2,F)$ such that and that if $\mathfrak{b}=O_K\alpha+\mathfrak{a}\beta$, then $O_K\beta^*+\mathfrak{a}\alpha^*=\mathfrak{b}^{-1}$. Clearly, $A\infty=\eta$, and moreover,

$$A^{-1}\Gamma'_\eta A=\left\{\begin{pmatrix}1 & \omega\\ 0 & 1\end{pmatrix}:\omega\in\mathfrak{a}\mathfrak{b}^{-2}\right\}.$$

Since $\widehat{E}_{2,\mathfrak{a}}(\tau,s)$, and hence $E_{2,\mathfrak{a}}(\tau,s)$, is $SL(O_K\oplus\mathfrak{a})$-invariant, $E_{2,\mathfrak{a}}(\tau,s)$ is $\Gamma'_\eta\subset SL(O_K\oplus\mathfrak{a})$-invariant. Therefore $E_{2,\mathfrak{a}}(A\tau,s)$ is $\mathfrak{a}\mathfrak{b}^{-2}$-invariant, that is, $E_{2,\mathfrak{a}}(A\tau,s)$ is invariant under parallel transforms by elements of $\mathfrak{a}\mathfrak{b}^{-2}$. Consequently, we have the Fourier expansion

$$E_{2,\mathfrak{a}}(A\tau,s)=\sum_{\omega'\in(\mathfrak{a}\mathfrak{b}^{-2})^\vee}a_{\omega'}\Big(\mathrm{ImJ}(\tau),s\Big)\cdot e^{2\pi i\langle\omega',\mathrm{ReZ}(\tau)\rangle},$$

where $(\mathfrak{a}\mathfrak{b}^{-2})^\vee$ denotes the dual lattice of $\mathfrak{a}\mathfrak{b}^{-2}$. Thus, if we use Q to denote a fundamental parallolgram of $\mathfrak{a}\mathfrak{b}^{-2}$ in $\mathbb{R}^{r_1}\times\mathbb{C}^{r_2}$, then

$$a_{\omega'}(\mathrm{ImJ}(\tau),s):=\frac{1}{\mathrm{Vol}(\mathfrak{a}\mathfrak{b}^{-2})}\sum_{(c,d)\in(O_K\oplus\mathfrak{a})A/U_K,(c,d)\neq(0,0)}$$
$$\int_Q\left(\frac{N(\mathrm{ImJ}(\tau))}{\|c\tau+d\|^2}\right)^s\cdot e^{-2\pi i\langle\omega',\mathrm{ReZ}(\tau)\rangle}\prod_{\sigma:\mathbb{R}}dx_\sigma\cdot\prod_{\tau:\mathbb{C}}dx_\tau dy_\tau.$$

Theorem. ([We6,8]) *We have the following Fourier expansion*

$$\begin{aligned}E_{2,\mathfrak{a}}(A\tau,s)=&\zeta([\mathfrak{a}^{-1}\mathfrak{b}],2s)\cdot N(\mathfrak{a}\mathfrak{b}^{-1})^{-2s}\cdot N(\mathrm{ImJ}(\tau))^s\\ &+\frac{1}{\mathrm{Vol}(\mathfrak{a}\mathfrak{b}^{-2})}\sum_{\begin{pmatrix}* & *\\ c & c\omega+d\end{pmatrix}\in\mathcal{R}}\frac{1}{N(c)^{2s}}\cdot(\pi^{\frac12})^{r_1}\\ &\cdot\left(\frac{\Gamma\big(s-\frac12\big)}{\Gamma(s)}\right)^{r_1}\left(\frac{\pi}{2s-1}\right)^{r_2}\cdot N(\mathrm{ImJ}(\tau))^{1-s}\end{aligned}$$

$$+\frac{1}{\mathrm{Vol}(\mathfrak{a}\mathfrak{b}^{-2})}\sum_{\begin{pmatrix}* & *\\ c & d\end{pmatrix}\in\mathcal{R}}\frac{e^{2\pi i\langle\omega',\frac{d}{c}\rangle}}{N(c)^{2s}}\cdot N(\mathrm{ImJ}(\tau))^{\frac{1}{2}}\cdot N(\omega')^{s-\frac{1}{2}}\Big(\frac{2\pi^s}{\Gamma(s)}\Big)^{r_1}$$

$$\times\prod_{\sigma:\mathbb{R}}K_{s-\frac{1}{2}}(2\pi|\omega'|_\sigma y_\sigma)\cdot\Big(\frac{2\pi^{2s}|\omega'|^{2s-1}}{\Gamma(2s)}\Big)^{r_2}\cdot\prod_{\tau:\mathbb{C}}K_{2s-1}(2\pi|\omega'|_\tau r_\tau).$$

7.3 Explicit Formula for Rank Two Zetas

The original Rankin-Selberg method gives a way to express the Mellin transfrom of the constant term in the Fourier expansion of an automorphic function as the scalar product of the automorphic function with an Eisenstein series if the automorphic function is very small when approach to the cusp. In a paper of Zagier [Z], this method is extended to a much broad type of automorphic functions. Here, we use a generalization of the Rankin-Selberg & Zagier method to give an explicit expression for rank two zeta functions of number fields in terms of Dedekind zeta functions.

In the sequel, T is assumed to be a positive real number ≥ 1. Our aim is to compute the integration

$$\iiint_{\mathcal{D}_T}\widehat{E}_{2,\mathfrak{a}}(\tau,s)d\mu(\tau).$$

Here, as in 6.8, $\mathcal{D}_T$ is the compact part obtained from the fundamental domain $\mathcal{D}$ for $SL(O_K\oplus\mathfrak{a})\backslash\big(\mathcal{H}^{r_1}\times\mathbb{H}^{r_2}\big)$ by cutting off the cusp neighborhoods defined by the conditions that the distance to cusps is less than T^{-1}. (Recall that, as such, $\mathcal{D}_1$ is simply the part corresponding to semi-stable lattices.)

For doing so, first formulate the integration

$$\iiint_{\mathcal{D}_T}\Big(\Delta_K\,\widehat{E}_{2,\mathfrak{a}}(\tau,s)\Big)\,d\mu(\tau)$$

where

$$\Delta_K:=\sum_{\sigma:\mathbb{R}}\Delta_\sigma+\sum_{\tau:\mathbb{C}}\Delta_\tau$$

with

$$\Delta_\sigma:=y_\sigma^2\Big(\frac{\partial^2}{\partial x_\sigma^2}+\frac{\partial^2}{\partial y_\sigma^2}\Big),\quad \Delta_\tau:=r_\tau^2\Big(\frac{\partial^2}{\partial x_\tau^2}+\frac{\partial^2}{\partial y_\tau^2}+\frac{\partial^2}{\partial r_\tau^2}\Big)-r\frac{\partial}{\partial r_\tau}.$$

(For the time being, by an abuse of notation, we use Δ_K to denote the hyperbolic Laplace operator for the space $\mathcal{H}^{r_1} \times \mathbb{H}^{r_2}$, not the absolute value of the discriminant of K which accordingly is changed to D_K.)

Note that

$$\Delta_\sigma\left(y_\sigma^s\right) = s(s-1) \cdot y_\sigma^s, \qquad \text{and} \qquad \Delta_\tau\left(r_\tau^{2s}\right) = 2s(2s-2) \cdot r_\tau^{2s}$$

by the SL-invariance of the metrics, we conclude that

$$\Delta_K\left(\widehat{E}_{2,\mathfrak{a}}(\tau, s)\right) = \left(r_1 \cdot \left(s(s-1)\right) + r_2 \cdot \left(2s(2s-2)\right)\right) \cdot \widehat{E}_{2,\mathfrak{a}}(\tau, s).$$

Hence

$$\iiint_{\mathcal{D}_T} \widehat{E}_{2,\mathfrak{a}}(\tau, s)\, d\mu(tau) = \frac{r_1 + 4r_2}{s(s-1)} \iiint_{\mathcal{D}_T} \Delta_K \widehat{E}_{2,\mathfrak{a}}(\tau, s)\, d\mu(\tau).$$

On the other hand, using Stokes' Formula, we have

$$\begin{aligned}
&\iiint_{\mathcal{D}_T} \Delta_K \widehat{E}_{2,\mathfrak{a}}(\tau, s) d\mu(\tau) \\
&= \iiint_{\mathcal{D}_T} \left(\Delta_K \widehat{E}_{2,\mathfrak{a}}(\tau, s)\right) \cdot 1 d\mu(\tau) \\
&\qquad - \iiint_{\mathcal{D}_T} \widehat{E}_{2,\mathfrak{a}}(\tau, s) \cdot \left(\Delta_K 1\right) d\mu(\tau) \\
&= \iiint_{\mathcal{D}_T} \left(\left(\Delta_K \widehat{E}_{2,\mathfrak{a}}(\tau, s)\right) \cdot 1 - \widehat{E}_{2,\mathfrak{a}}(\tau, s) \cdot \left(\Delta_K 1\right)\right) d\mu(\tau) \\
&= \iint_{\partial D(T)} \left(\frac{\partial \widehat{E}_{2,\mathfrak{a}}(\tau, s)}{\partial \nu} \cdot 1 - \widehat{E}_{2,\mathfrak{a}}(\tau, s) \cdot \frac{\partial 1}{\partial \nu}\right) d\mu \\
&= \iint_{\partial D(T)} \frac{\partial \widehat{E}_{2,\mathfrak{a}}(\tau, s)}{\partial \nu} d\mu
\end{aligned}$$

where $\frac{\partial}{\partial \nu}$ is the outer normal derivative, $d\mu$ is the volume element of the boundary $\partial \mathcal{D}_T$.

Theorem. ([We6,8]) *Up to a constant factor depending only on K,*

$$\iiint_{\mathcal{D}_T} \widehat{E}_{2,\mathfrak{a}}(\tau, s) d\mu(\tau) = \frac{\xi_K(2s)}{s-1} \left(\Delta_K \cdot T\right)^{s-1} - \frac{\xi_K(2-2s)}{s} \left(\Delta_K \cdot T\right)^{-s}.$$

Consequently, we have the following

Theorem′. ([We6,8]) *The rank two zeta function $\xi_{K,2}(s)$ of K is given by*

$$\xi_{K,2}(s) = \frac{\xi_K(2s)}{s-1} - \frac{\xi_K(2s-1)}{s} \qquad \mathrm{Re}\,(s) > 1.$$

8 Zeros of Rank Two Zetas

8.1 Zeros of Rank Two Zeta of $\mathbb{Q}$

Following what was happened in history, let me first start with Suzuki's weak result [Su] and then give Lagarias' unconditional result. Meanwhile, for an independent parallel work, please go to Ki's beautiful paper [Ki].

Theorem. *If the Riemann Hypothesis for the Riemann zeta function holds, then all zeros of $\xi_{\mathbb{Q},2}(s)$ lie on the critical line* $\mathrm{Re}\,(s) = \frac{1}{2}$.

This is a very clever observation, rooted back to Titchimashi's book on Riemann Zeta Functions.

8.1.1 Product Formula for Entire Function of Order 1

Let $f(z)$ be an entire function of order one on $\mathbb{C}$, that is,
(i) $f(z)$ is analytic over $\mathbb{C}$;
(ii) $f(z) = O\Big(\exp(|z|^{1+\varepsilon})\Big),\ \ \forall\varepsilon > 0.$
Let $n(R)$ denote the number of zeros of $f(z)$ inside C_R, the circle of radius R centered at the origin. Then
(1) $n(R) = O(R^\alpha),\ \ \forall\alpha > 1;$
(2) $\sum_{\rho_n, f(\rho_n)=0} |\rho_n|^{-\alpha}$ converges. In particular, $(1-\frac{z}{\rho_n})\exp(\frac{z}{\rho_n}) = 1+O\Big((\frac{z}{\rho_n})^2\Big)$ as $n \to \infty$. As a direct consequnce,
(3) $P(z) := \prod_{\rho_n}\Big(1-\frac{z}{\rho_n}\Big)\cdot\exp\Big(\frac{z}{\rho_n}\Big)$ converges and $\frac{f(z)}{P(z)}$ is an entire function of order 1 without zeros, hence should be in the form $\exp(A + Bz)$ for certain constants A, B.

That is to say, we have the following

Hadamard Product Theorem. *Let $f(z)$ be an entire function of order one on $\mathbb{C}$, then there exist constants A, B such that*

$$f(z) = e^{A+Bz}\cdot\prod_\rho\Big(1-\frac{z}{\rho}\Big)\cdot\exp\Big(\frac{z}{\rho}\Big).$$

Example. (See e.g. [Ed]) We have

$$\frac{1}{2}s(s-1)\cdot\xi(s) = e^{A+Bz}\cdot\prod_\rho\Big(1-\frac{z}{\rho}\Big)\cdot\exp\Big(\frac{z}{\rho}\Big)$$

with $A = -\log 2$, $B = -\frac{\gamma}{2} - 1 + \frac{1}{2}\log 4\pi$, where $\gamma = \lim_{n\to\infty}\left(1 + \frac{1}{2} + \cdots + \frac{1}{n} - \log n\right)$ denotes the Euler constant.

8.1.2 Proof

Let

$$F(z) = -Z(\frac{1}{2} + 2iz) \qquad \text{with} \qquad Z(s) = s(1-s)\xi(s).$$

Proposition. (Suzuki) (1) $F(z+\frac{i}{4}) - F(z-\frac{i}{4}) = iz(1+4z^2)\,\xi_{\mathbb{Q},2}(\frac{1}{2}+iz)$.
(2) *Assume the RH, then all zeros of* $F(z+\frac{i}{4}) - F(z-\frac{i}{4})$ *are real.*
In particular, then the RH implies that $\xi_{\mathbb{Q},2}(\frac{1}{2}+zi)$ *admits only real zeros.*

Proof. (1) Simple calculation. Indeed,

$$F\Big(z+\frac{i}{4}\Big) = -Z\Big(\frac{1}{2} + 2i(\frac{i}{4}+z)\Big) = -Z\Big(\frac{1}{2} - \frac{1}{2} + 2iz\Big) = -Z(2iz).$$

So

$$F\Big(z-\frac{i}{4}\Big) = -Z\Big(1+2iz\Big)$$

and

$$\begin{aligned}
&F(z+\frac{i}{4}) - F(z-\frac{i}{4})\\
&= (1+2iz)(-2iz)\xi(1+2iz) - 2iz(1-2iz)\xi(2iz)\\
&= 2iz(1-2iz)(1+2iz)\cdot\Big(\frac{\xi(1+2iz)}{2iz-1} - \frac{\xi(2iz)}{1+2iz}\Big)\\
&= iz(1+4z^2)\cdot\Big(\frac{\xi(2(\frac{1}{2}+iz))}{(\frac{1}{2}+iz)-1} - \frac{\xi(2(\frac{1}{2}+iz)-1)}{\frac{1}{2}+iz}\Big)\\
&= iz(1+4z^2)\,\xi_{\mathbb{Q},2}(\frac{1}{2}+iz).
\end{aligned}$$

(2) Clearly, $F(z)$ is an entire function of order 1, so there are constants A, B such that

$$F(z) = e^{A+Bz}\cdot\prod_{\rho:F(\rho)=0}\Big(1-\frac{z}{\rho}\Big)\cdot\exp\Big(\frac{z}{\rho}\Big).$$

Note that essentially, ρ are zeros of the completed Riemann zeta but transformed from z to $\frac{1}{2}+2iz$. Hence, *by the RH*, all ρ are real.

Moreover, since $F(z) = -Z(\frac{1}{2} + 2iz)$ with $Z(s) = s(1-s)\,\xi(s)$, we have for $x \in \mathbb{R}$,

$$\overline{F(x)} = \overline{-Z\big(1/2 + 2ix\big)} = -Z\big(\overline{1/2 + 2ix}\big) = -Z\Big(\frac{1}{2} - 2ix\Big)$$

which by the functional equation is simply

$$-Z\Big(1 - (\frac{1}{2} - 2ix)\Big) = -Z\Big(\frac{1}{2} + 2ix\Big) = F(x).$$

That is to say, for $x \in \mathbb{R}$, $F(x)$ takes only real values. Hence, constants A and B are both real.

Now let $z_0 = x_0 + iy_0$ be a zero of

$$F(z + \frac{i}{4}) - F(z - \frac{i}{4}) = iz(1 + 4z^2) \cdot \xi_{\mathbb{Q},2}(\frac{1}{2} + iz).$$

Then $z_0 = 0$ and/or z_0 is a zero of $\xi_{\mathbb{Q},2}(\frac{1}{2} + iz)$ since $\xi_{\mathbb{Q},2}(\frac{1}{2} + iz)$ admits simple poles at $z = \pm\frac{1}{2}i$.

In any case,

$$F(z_0 + \frac{i}{4}) = F(z_0 - \frac{i}{4}).$$

By taking absolute values on both sides,

$$\Big|e^{A+B(z_0+\frac{i}{4})} \cdot \prod\Big(1 - \frac{z_0 + \frac{i}{4}}{\rho}\Big) \cdot \exp\Big(\frac{z_0 + \frac{i}{4}}{\rho}\Big)\Big|$$
$$= \Big|e^{A+B(z_0-\frac{i}{4})} \cdot \prod\Big(1 - \frac{z_0 - \frac{i}{4}}{\rho}\Big) \cdot \exp\Big(\frac{z_0 - \frac{i}{4}}{\rho}\Big)\Big|.$$

Since $B \in \mathbb{R}$ and $\rho_n \in \mathbb{R}$ (which is obtained by the RH as said above), we hence get

$$1 = \prod_{n=1}^{\infty} \frac{(x_0 - \rho_n)^2 + (y_0 - \frac{1}{4})^2}{(x_0 - \rho_n)^2 + (y_0 + \frac{1}{4})^2}.$$

Thus if $y_0 > 0$, then the right hand side is < 1, while if $y_0 < 0$, then the right hand side is > 1. Contradiction. This leads then $y_0 = 0$. □

With this in mind, note that in the proof above, the RH was used to ensure that ρ are real, which have the effect that then in the calculation for the exponential factor $\exp\left(\frac{z_0+\frac{i}{4}}{\rho}\right)$, the ratio $\frac{\left|\exp\left(\frac{z_0+\frac{i}{4}}{\rho}\right)\right|}{\left|\exp\left(\frac{z_0-\frac{i}{4}}{\rho}\right)\right|}$ gives us the exact value 1. That is to say, this factor of ratio of exp's does not contribute.

However, one does not need such an argument from the very beginning to eliminate the factors $\exp\left(\frac{z_0\pm\frac{i}{4}}{\rho}\right)$. In fact, this is the improvement of Lagarias, who gets his own unconditional result totally independently, as a part of his understanding of de Branges's work. The trick is very simple: Use the functional equation. So instead of working on individual ρ_n in the product, we may equally use the functional equation to pair ρ and $1-\rho$ for the zeros of the completed Riemann zeta function, or even to group ρ, $1-\rho$, $\overline{\rho}$ and $1-\overline{\rho}$ together. Consequently, the exponential factor appeared inside the infinite product may be totally omitted. That is to say, from the very beginning, we may simply assume that the Hadamard product involved takes the form

$$F(z) = e^{A+Bz}\cdot \prod_{\rho:F(\rho)=0}{}' \left(1-\frac{z}{\rho}\right)$$

where $\prod'$ means that ρ's are paired or grouped as above. Form here, it is an easy exercise to deduce the following result of (Suzuki and) Lagarias.

Theorem$_{\mathbb{Q}}$. *All zeros of* $\xi_{\mathbb{Q},2}(s)$ *lie on the line* $\mathrm{Re}\,(s) = \frac{1}{2}$.

Proof. Alternatively, as above, we have

$$\begin{aligned}
&\left|e^{A+B(z_0+\frac{i}{4})}\cdot \prod\left(1-\frac{z_0+\frac{i}{4}}{\rho}\right)\cdot\exp\left(\frac{z_0+\frac{i}{4}}{\rho}\right)\right|\\
&= \left|e^{A+B(z_0-\frac{i}{4})}\cdot \prod\left(1-\frac{z_0-\frac{i}{4}}{\rho}\right)\cdot\exp\left(\frac{z_0-\frac{i}{4}}{\rho}\right)\right|.
\end{aligned}$$

Since $B\in\mathbb{R}$, so if we can take care of the factors $\exp\left(\frac{z_0+\frac{i}{4}}{\rho}\right)$ and $\exp\left(\frac{z_0-\frac{i}{4}}{\rho}\right)$ in a nice way, we are done. For this, as said above, let us group ρ, $\bar{\rho}$, $1-\rho$, $1-\bar{\rho}$ together, we see that $\frac{1}{\rho}+\frac{1}{\bar{\rho}}=\frac{2\,\mathrm{Re}\,(\rho)}{|\rho|^2}$ and $\frac{1}{1-\rho}+\frac{1}{1-\bar{\rho}}=\frac{2-2\,\mathrm{Re}\,(\rho)}{|1-\rho|^2}$ are all reals, hence, the same prove as above works. □

8.1.3 A Simple Generalization

The above method works for the functions $\xi_{\mathbb{Q},2}^{T}(s)$ as well, provided that $T \geq 1$. Indeed, first, recall that we have the precise relation

$$\xi_{\mathbb{Q},2}^{T}(s) = \frac{\xi(2s)}{s-1} \cdot T^{s-1} - \frac{\xi(2s-1)}{s} \cdot T^{-s}.$$

Consequently,

$$F(z+\frac{i}{4}) \cdot T^{-\frac{1}{2}-iz} - F(z-\frac{i}{4}) \cdot T^{-\frac{1}{2}+iz} = iz(1+4z^2)\,\xi_{\mathbb{Q},2}^{T}(\frac{1}{2}+zi).$$

Therefore, using the same proof, we arrive at the relation

$$\begin{aligned}&\left|e^{A+B(z_0+\frac{i}{4})} \cdot \prod\left(1-\frac{z_0+\frac{i}{4}}{\rho}\right) \cdot \exp\left(\frac{z_0+\frac{i}{4}}{\rho}\right)\right| \cdot \left|T^{-iz-\frac{1}{2}}\right| \\ &\quad = \left|e^{A+B(z_0-\frac{i}{4})} \cdot \prod\left(1-\frac{z_0-\frac{i}{4}}{\rho}\right) \cdot \exp\left(\frac{z_0-\frac{i}{4}}{\rho}\right)\right| \cdot \left|T^{iz-\frac{1}{2}}\right|.\end{aligned}$$

That is to say,

$$1 = \prod_{n=1}^{\infty} \frac{(x_0-\rho_n)^2 + (y_0-\frac{1}{4})^2}{(x_0-\rho_n)^2 + (y_0+\frac{1}{4})^2} \cdot \frac{T^{-y_0}}{T^{y_0}}.$$

Or equivalently,

$$T^{2y_0} = \prod_{n=1}^{\infty} \frac{(x_0-\rho_n)^2 + (y_0-\frac{1}{4})^2}{(x_0-\rho_n)^2 + (y_0+\frac{1}{4})^2}.$$

Thus with $T \geq 1$, we have
(i) if $y_0 > 0$, the left hand side is > 1, while the right hand side is < 1, contradiction; while
(ii) if $y_0 < 0$, the left hand side is < 1, while the right hand side is > 1, contradiction. That is to say, we obtain the following

Theorem$'_{\mathbb{Q}}$. *For $T \geq 1$, all zeros of $\xi_{\mathbb{Q},2}^{T}(s)$ lie on the critical line* $\mathrm{Re}\,(s) = \frac{1}{2}$.

Recall that

$$\xi_{\mathbb{Q},2}^{T}(s) = \frac{\xi(2s)}{s-1} \cdot T^{s-1} - \frac{\xi(2s-1)}{s} \cdot T^{-s}.$$

Clearly

$$\lim_{T\to 1^+} \xi^T_{\mathbb{Q},2}(s) = \xi_{\mathbb{Q},2}(s),$$

while $\lim_{T\to +\infty} \xi^T_{\mathbb{Q},2}(s)$ does not really make any sense. This says that even when we have a family of natural functions whose zeros all lie on the critical line, in general, we cannot take limit for this family to preserve this property, no matter how careful we are.

8.2 Zeros of Rank Two Zetas

Finally, we are ready to state the following

Theorem. *All zeros of rank two zeta functions for number fields are on the critical line* $\mathrm{Re}\,(s) = \frac{1}{2}$.

Proof. This is a direct consequence of the following two facts:
First, as stated in Thm 7.2, we know that, by the Rankin-Selberg & Zagier method, up to a constant factor depending only on K,

$$\xi_{K,2}(s) = \frac{\xi_K(2s)}{s-1} - \frac{\xi_K(2s-1)}{s}.$$

Secondly, $s(s-1)\cdot\xi_K(s)$ is also an entire function of order one [L1]. Consequently, the proof in the previous section, more precisely, that of 8.1.2, on the zeros works here as well by a simple change from the Riemann ξ for the field of rationals $\mathbb{Q}$ to the Dedekind ξ_K for the number field K. □

9 A Rank Three Zeta and Its Zeros

In this chapter, we evaluate rank 3 zeta functions, using Langlands' theory of Eisenstein series, Arthur's theory of analytic truncations, and an advance version of Rankin-Selberg & Zagier method established in 4.7.

The starting point is a general relation between Arthur's analytic truncation and the geometric truncation. Simply put, geometric truncation using stability is the same as Arthur's analytic truncation (taking effect on constant function **1**) for suitable choices of parameters. But then, an essential difficulty appears: In the theory of Arthur, parameters involved are all supposed to be very large, i.e., the truncation is taken to be very near cusps. With such sufficiently large parameters, because the cut-off part is far away, basic properties that the truncation operator is indeed an orthogonal projection and that the truncation for constant function **1** becomes a characteristic function do hold. On the other hand, in our geometric truncation, say, the one for semi-stable lattices, parameters are relatively small. And, it is by no means clear that the above mentioned properties hold for such parameters. Fortunately, geometric truncation using stability is intrinsic and hence has to be understood at each and hence all levels of parabolic subgroups. Consequently, the corresponding analytic truncation for **1** is still a characteristic function and the associated truncation operator is an orthogonal projection too. With this cleared, for the calculation of rank 3 zeta, it suffices to find out the so-called constant terms of certain Eisenstein series with the help of a residue argument, providing that all the integrals appeared are convergent.

However, this latest assumption turns out to be a very difficult one: The convergence argument is really the heart of Arthur's analytic truncation theory, in which, as mentioned above, the parameters are sufficiently large, not just 'normal' in our current situation. Luckily, this can be solved in due course as well, thanks to the advanced studies for trace formula. The notable points are 1) for large parameters, by the original argument of Arthur, all integrations appeared are convergent; and 2) the integrands, varying over difference of regions obtained between large parameters and for smaller parameters are compactly supported. (We here should mention that while 1) is quite hard, 2) is quite obvious. But still it is this easy new input that solves our problem instantly.)

With the convergence problem being fixed, we are now ready to come to the point of constant terms of Eisenstein series. As stated in Ch. 1, Eisenstein series used in high rank zetas are Epstein type zeta functions. Being a special kind of Eisenstein series, they come from suitable L^2-functions on maximal parabolic subgroups. Hence, according to general theory of Langlands on Eisenstein series, they can be obtained as residues of Eisenstein series associated to cusp forms on lower rank parabolic subgroups. As such, the problem is changed to the one of writing down concretely such a residue argument. In the case at hand, by a classical result, we know that the cusp form can be taken to be the constant function **1** on minimal parabolic subgroup. (Recall that being abelian, as a minimal parabolic of SL, the L^2-function **1** is indeed cuspidal on the Levi.) To be more precise, we will use classical Eisenstein series such as Koecher zeta functions and Selberg's Eisenstein series associated to parabolic subgroups of SL_n.

With this important matter being settled again, to go further we need to recall the theory of constant terms of Eisenstein series associated to cusp forms. According to Langlands, they are closed related and are indeed given by the so-called intertwining operators, which, following Langlands as well, can be naturally given by the so-called Gindikin-Karpelevich formula. So a calculation about the action of Weyl group on the roots system must be worked out in details. After that, by applying our generalization in 4.1 of a result of Jacquet-Lapid-Rogowski, an advance version of Rankin-Selberg & Zagier method, final expression of rank 3 zeta can be obtained by taking residues along suitable singular hyperplanes.

Let us now briefly review the related works on constant terms of Eisenstein series for SL_3. In fact, a right expression first appeared in Venkov's basic example of trace formulas for SL_3. However Venkov only briefly states the final result and gives no details. (He justifies this by merely saying that what he gets comes from Langlands' general theory.) A few years later, in an effort of giving all Fourier coefficients of the Eisenstein series for SL(3,Z), Vinagradov-Takhtajan obtained the same result, from which in particular, they proved the functional equation for the Eisenstein series. Their derivation is completely elementary and is independent on anything except Epstein zeta result for SL_2.

Sure, the story does not stop here. Immediately after obtaining the precise formula for the rank 3 zeta, I sent it to Masatoshi Suzuki, and asked him to check whether the Riemann Hypothesis works for this rank 3 zeta as well. Suzuki first did some numerical calculations from which he was convinced that all zeros do lie on the central line. So he started working more carefully. Now we have the following theorem of him: *All zeros of rank 3 zeta for the field of rationals lie on the central line* $\mathrm{Re}\, s = \frac{1}{2}$. As such then it appears that the time is ripe to openly make the following generalization of the Riemann Hypothesis:

The RH. *All zeros of zetas, rank one or higher, lie on the central line* $\mathrm{Re}\, s = \frac{1}{2}$.

9.1 Various Eisenstein Series

9.1.1 Eisenstein Series Associated to 1

Let G be a reductive group defined over a number field F. Fix a minimal parabolic subgroup P_0 and its Levi subgroup M_0, both are supposed to be defined over F as well. Take a standard parabolic subgroup M with Levi decomposition $P = MU$ where M is the corresponding standard Levi subgroup and U is the unipotent radical. Then for an automorphic form $\psi \in A\big(U(\mathbb{A})M(F)\backslash G(\mathbb{A})\big)$ of P-level, by definition, the associated *Eisenstein series* is given by the convergence series

$$E\big(\psi, \lambda, g\big) := \sum_{\delta \in P(F)\backslash G(F)} m_P(\delta g)^{\lambda+\rho} \cdot \psi(\delta g)$$

where $\lambda \in \mathfrak{P}$ satisfying $\mathrm{Re}\lambda \in C$ a certain positive cone. (For unknown notions, please refer to 2.1.) In particular, if ψ is a cusp form, then C may be chosen to be the positive Weyl chamber of $\mathfrak{P}$.

For cusp forms, by a result of Langlands, the constant term of E with respect to a parabolic subgroup $P' = M'U'$ is given by

$$E_{P'}(\psi, \lambda; g) := \sum_{w \in W(M,M')} E^{M'/M}\big(M(w, \lambda)\psi, w\lambda\big)(g)$$

where

$$\begin{aligned} W(M, M') := \Big\{ w \in W \,\Big|\, & w^{-1}(\alpha) > 0 \\ & \forall \alpha \in T^+(T_0, M'),\ wMw^{-1} \hookrightarrow M'\text{standard Levi}\Big\}, \end{aligned}$$

and $E^{M'/M}$ denotes the relative Eisenstein series. As a special case, if M' is conjuagate to M, then

$$E_{P'}(\psi,\lambda;g) := \sum_{w\in W(M)} \Big(M(w,\lambda)\,\psi\Big)(g).$$

For our purpose, we take G to be SL_3, $P = P_0 = M_0U_0$ the minimal parabolic subgroup consisting of upper triangular matrices with M_0 diagonal matrices in P_0 (so it is a torus, in particular, is commutative) and U_0 the upper triangular matrices with all entries on the diagonal being 1. As such, then the constant function $\mathbf{1}$ on M_0 is cuspidal and its associated Eisenstein series is simply

$$E(\mathbf{1},\lambda)(g) = \sum_{\delta\in P_0(F)\backslash SL_3(F)} m_{P_0}(\delta g)^{\lambda+\rho_0}, \qquad \forall \mathrm{Re}\,\lambda \in C_0$$

and

$$E_{P_0}(\mathbf{1},\lambda)(g) = \sum_{w\in W(M_0)} M(w,\lambda)\cdot m_{P_0}(g)^{w(\lambda+\rho_0)}$$

while $E_{P'}(\mathbf{1},\lambda)(g) \equiv 0$ since $wM_0w^{-1} \not\leftrightarrow M'$ if $M' \simeq M_0$ (recall that while P_0 is minimal, its Levi M_0 is maximal).

9.1.2 Koecher Zeta Functions

Defined the so-called *Koecher zeta function* by

$$Z_{m,n-m}(X,s) := \sum_{A\in\mathbb{Z}^{m\times n}\backslash GL(m,\mathbb{Z}),\mathrm{rk}A=m} \Big|X[A]\Big|^{-s}, \qquad \mathrm{Re}(s) > \frac{n}{2}$$

where $X[A] := A^t\cdot X\cdot A$, and $|\ |$ denotes the determinant. It is convergent hence well-defined, and satisfies many interesting properties. (For a discussion related to classical Eisenstein series, a useful reference is [Te]. In fact we here adopt her notation as well.) In particular, when $m = 1$, we get

$$
\begin{aligned}
Z_{1,n-1}(X,s) :=& \sum_{A\in\mathbb{Z}^{1\times n}/GL(1,\mathbb{Z}),\mathrm{rk}A=1} \Big|X[A]\Big|^{-s} \\
=& \sum_{A\in\mathbb{Z}^{1\times n}/\pm 1,\mathrm{rk}A=1} \Big|X[A]\Big|^{-s} \\
=& \sum_{A\in\mathbb{Z}^{n}\backslash 0, A=-A} \Big|X[A]\Big|^{-s} \\
=& \frac{1}{2}\sum_{\nu\in\mathbb{Z}^n\backslash 0} \frac{1}{|\nu^t\cdot X\cdot \nu|^s}
\end{aligned}
$$

which is indeed the standard Epstein zeta functions. For example,

$$Z_{1,0}(X,s) = \frac{1}{2}\sum_{\nu\in\mathbb{Z}\backslash 0}\frac{1}{|\nu^t\cdot X\cdot\nu|^s} = \frac{1}{X^s}\zeta(2s).$$

9.1.3 A Special Eisenstein Series

Suppose that ψ is an automorphic form on a lower rank parabolic subgroups, $1 \le m < n$, $\mathrm{Re}\, s > \frac{n}{2}$, $Y \in \mathcal{P}_n$, *the space of positive* $n\times n$ *real matrices.* Define then the associated Eisenstein series $E_{m,n-m}(\psi, s|Y) := E(\psi, s|Y)$ by

$$E(\psi, s|Y) = \sum_{A=(A_1,*)\in\Gamma_n/P(m,n-m)} \Big|Y[A_1]\Big|^{-s}\cdot\psi(Y[A_1]^o).$$

Here if $Y \in \mathcal{P}_n$ we write $Y = t^{\frac{1}{n}}W$ with $t = |Y| > 0$ and $W =: Y^o \in S\mathcal{P}_n := \{Y \in \mathcal{P}_n \,|\, |Y| = 1\}$; $A_1 \in \mathbb{Z}^{n\times m}$, $\Gamma_n = GL(n,\mathbb{Z})$, $P(m, n-m) :=$ the subgroup consisting of elements of the form $\begin{pmatrix} *_1 & 0 \\ 0 & *_2\end{pmatrix}$ where $*_1$ (resp. $*_2$) is a matrix of size m (resp. n).

9.1.4 Relation Between Z and E

Lemma 1. *The lattice version and the group version of Eisenstein series are related by the following relations*

$$Z_{m,n-m}(Y,s) = Z_{m,0}(I,s)\cdot E_{m,n-m}(\mathbf{1}, s|Y)$$

and

$$Z_{m,0}(X,s) = |X|^{-s} \cdot \prod_{j=0}^{m-1} \zeta(2s-j).$$

In particular,

$$\begin{aligned} & Z_{1,n-1}(Y,s) \\ &= |Y|^{-s} \cdot \zeta(2s) \cdot E_{1,n-1}(\mathbf{1}, s|Y) \\ &= |Y|^{-s} \cdot \zeta(2s) \cdot \sum_{A=(A_1,*)\in\Gamma_n/P(1,n-1), A_1\in\mathbb{Z}^{n\times 1}} |Y[A_1]|^{-s}. \end{aligned}$$

Proof. We need the following

Lemma 2. *The quotient $\mathbb{Z}^{n\times m}\,\mathrm{rk}\, m/\Gamma_m$, which means the $n \times m$, rank m integral matrices in a complete set of matrices inequivalent under right multiplication by matrices in $\Gamma_m = GL(m,\mathbb{Z})$, for $1 \leq m < n$, can be expressed by matrices $A = B{\cdot}C$ where $B \in \mathbb{Z}^{n\times m}$, $(B,*) \in GL(n,\mathbb{Z})/P(m,n-m)$, $C \in \mathbb{Z}^{m\times m}\mathrm{rk}\, m/GL(m,\mathbb{Z})$.*

For a proof of this lemma, see e.g., [Tr].

Indeed, then by definition,

$$\begin{aligned} & Z_{m,n-m}(X,s) \\ &= \sum_{A\in\mathbb{Z}^{n\times m}\mathrm{rk}\, m/GL(m,\mathbb{Z})} |X[A]|^{-s} \\ &= \sum_{BC\in\mathbb{Z}^{n\times m},(B*)\in GL(n,\mathbb{Z})/P(m,n-m),C\in\mathbb{Z}^{m\times m}\,\mathrm{rk} m/GL(m,\mathbb{Z})} |C^tB^tXBC|^{-s} \end{aligned}$$

by the lemma since $|AB| = |A| \cdot |B|$, while

$$\begin{aligned} Z_{m,0}(I,s) \cdot E_{m,n-m}(\mathbf{1}, s|X) = & \sum_{C\in\mathbb{Z}^{m\times m}/GL(m,\mathbb{Z}),\mathrm{rk} C=m} |C^t \cdot C|^{-s} \\ & \times \sum_{(B,*)\in GL(n,\mathbb{Z})/P(m,n-m)} |B^tXB|^{-s}. \end{aligned}$$

□

With this, for high rank zetas, we shift our attention to $E_{1,n-1}$.

Remark. In fact what we really use for rank 3 zeta function is $E_{1,2}$. For this, to see the relation between $Z_{1,2}$ and $E_{1,2}$ stated above, we may also use the following more direct

Lemma 2′. *Let* $(x, y, z) \in \mathbb{Z}^3 \backslash \{0\}$ *such that* $\gcd(x, y, z) = 1$, *then*

(i) *There exists a matrix* $A \in SL(3, \mathbb{Z})$ *with the form* $A = \begin{pmatrix} a_{11} & a_{12} & a_{13} \\ a_{21} & a_{22} & a_{23} \\ x & y & z \end{pmatrix}$;

(ii) *The above matrix* A *is unique up to a left multiplication of a matrix in* $P(1, 2) \cap SL(3, \mathbb{Z})$.

Proof. First let us prove (ii). If A and A' are as in (i). Then consider the product $A' \cdot A^{-1}$ and pay attention to the last row of the resulting matrix. Since the last rows of A and A' are the same, namely (x, y, z), the last row of the matrix $A' \cdot A^{-1}$ equals to $(0\,0\,1)$. This proves (ii).

Next let us prove (i). The matrix $A = \begin{pmatrix} a_{11} & a_{12} & a_{13} \\ a_{21} & a_{22} & a_{23} \\ x & y & z \end{pmatrix} \in SL(3, \mathbb{Z})$ we want to construct will take the form $A = \begin{pmatrix} a_{11} & a_{12} & a_{13} \\ 0 & a_{22} & a_{23} \\ x & y & z \end{pmatrix}$, that is, the entry $a_{12} = 0$. Then the condition that $A \in SL(3, \mathbb{Z})$ means that

$$x \cdot \begin{vmatrix} a_{12} & a_{13} \\ a_{22} & a_{23} \end{vmatrix} - a_{11} \cdot \left(a_{23} y - a_{22} z\right) = 1.$$

Let $\gcd(y, z) = \delta$. There exist $a, b \in \mathbb{Z}$ such that $ay + bz = \delta$ and $\gcd(a, b) = 1$. Set now $a_{23} = a, a_{22} = -b$. Then the above equation becomes

$$x \cdot \begin{vmatrix} a_{12} & a_{13} \\ -b & a \end{vmatrix} - a_{11} \cdot \left(ay + bz\right) = 1,$$

or equivalently,

$$x \cdot \begin{vmatrix} a_{12} & a_{13} \\ -b & a \end{vmatrix} - a_{11} \cdot \delta = 1.$$

Now note that $\gcd(x, \delta) = \gcd(x, y, z) = 1$, there exist $X,\ Y \in \mathbb{Z}$ such that $x \cdot X + \delta \cdot Y = 1$. Set $a_{11} = -Y$. Then the problem becomes that whether there exist $a_{12}, a_{13} \in \mathbb{Z}$ such that $X = \begin{vmatrix} a_{12} & a_{13} \\ -b & a \end{vmatrix}$. But this is clear, since gcd $(a, b) = 1$. Indeed, there exist $\alpha, \beta \in \mathbb{Z}$ such that $\alpha \cdot a + \beta \cdot b = 1$. As such, set $a_{12} := \alpha \cdot X$, $a_{13} := \beta \cdot X$ will do the job. □

9.1.5 Selberg Eisenstein Series

Let $\mathbf{s} = (s_1, \ldots, s_n) \in \mathbb{C}^n$ and $Y \in \mathcal{P}_n$. Define the so-called *Selberg Eisenstein series* by

$$E_{(n)}(\mathbf{s}|Y) := \sum_{\nu \in \Gamma_n / P_{(n)}} p_{-\mathbf{s}}(Y[\nu])$$

where for a matrix $A \in \mathcal{P}_n$, $p_{\mathbf{s}}(A)$ is the *power function* defined by $p_s(A) := \prod_{j=1}^{n} |A_j|^{s_j}$ with $A_j \in \mathcal{P}_j$, the $j \times j$ upper left hand corner in Y, $j = 1, \ldots, n$. (Note that when $n = 2$ and $Y \in S\mathcal{P}_2$, the power function can be identified with the function on the upper half plane defined by $p_s(x + iy) = y^s$ for $y > 0$.)

As usual, let $P = P(n_1, n_2, \ldots, n_q)$ be the parabolic subgroup of SL_n consisting of blocked upper triangular matrices of sizes $n_1, n_2, \ldots, n_q$ corresponding to the partition $n = n_1 + n_2 + \cdots + n_q$. Then for Selberg Eisenstein series, we have the following

Lemma 3. *For each* $i = 1, 2, \ldots, n-1$,

$$E_{(n)}(\mathbf{s}|Y) = \sum_{\nu=(\nu_1 *)=(\nu_2 *)=\cdots=(\nu_n)\in\Gamma_n/P_i^*, \nu_1\in\mathbb{Z}^{n\times 1}, \nu_2\in\mathbb{Z}^{n\times 2}, \ldots, \nu_n\in\mathbb{Z}^{n\times n}} E_{(2)}(s_i|T_i) \cdot \prod_{j=1, j\neq i}^{n} |Y[\nu_j]|^{-s_j}$$

where $P_i^* := P(1, \ldots, 1, 2, 1, \ldots, 1)$ *with 2 the i-th entry, and* T_i *is defined as follows: First write* $Y[\nu_{i+1}] = \begin{pmatrix} Y[\nu_{i-1}] & 0 \\ 0 & R_i \end{pmatrix} \begin{bmatrix} I_{i-1} & Q \\ 0 & I_2 \end{bmatrix}, i = 2, \ldots, n-1$ *and* $Y[\nu_2] = R_1$. *We see that all* $R_i \in \mathcal{P}_2$; *then introduce* $T_i \in \mathcal{P}_2$ *by setting* $T_1 = R_1 = Y[\nu_2]$ *and* $T_i := |Y[\nu_{i-1}]| \cdot R_i$ *for* $i = 2, \ldots, n-1$.

(In particular, $|T_1| = |Y[\nu_2]|$, $T_i = |Y[\nu_{i-1}]| \cdot |Y[\nu_{i+1}]|$ for $i = 2, \ldots, n-2$ and $T_{n-1} = |Y[\nu_{n-2}]| \cdot |Y|$.) For a proof, see e.g., [Te].

Example. For $n = 3$ with $i = 1, 2$, we obtain the following formulae

$$E_{(3)}(\mathbf{s}|Y) \overset{i=1}{=} \sum_{\nu=(\nu_2*)=\nu_3\in\Gamma_3/P(2,1),\nu_2\in\mathbb{Z}^{3\times 2},\nu_3\in\mathbb{Z}^{3\times 3}} E_{(2)}(s_1|T_1)\cdot|Y[\nu_2]|^{-s_2}\cdot|Y[\nu_3]|^{-s_3}$$

$$\overset{i=2}{=} \sum_{\nu=(\nu_1*)=\nu_3\in\Gamma_3/P(1,2),\nu_1\in\mathbb{Z}^{3\times 1},\nu_3\in\mathbb{Z}^{3\times 3}} E_{(2)}(s_2|T_2)\cdot|Y[\nu_1]|^{-s_1}\cdot|Y[\nu_3]|^{-s_3}$$

where $T_1 = Y[\nu_2], T_2 = |Y[\nu_1]|\cdot R_2$ and $Y[\nu] = \begin{pmatrix} Y[\nu_1] & 0 \\ 0 & R_2 \end{pmatrix}\begin{bmatrix} I_1 & Q \\ 0 & I_2 \end{bmatrix}$. Consequently,

$$\begin{aligned} &E_{(3)}(s_1, s_2, 0|Y) \\ &= \sum_{\nu\in\Gamma_3/P(1,2),\nu=(\nu_1*),\nu_1\in\mathbb{Z}^{3,1}} E_{(2)}(s_2|T_2)\cdot Y[\nu_1]^{-s_1} \\ &= \sum_{\nu\in\Gamma_3/P(1,2),\nu=(\nu_1*),\nu_1\in\mathbb{Z}^{3,1}} E_{(2)}(s_2|Y[\nu_1]\cdot R_2)\cdot Y[\nu_1]^{-s_1} \\ &= \sum_{\nu\in\Gamma_3/P(1,2),\nu=(\nu_1*),\nu_1\in\mathbb{Z}^{3,1}} E_{(2)}(s_2|R_2)\cdot Y[\nu_1]^{-s_1+\frac{s_2}{2}}. \end{aligned}$$

More generally, Siegel introduced the following Eisenstein series for the parabolic subgroup $P = P(n_1, n_2, \ldots, n_q)$:

$$E_P(\mathbf{s}|Y) := \sum_{A\in\Gamma_n/P,A=(A_j*),A_j\in\mathbb{Z}^{n\times N_j},j=1,\ldots,q} \prod_{j=1}^{q} |Y[A_j]|^{-s_j}$$

where $N_j := n_1 + n_2 + \cdots + n_j$.

For example, $E_{P_2^*}(s_1, s_3|Y)$

$$= \sum_{A\in\Gamma_3/P(1,2),A=(A_1*)=A_3,A_1\in\mathbb{Z}^{3\times 1},A_3\in\mathbb{Z}^{3\times 3}} |Y[A_1]|^{-s_1}\cdot|Y[A_3]|^{-s_3}.$$

In particular, $E_{P_2^*}(s, 0|Y)$

$$= \sum_{A\in\Gamma_3/P(1,2),A=(A_1*)A_1\in\mathbb{Z}^{3\times 1}} |Y[A_1]|^{-s} = \frac{|Y|^s}{\zeta(2s)}\cdot Z_{1,2}(Y, s).$$

9.1.6 Taking Residues

It is also known that

$$\begin{aligned} E_{(n)}(\mathbf{s}|Y)|_{s_i=0} =& E_{P_i^*}(s_1,\ldots,s_{i-1},s_{i+1},\ldots,s_n|Y),\\ \mathrm{Res}_{s_i=1}E_{(n)}(\mathbf{s}|Y) =& \frac{3}{2\pi}\cdot E_{P_i^*}(s_1^*,\ldots,s_{i-1}^*,s_{i+1}^*,\ldots,s_n^*|Y) \end{aligned}$$

with $s_j^* = s_j$ for $j \neq i \pm 1$, $s_{i\pm1}^* = s_{i\pm1} + \frac{1}{2}$. (See e.g. [Te].) In particular,

$$\begin{aligned} E_{(3)}(\mathbf{s}|Y)|_{s_2=0} =& E_{P_2^*}(s_1,s_3|Y),\\ \mathrm{Res}_{s_2=1}E_{(3)}(\mathbf{s}|Y) =& \frac{3}{2\pi}\cdot E_{P_2^*}(s_1+\frac{1}{2},s_3+\frac{1}{2}|Y). \end{aligned}$$

Or put it in a better form,

$$\begin{aligned} E_{(3)}(s,t,0|Y)|_{t=0} =& E_{P_2^*}(s,0|Y),\\ \mathrm{Res}_{t=1}E_{(3)}(s-\frac{1}{2},t,-\frac{1}{2}|Y) =& \frac{3}{2\pi}\cdot E_{P_2^*}(s,0|Y). \end{aligned}$$

To go further, note that, by definition,

$$E_{(3)}(s_1,s_2,s_3|Y) = \prod_{\gamma\in\Gamma_3/P_{(3)}} p_{-s}(Y[\gamma]) = |Y|^{-s_3}\cdot E_{(3)}(s_1,s_2,0|Y)$$

since $|Y[\gamma]|^{-s_3} = |\gamma^tY\gamma|^{-s_3} = |\gamma|^2\cdot|Y|^{-s_3}$. Consequently, to understand all $E_{(3)}(s_1,s_2,s_3|Y)$, we only need to understand $E_{(3)}(s,t,0|Y)$.

9.1.7 $\mathbf{E}_{2,1}$: An Example

As an example, we have in particular,

$$E_{2,3-2}(E_u,s|Y) = \sum_{(C_i,*)=C\in\Gamma_3/P_{(3)},C_i\in\mathbb{Z}^{3\times i}} Y[C_1]^{-u}\cdot|Y[C_2]|^{-s+u/2}$$

where $E_u(z) := \sum_{\gamma\in\Gamma_2/P(1,1)} y^u(\gamma z)$, and $P_{(3)} = P(1,1,1)$ is the minimal subgroup P_0.

Indeed, the right hand side is equal to

$$\begin{aligned} &\sum_{C\in\Gamma_3/P(1,1,1),C=(C_1*)=(C_2*),C_1\in\mathbb{Z}^{3\times1},C_2\in\mathbb{Z}^{3\times2}} |Y[C_1]|^{-u}\cdot|Y[C_2]|^{-s+u/2}\\ =&\sum_{C=(C_2*)\in\Gamma_3/P(2,1)}\big(Y[C_2]\big)^{-s+u/2}\cdot\sum_{(b*)\in\Gamma_2/P(1,1)}\big(Y[C_2][b]\big)^{-u} \end{aligned}$$

since $\Gamma_3/P(1,1,1) = \Gamma_3/P(2,1) \times P(2,1)/P(1,1,1)$ and

$$P(2,1)/P(1,1,1) \simeq \begin{pmatrix} \Gamma_2/P(1,1) & 0 \\ 0 & \pm 1 \end{pmatrix}.$$

Also by definition

$$\begin{aligned} E(E_u, s|Y) :=& \sum_{A=(A_1 *)\in\Gamma_3/P(2,1)} |Y[A_1]|^{-s} \cdot E_u(Y[A_1]^o) \\ =& E_{(3)}(u, s - \frac{u}{2}, 0|Y). \end{aligned}$$

(Note that in E_u, $*^o$ is used, consequently, in $E_{(3)}$, the variable becomes $s - \frac{u}{2}$ not simply s).

9.2 What Matters to Us

9.2.1 The Function $p_s(Y)$

In this section, for SL_3, we give a precise relation between classical and current expressions of Eisenstein series. The key then is that for SL_3, any element g can be written as

$$g = \begin{pmatrix} \sqrt{u} & \frac{v}{\sqrt{u}} & 0 \\ 0 & \frac{1}{\sqrt{u}} & 0 \\ 0 & 0 & 1 \end{pmatrix} \cdot \begin{pmatrix} \alpha & 0 & 0 \\ 0 & \alpha & 0 \\ 0 & 0 & \alpha^{-2} \end{pmatrix} \cdot \begin{pmatrix} 1 & 0 & x \\ 0 & 1 & t \\ 0 & 0 & 1 \end{pmatrix},$$

where $\alpha > 0, u > 0, v, x, t \in \mathbb{R}$, so

$$g^t = \begin{pmatrix} 1 & 0 & 0 \\ 0 & 1 & 0 \\ x & t & 1 \end{pmatrix} \cdot \begin{pmatrix} \alpha & 0 & 0 \\ 0 & \alpha & 0 \\ 0 & 0 & \alpha^{-2} \end{pmatrix} \cdot \begin{pmatrix} \sqrt{u} & 0 & 0 \\ \frac{v}{\sqrt{u}} & \frac{1}{\sqrt{u}} & 0 \\ 0 & 0 & 1 \end{pmatrix}.$$

Consequently,

$$Y := g^t g =: \begin{pmatrix} A & * \\ * & * \end{pmatrix}$$

with

$$A = \alpha^2 \cdot \begin{pmatrix} \sqrt{u} & 0 \\ \frac{v}{\sqrt{u}} & \frac{1}{\sqrt{u}} \end{pmatrix} \cdot \begin{pmatrix} \sqrt{u} & \frac{v}{\sqrt{u}} \\ 0 & \frac{1}{\sqrt{u}} \end{pmatrix} = \alpha^2 \cdot \begin{pmatrix} u & v \\ v & \frac{v^2+1}{u} \end{pmatrix}.$$

In particular, $|Y_1| = \alpha^2 u$, $|Y_2| = \alpha^4\Big(u \cdot \frac{v^2+1}{u} - v^2\Big) = \alpha^4$. This then leads to, for $y := \alpha^6$,

$$\begin{aligned} p_s(Y) &:= |Y_1|^{s_1} \cdot |Y_2|^{s_2} = (\alpha^2 u)^{s_1} (\alpha^4)^{s_2} \\ &= u^{s_1} \cdot (\alpha^6)^{\frac{s_1+2s_2}{3}} = u^{s_1} \cdot y^{\frac{s_1+2s_2}{3}}. \end{aligned}$$

9.2.2 Root System for SL_3

In $\mathbb{R}^3$, set $e_1 := (1,0,0), e_2 = (0,1,0), e_3 = (0,0,1)$ be the standard orthonormal basis. Then with $\alpha_1 := e_1 - e_2, \alpha_2 := e_2 - e_3$, we get $\|\alpha_1\|^2 = \|\alpha_2\|^2 = 2$. And from $-1 = \langle \alpha_1, \alpha_2 \rangle = \|\alpha_1\| \cdot \|\alpha_2\| \cdot \cos\theta = 2\cos\theta$, we get $\theta = \frac{2}{3}\pi$. As usual, introduce the so-called co-root by $\alpha^\vee := \frac{2}{\langle \alpha, \alpha \rangle} \cdot \alpha$. Then

$$\alpha_1^\vee = \frac{2}{2}\alpha_1 = \alpha_1, \qquad \alpha_2^\vee = \frac{2}{2}\alpha_2 = \alpha_2.$$

Set also $\Delta_0^\vee := \{\alpha_1^\vee, \alpha_2^\vee\}$ and denote its dual basis by $\widehat{\Delta}_0 := \{\varpi_1, \varpi_2\}$ (so that $\varpi_i(\alpha_j^\vee) = \delta_{ij}$). Note that

$$\begin{aligned} (a\alpha_1 + b\alpha_2)(u_1, u_2, u_3) &= a(u_1 - u_2) + b(u_2 - u_3) \\ &= au_1 + (b-a)u_2 - bu_3 \end{aligned}$$

from $\varpi_1(\alpha_1) = 1, \varpi_1(\alpha_2) = 0$, we get $a - (b-a) = 1, (b-a) + b = 0$, that is, $a = \frac{2}{3}, b = \frac{1}{3}$. Or better,

$$\varpi_1 = \frac{2}{3}\alpha_1 + \frac{1}{3}\alpha_2 = (\frac{2}{3}, -\frac{1}{3}, -\frac{1}{3}).$$

Similarly from $\varpi_2(\alpha_1) = 0, \varpi_2(\alpha_2) = 1$, we get $a-(b-a) = 0, (b-a)+b = 1$, that is, $a = \frac{1}{3}, b = \frac{2}{3}$. Or better,

$$\varpi_2 = \frac{1}{3}\alpha_1 + \frac{2}{3}\alpha_2 = (\frac{1}{3}, \frac{1}{3}, -\frac{2}{3}).$$

Consequently, $\|\varpi_1\|^2 = \frac{4}{9} + \frac{1}{9} + \frac{1}{9} = \frac{2}{3}$ and $\|\varpi_2\|^2 = \frac{1}{9} + \frac{1}{9} + \frac{4}{9} = \frac{2}{3}$.

Also by definition, $\alpha_1 \perp \varpi_2$, that is, α_1 and ϖ_2 are orthogonal to each other. So we may use them, or better use $\{3\varpi_2 = \alpha_1 + 2\alpha_2, \alpha_1\}$ as a new coordinate system so that every point λ on the roots plane can be written as

$$\lambda = s \cdot (3\varpi_2) + t \cdot \alpha_1 = s \cdot (\alpha_1 + 2\alpha_2) + t \cdot \alpha_1 = (s+t)\alpha_1 + 2s\alpha_2.$$

In particular, we have

$$e^{\langle\lambda,H(g)\rangle} := g^\lambda := \begin{pmatrix} \alpha\sqrt{u} & 0 & 0 \\ 0 & \frac{\alpha}{\sqrt{u}} & 0 \\ 0 & 0 & \alpha^{-2} \end{pmatrix}^\lambda = \left(\frac{\alpha\sqrt{u}}{\frac{\alpha}{\sqrt{u}}}\right)^{s+t} \cdot \left(\frac{\frac{\alpha}{\sqrt{u}}}{\alpha^{-2}}\right)^{2s}$$
$$= u^{s+t}\cdot\alpha^{6s}\cdot u^{-s} = u^t\cdot(\alpha^6)^s = u^t\cdot y^s.$$

Consequently, $e^{\langle\lambda,H_0(g)\rangle} = u^ty^s = p_{(s_1,s_2)}(Y)$ when $s_1 = t$, $\frac{s_1+2s_2}{3} = s$ or better $s_1 = t$, $s_2 = \frac{3s-t}{2}$. In particular, when $\lambda = s\cdot(3\varpi_2)+t\cdot\alpha_1$,

$$E(\mathbf{1},\lambda)(g) = E_{(3)}(t+\frac{1}{2},\frac{3s-t}{2}+\frac{1}{2},0|g^tg)$$

since $\rho_0 = \alpha_1+\alpha_2$, $E(\mathbf{1},\lambda)(g) = E_{2,1}(E_{t+\frac{1}{2}}, s+\frac{1}{2}|Y)$, and

$$E_{(3)}(s_1,s_2,-\frac{1}{2}|Y) = E_{(3)}(t,\frac{3s-t}{2},-\frac{1}{2}|Y) = E_{(3)}(t,\frac{3s-t}{2},0|Y).$$

Remark. It is the function $E(\mathbf{1},\lambda-\rho_0)(g) = E^0(Y;s,t)$ that is used by Venkov ([Ve]). This explains the above change of variables $s\mapsto s+\frac{1}{2}, t\mapsto t+\frac{1}{2}$.

9.3 Relation with the Rank Three Zeta

Define the *completed Eisenstein series* associated to $E^0(Y;s,t)$ by

$$\widehat{E}^0(Y;s,t) := \widehat{E}(\mathbf{1},\lambda-\rho_0)(g) := \xi(2t)\xi(3s-t)\xi(3s+t-1)\cdot E^0(Y;s,t).$$

Then we have the following very neat

Functional Equations. *With the same notations as above,*

(i) $\widehat{E}^0(Y;s,1-t) = \widehat{E}^0(Y;s,t)$;

(ii) $\widehat{E}^0(Y;\frac{1}{2}(1-s+t),\frac{1}{2}(-1+3s+t)) = \widehat{E}^0(Y;s,t)$;

(iii) $\widehat{E}^0(Y;\frac{1}{2}(2-s-t),\frac{1}{2}(2-3s+t)) = \widehat{E}^0(Y;s,t)$;

(iv) $\widehat{E}^0(Y;\frac{1}{2}(1-s+t),\frac{1}{2}(3-3s-t)) = \widehat{E}^0(Y;s,t)$;

(v) $\widehat{E}^0(Y;\frac{1}{2}(2-s-t),\frac{1}{2}(3s-t)) = \widehat{E}^0(Y;s,t)$.

Proof. Replacing $\widehat{E}$ by E, this is a precise realization of Langlands' result on functional equations for the Eisenstein series involved: The totality should be 6, the cardinal number of the Weyl group. Here we omit the trivial one corresponding to the identity. For details, please also refer the discussion in the next section. □

Moreover, by 3.3, 5.3 and 9.1-2, we have, up to a constant factor,

$$\begin{aligned}
\xi_{F,3}(s) :=& \int_{\mathcal{M}_{F,3}} \widehat{E}_{\mathbb{Q},3}(\Lambda, s)\, d\mu(\Lambda)\\
=& \int_{\mathcal{M}_{F,3}} \mathrm{Res}_{\frac{3s-t}{2}+\frac{1}{2}=1} \widehat{E}_{(3)}\Big(t+\frac{1}{2}, \frac{3s-t}{2}+\frac{1}{2}, 0|Y(\Lambda)\Big)\, d\mu(\Lambda(Y))\\
=& \mathrm{Res}_{3s-t=1} \int_{\mathcal{M}_{F,3}} \widehat{E}_{(3)}\Big(t+\frac{1}{2}, \frac{3s-t}{2}+\frac{1}{2}, 0|Y(\Lambda)\Big)\, d\mu(\Lambda(Y))\\
=& \mathrm{Res}_{3s-t=1} \int_{\Lambda(G(F)\backslash G(\mathbb{A}))} \Lambda^{(u=1,y=1)} \widehat{E}(\mathbf{1}, \lambda)(g)\, d\mu(g).
\end{aligned}$$

This latest integration can be evaluated, since by Theorem 4.7,

$$\begin{aligned}
&\int_{\Lambda(G(F)\backslash G(\mathbb{A}))} \Lambda^{(u=1,y=1)} E(\mathbf{1}, \lambda)(g)\, d\mu(g)\\
=&v \cdot \sum_{w\in W} \frac{M(w,\lambda)}{\prod_{\alpha\in\Delta_0} \langle w\lambda - \rho_0, \alpha^\vee\rangle} \cdot \mathrm{Vol}\big(A_0(F)\backslash A_0(\mathbb{A})\big).
\end{aligned}$$

This then leads to the calculation of next section.

Remark. While it is the completed Eisenstein series that enters into the final discussion of our rank 3 zeta, but to follow notations used by most of current researchers, we will use $E^0(Y; s+\frac{1}{2}, t+\frac{1}{2}) = E(\mathbf{1},\lambda)(g)$, in other words, without clearing the denominators for the ξ's appeared in the constant term along with the minimal parabolic subgroup $P(1,1,1)$, to do the calculation.

9.4 Elementary Calculations

9.4.1 Action of Weyl Group on Root System

This is rather elementary, but for inexperienced reader, we give all the details.

For any root α, we have the associated reflection

$$s_\alpha : x \mapsto x - \langle \alpha^\vee, x\rangle \cdot \alpha = x - \frac{\langle 2\alpha, x\rangle}{\langle \alpha, \alpha\rangle} \cdot \alpha.$$

In particular, if $x = k\alpha$, then

$$k\alpha \mapsto k\alpha - 2\frac{\langle \alpha, k\alpha\rangle}{\langle \alpha, \alpha\rangle} \cdot \alpha = k\alpha - 2k\alpha = -k\alpha.$$

So, $\alpha \mapsto -\alpha$ while for any x such $x \perp \alpha$, i.e., $\langle x, \alpha \rangle = 0$, we have $x \mapsto x - 0 \cdot \alpha = x$ remains unchanged.

It is well-known that the Weyl group W of SL_3 is isomorphic to S_3, consists of 6 elements and is generated by

$$\tau_1 = s_{\alpha_1} : \alpha_1 \mapsto -\alpha_1, \alpha_2 \mapsto \alpha_2 - \langle \alpha_1, \alpha_2 \rangle \alpha_1 = \alpha_2 - (-1) \cdot \alpha_1 = \alpha_1 + \alpha_2$$

and

$$\tau_2 = s_{\alpha_2} : \alpha_1 \mapsto \alpha_1 - \langle \alpha_2, \alpha_1 \rangle \alpha_2 = \alpha_1 - (-1)\alpha_2 = \alpha_1 + \alpha_2, \alpha_2 \mapsto -\alpha_2.$$

Other elements of W are Id,

$$\tau_2 \circ \tau_1 : \alpha_1 \mapsto -\alpha_1 \mapsto -(\alpha_1 + \alpha_2), \alpha_2 \mapsto (\alpha_1 + \alpha_2) \mapsto (\alpha_1 + \alpha_2) - \alpha_2 = \alpha_1,$$

$$\tau_1 \circ \tau_2 : \alpha_1 \mapsto \alpha_1 + \alpha_2 \mapsto (-\alpha_1) + \alpha_1 + \alpha_2 = \alpha_2, \alpha_2 \mapsto -\alpha_2 \mapsto -(\alpha_1 + \alpha_2)$$

and

$$\tau_1 \circ (\tau_2 \circ \tau_1) : \alpha_1 \mapsto -(\alpha_1 + \alpha_2) \mapsto \alpha_1 - (\alpha_1 + \alpha_2) = -\alpha_2, \alpha_2 \mapsto \alpha_1 \mapsto -\alpha_1.$$

That is to say, we have

i) the identity map

$$\text{Id} : \alpha_1 \mapsto \alpha_1, \alpha_2 \mapsto \alpha_2;$$

ii) the reflection with respect to ϖ_2

$$\tau_1 : \alpha_1 \mapsto -\alpha_1, \alpha_2 \mapsto \alpha_1 + \alpha_2;$$

iii) the reflection with respect to ϖ_1

$$\tau_2 : \alpha_1 \mapsto \alpha_1 + \alpha_2, \alpha_2 \mapsto -\alpha_2;$$

iv) the rotation with the center at origin and the angle $\frac{2}{3}\pi$

$$\tau_2 \circ \tau_1 : \alpha_1 \mapsto -(\alpha_1 + \alpha_2), \alpha_2 \mapsto \alpha_1;$$

v) the rotation with the center at origin and the angle $\frac{4}{3}\pi$

$$\tau_1 \circ \tau_2 : \alpha_1 \mapsto \alpha_2, \alpha_2 \mapsto -(\alpha_1 + \alpha_2);$$

vi) the reflection with respect to the line $l_{-\frac{\pi}{6}}$.

$$\tau_1 \circ \tau_2 \circ \tau_1 = \tau_2 \circ \tau_1 \circ \tau_2 : \alpha_1 \mapsto -\alpha_2, \alpha_2 \mapsto -\alpha_1.$$

As such, then if $\lambda = s \cdot (3\varpi_2) + t\alpha_1 = (s+t)\alpha_1 + 2s\alpha_2$, then $g^\lambda = u^t y^s$, while the action of other elements in the Weyl group is given by

$$\begin{aligned}\tau_1\lambda &= \tau_1\big((s+t)\alpha_1 + 2s\alpha_2\big)\\ &= (s+t)(-\alpha_1) + 2s(\alpha_1+\alpha_2) = (s-t)\alpha_1 + 2s\alpha_2\end{aligned}$$

hence

$$g^{\tau_1\lambda} = u^{-t}y^s;$$

$$\begin{aligned}\tau_2\lambda &= \tau_2\big((s+t)\alpha_1 + 2s\alpha_2\big)\\ &= (s+t)(\alpha_1+\alpha_2) + 2s(-\alpha_1) = (s+t)\alpha_1 + (t-s)\alpha_2,\end{aligned}$$

hence

$$g^{\tau_2\lambda} = u^{s+t}y^{\frac{t-s}{2}}u^{-\frac{t-s}{2}} = u^{\frac{3s+t}{2}}y^{\frac{t-s}{2}};$$

$$\begin{aligned}\tau_2\circ\tau_1\lambda &= \tau_2\circ\tau_1\big((s+t)\alpha_1 + 2s\alpha_2\big)\\ &= (s+t)[-(\alpha_1+\alpha_2)] + 2s\alpha_1 = (s-t)\alpha_1 - (s+t)\alpha_2,\end{aligned}$$

hence

$$g^{\tau_2\circ\tau_1\lambda} = u^{s-t}y^{-\frac{s+t}{2}}u^{\frac{s+t}{2}} = u^{\frac{3s-t}{2}}y^{-\frac{s+t}{2}};$$

$$\begin{aligned}\tau_1\circ\tau_2\lambda &= \tau_1\circ\tau_2\big((s+t)\alpha_1 + 2s\alpha_2\big)\\ &= (s+t)\alpha_2 + 2s[-(\alpha_1+\alpha_2)] = -2s\alpha_1 + (-s+t)\alpha_2,\end{aligned}$$

hence

$$g^{\tau_1\circ\tau_2\lambda} = u^{-2s}y^{\frac{-s+t}{2}}u^{\frac{s-t}{2}} = u^{\frac{-3s-t}{2}}y^{\frac{-s+t}{2}};$$

$$\begin{aligned}\tau_1\circ\tau_2\circ\tau_1\lambda &= \tau_1\circ\tau_2\circ\tau_1\big((s+t)\alpha_1 + 2s\alpha_2\big)\\ &= (s+t)(-\alpha_2) + 2s(-\alpha_1) = -2s\alpha_1 - (s+t)\alpha_2,\end{aligned}$$

a change from the previous one via $s \mapsto s, t \mapsto -t$, hence

$$g^{\tau_1\circ\tau_2\circ\tau_1\lambda} = u^{\frac{-3s+t}{2}} y^{\frac{-s-t}{2}}.$$

In particular, with

$$\rho_0 := \frac{1}{2}\sum_{\alpha>0}\alpha = \frac{1}{2}\Big(\alpha_1 + \alpha_2 + (\alpha_1 + \alpha_2)\Big) = \alpha_1 + \alpha_2,$$

we have by taking $s = t = \frac{1}{2}$ in the above discussion,

$$g^{\rho_0} = u^{\frac{1}{2}}y^{\frac{1}{2}}, \quad g^{\tau_1\rho_0} = u^{-\frac{1}{2}}y^{\frac{1}{2}}, \quad g^{\tau_2\rho_0} = u^1y^0,$$

$$g^{\tau_2\circ\tau_1\rho_0} = u^{\frac{1}{2}}y^{-\frac{1}{2}},\ g^{\tau_1\circ\tau_2\rho_0} = u^{-1}y^0,\ g^{\tau_1\circ\tau_2\circ\tau_1\rho_0} = u^{-\frac{1}{2}}y^{-\frac{1}{2}},$$

while $g^{w\rho_0-\rho_0}$ is given by

$$g^{\rho_0-\rho_0} = u^0y^0, \quad g^{\tau_1\rho_0-\rho_0} = u^1y^0, \quad g^{\tau_2\rho_0-\rho_0} = u^{-\frac{1}{2}}y^{\frac{1}{2}},$$

$$g^{\tau_2\circ\tau_1\rho_0-\rho_0} = u^0y^1,\ g^{\tau_1\circ\tau_2\rho_0-\rho_0} = u^{\frac{3}{2}}y^{\frac{1}{2}},\ g^{\tau_1\circ\tau_2\circ\tau_1\rho_0-\rho_0} = u^1y^1.$$

9.4.2 Intertwining Operator

By the Gindikin-Karpelerich formua, the action of the intertwining operator $M(w,\lambda)$ on constant function **1** is given by

$$M(w,\lambda) = \prod_{\alpha>0, w\alpha<0} \frac{\xi_F(\langle\lambda,\alpha^\vee\rangle)}{\xi_F(\langle\lambda,\alpha^\vee\rangle + 1)}.$$

So we need to calculate the corresponding product.

First recall the following table

α	α_1	α_2	$\alpha_1+\alpha_2$
$\tau_1\alpha$	$-\alpha_1$	$\alpha_1+\alpha_2$	α_2
$\tau_2\alpha$	$\alpha_1+\alpha_2$	$-\alpha_2$	α_1
$\tau_2\circ\tau_1\alpha$	$-(\alpha_1+\alpha_2)$	α_1	$-\alpha_2$
$\tau_1\circ\tau_2\alpha$	α_2	$-(\alpha_1+\alpha_2)$	$-\alpha_1$
$\tau_1\circ\tau_2\circ\tau_1\alpha$	$-\alpha_2$	$-\alpha_1$	$-(\alpha_1+\alpha_2)$

from which we see that the following cases must be taken into the product appeared in the above formula:

$$\begin{array}{cccc}
\alpha & * & * & * \\
\tau_1\alpha & -\alpha_1 & * & * \\
\tau_2\alpha & * & -\alpha_2 & * \\
\tau_2\circ\tau_1\alpha & -(\alpha_1+\alpha_2) & * & -\alpha_2 \\
\tau_1\circ\tau_2\alpha & * & -(\alpha_1+\alpha_2) & -\alpha_1 \\
\tau_1\circ\tau_2\circ\tau_1\alpha & -\alpha_2 & -\alpha_1 & -(\alpha_1+\alpha_2).
\end{array}$$

Consequently,

$$\begin{aligned}
M(\mathrm{Id},\lambda) &= 1;\\
M(\tau_1,\lambda) &= \frac{\xi_F(\langle\lambda,\alpha_1\rangle)}{\xi_F(\langle\lambda,\alpha_1\rangle+1)} = \frac{\xi_F(2t)}{\xi_F(2t+1)} = c_F(t+\frac{1}{2});\\
M(\tau_2,\lambda) &= \frac{\xi_F(\langle\lambda,\alpha_2\rangle)}{\xi_F(\langle\lambda,\alpha_2\rangle+1)} = \frac{\xi_F(3s-t)}{\xi_F(3s-t+1)} = c_F(\frac{3s-t}{2}+\frac{1}{2});\\
M(\tau_2\circ\tau_1,\lambda) &= \frac{\xi_F(\langle\lambda,\alpha_1\rangle)}{\xi_F(\langle\lambda,\alpha_1\rangle+1)}\cdot\frac{\xi_F(\langle\lambda,\alpha_1+\alpha_2\rangle)}{\xi_F(\langle\lambda,\alpha_1+\alpha_2\rangle+1)}\\
&= \frac{\xi_F(2t)}{\xi_F(2t+1)}\cdot\frac{\xi_F(3s+t)}{\xi_F(3s+t+1)}\\
&= c_F(t+\frac{1}{2})\cdot c_F(\frac{3s+t}{2}+\frac{1}{2});\\
M(\tau_1\circ\tau_2,\lambda) &= \frac{\xi_F(\langle\lambda,\alpha_2\rangle)}{\xi_F(\langle\lambda,\alpha_2\rangle+1)}\cdot\frac{\xi_F(\langle\lambda,\alpha_1+\alpha_2\rangle)}{\xi_F(\langle\lambda,\alpha_1+\alpha_2\rangle+1)}\\
&= \frac{\xi_F(3s-t)}{\xi_F(3s-t+1)}\cdot\frac{\xi_F(3s+t)}{\xi_F(3s+t+1)}\\
&= c_F(\frac{3s-t}{2}+\frac{1}{2})\cdot c_F(\frac{3s+t}{2}+\frac{1}{2});\\
M(\tau_1\circ\tau_2\circ\tau_1,\lambda) &= \frac{\xi_F(\langle\lambda,\alpha_1\rangle)}{\xi_F(\langle\lambda,\alpha_1\rangle+1)}\cdot\frac{\xi_F(\langle\lambda,\alpha_2\rangle)}{\xi_F(\langle\lambda,\alpha_2\rangle+1)}\\
&\qquad\cdot\frac{\xi_F(\langle\lambda,\alpha_1+\alpha_2\rangle)}{\xi_F(\langle\lambda,\alpha_1+\alpha_2\rangle+1)}\\
&= \frac{\xi_F(2t)}{\xi_F(2t+1)}\cdot\frac{\xi_F(3s-t)}{\xi_F(3s-t+1)}\cdot\frac{\xi_F(3s+t)}{\xi_F(3s+t+1)}\\
&= c_F(t+\frac{1}{2})\cdot c_F(\frac{3s-t}{2}+\frac{1}{2})\cdot c_F(\frac{3s+t}{2}+\frac{1}{2}).
\end{aligned}$$

since $\langle \lambda, \alpha_1 \rangle = 2(s+t) - 2s = 2t$, $\langle \lambda, \alpha_2 \rangle = -(s+t) + 4s = 3s - t$, $\langle \lambda, \alpha_1 + \alpha_2 \rangle = 3s + t$. Here $c_F(t) := \frac{\xi_F(2t-1)}{\xi_F(2t)}$ used in [Ve].

Remark. If we make the variable changes $t \mapsto t + \frac{1}{2}$, $s \mapsto s + \frac{1}{2}$ from Venkov to our current situation, we see that

$$\frac{3s-t}{2} \mapsto \frac{3s-t+\frac{3}{2}-\frac{1}{2}}{2} = \frac{3s-t}{2} + \frac{1}{2},$$
$$\frac{3s+t-1}{2} \mapsto \frac{3s+t-1+\frac{3}{2}+\frac{1}{2}}{2} = \frac{3s+t}{2} + \frac{1}{2}.$$

So our result coincides with that of Venkov [Ve].

9.4.3 Calculation of $\prod_{\alpha\in\Delta_0}\langle w\lambda - \rho_0, \alpha^\vee \rangle$

We also need to calculate $\prod_{\alpha\in\Delta_0}\langle w\lambda - \rho_0, \alpha^\vee \rangle$. That is,

$$\langle w\lambda - \rho_0, \alpha_1^\vee \rangle \cdot \langle w\lambda - \rho_0, \alpha_2^\vee \rangle$$

or the same

$$\langle w\lambda - \rho_0, \alpha_1 \rangle \cdot \langle w\lambda - \rho_0, \alpha_2 \rangle.$$

This is given as follows:

$$\begin{aligned}
&\langle \lambda - \rho_0, \alpha_1 \rangle \cdot \langle \lambda - \rho_0, \alpha_2 \rangle \\
&= \langle (s+t-1)\alpha_1 + (2s-1)\alpha_2, \alpha_1 \rangle \\
&\qquad\qquad \cdot \langle (s+t-1)\alpha_1 + (2s-1)\alpha_2, \alpha_2 \rangle \\
&= \big(2(s+t-1) - (2s-1)\big) \cdot \big(-(s+t-1) + 2(2s-1)\big) \\
&= (2t-1)(3s-t-1);
\end{aligned}$$

$$\begin{aligned}
&\langle \tau_1\lambda - \rho_0, \alpha_1 \rangle \cdot \langle \tau_1\lambda - \rho_0, \alpha_2 \rangle \\
&= \langle (s-t-1)\alpha_1 + (2s-1)\alpha_2, \alpha_1 \rangle \\
&\qquad\qquad \cdot \langle (s-t-1)\alpha_1 + (2s-1)\alpha_2, \alpha_2 \rangle \\
&= \big(2(s-t-1) - (2s-1)\big) \cdot \big(-(s-t-1) + 2(2s-1)\big) \\
&= (-2t-1)(3s+t-1);
\end{aligned}$$

$$
\begin{aligned}
&\langle \tau_2\lambda - \rho_0, \alpha_1\rangle \cdot \langle \tau_2\lambda - \rho_0, \alpha_2\rangle \\
&= \langle (s+t-1)\alpha_1 + (t-s-1)\alpha_2, \alpha_1\rangle \\
&\qquad\qquad\qquad\qquad \cdot \langle (s+t-1)\alpha_1 + (t-s-1)\alpha_2, \alpha_2\rangle \\
&= \big(2(s+t-1) - (t-s-1)\big) \cdot \big(-(s+t-1) + 2(t-s-1)\big) \\
&= (3s+t-1)(-3s+t-1);
\end{aligned}
$$

$$
\begin{aligned}
&\langle \tau_2 \circ \tau_1\lambda - \rho_0, \alpha_1\rangle \cdot \langle \tau_2 \circ \tau_1\lambda - \rho_0, \alpha_2\rangle \\
&= \langle (s-t-1)\alpha_1 - (s+t+1)\alpha_2, \alpha_1\rangle \\
&\qquad\qquad\qquad\qquad \cdot \langle (s-t-1)\alpha_1 + (s+t+1)\alpha_2, \alpha_2\rangle \\
&= \big(2(s-t-1) + (s+t-1)\big) \cdot \big(-(s-t-1) - 2(s+t+1)\big) \\
&= (3s-t-1)(-3s-t-1);
\end{aligned}
$$

$$
\begin{aligned}
&\langle \tau_1 \circ \tau_2\lambda - \rho_0, \alpha_1\rangle \cdot \langle \tau_1 \circ \tau_2\lambda - \rho_0, \alpha_2\rangle \\
&= \langle (-2s-1)\alpha_1 + (-s+t-1)\alpha_2, \alpha_1\rangle \\
&\qquad\qquad\qquad\qquad \cdot \langle (-2s-1)\alpha_1 + (-s+t-1)\alpha_2, \alpha_2\rangle \\
&= \big(2(-2s-1) - (-s+t-1)\big) \cdot \big((2s+1) + 2(-s+t-1)\big) \\
&= (-3s-t-1)(2t-1);
\end{aligned}
$$

$$
\begin{aligned}
&\langle \tau_1 \circ \tau_2 \circ \tau_1\lambda - \rho_0, \alpha_1\rangle \cdot \langle \tau_1 \circ \tau_2 \circ \tau_1\lambda - \rho_0, \alpha_2\rangle \\
&= \langle (-2s-1)\alpha_1 - (s+t+1)\alpha_2, \alpha_1\rangle \\
&\qquad\qquad\qquad\qquad \cdot \langle (-2s-1)\alpha_1 - (s+t+1)\alpha_2, \alpha_2\rangle \\
&= \big(2(-2s-1) + (s+t+1)\big) \cdot \big((2s+1) - 2(s+t+1)\big) \\
&= (-3s+t-1)(-2t-1).
\end{aligned}
$$

9.5 Rank Three Zeta

By putting all the above calculation together, we have then obtain the following

$$
\begin{aligned}
&\int_{\Lambda(G(\mathbb{Q})\backslash G(\mathbb{A}))} \Lambda^{(u,y)} E(\mathbf{1},\lambda)(g)\, d\mu(g)\\
&=3\cdot \int_{A_0(\mathbb{Q})\backslash A_0(\mathbb{A})^1} da \cdot \Big[\frac{u^{t-\frac{1}{2}}y^{s-\frac{1}{2}}}{(2t-1)(3s-t-1)}\cdot 1\\
&\quad -\frac{u^{-t-\frac{1}{2}}y^{s-\frac{1}{2}}}{(2t+1)(3s+t-1)}\cdot\frac{\xi(2t)}{\xi(2t+1)}\\
&\quad -\frac{u^{\frac{3s+t}{2}-\frac{1}{2}}y^{\frac{t-s}{2}s-\frac{1}{2}}}{(3s+t-1)(3s-t+1)}\cdot\frac{\xi(3s-t)}{\xi(3s-t+1)}\\
&\quad -\frac{u^{\frac{3s-t}{2}-\frac{1}{2}}y^{-\frac{s+t}{2}-\frac{1}{2}}}{(3s-t-1)(3s+t+1)}\cdot\frac{\xi(2t)}{\xi(2t+1)}\cdot\frac{\xi(3s+t)}{\xi(3s+t+1)}\\
&\quad -\frac{u^{\frac{-3s-t}{2}-\frac{1}{2}}y^{\frac{-s+t}{2}-\frac{1}{2}}}{(3s+t+1)(2t-1)}\cdot\frac{\xi(3s-t)}{\xi(3s-t+1)}\cdot\frac{\xi(3s+t)}{\xi(3s+t+1)}\\
&\quad +\frac{u^{\frac{-3s+t}{2}-\frac{1}{2}}y^{\frac{-s-t}{2}-\frac{1}{2}}}{(3s-t+1)(2t+1)}\cdot\frac{\xi(2t)}{\xi(2t+1)}\cdot\frac{\xi(3s-t)}{\xi(3s-t+1)}\cdot\frac{\xi(3s+t)}{\xi(3s+t+1)}\Big].
\end{aligned}
$$

Thus in particular, by taking $u=y=1$, we get

$$
\begin{aligned}
&\mathrm{Res}_{3s-t=1}\int_{\Lambda(G(F)\backslash G(\mathbb{A}))} \Lambda^{(u=1,y=1)} E(\mathbf{1},\lambda)(g)\, d\mu(g)\\
&=\mathrm{Res}_{3s-t=1}\Big[\frac{1}{(2t-1)(3s-t-1)}\Big]\\
&\quad -\mathrm{Res}_{3s-t=1}\Big[\frac{1}{(3s+t-1)(3s-t+1)}\cdot\frac{\xi(3s-t)}{\xi(3s-t+1)}\Big]\\
&\quad -\mathrm{Res}_{3s-t=1}\Big[\frac{1}{(3s-t-1)(3s+t+1)}\cdot\frac{\xi(2t)}{\xi(2t+1)}\cdot\frac{\xi(3s+t)}{\xi(3s+t+1)}\Big]\\
&\quad -\mathrm{Res}_{3s-t=1}\Big[\frac{1}{(3s+t+1)(2t-1)}\cdot\frac{\xi(3s-t)}{\xi(3s-t+1)}\cdot\frac{\xi(3s+t)}{\xi(3s+t+1)}\Big]\\
&\quad +\mathrm{Res}_{3s-t=1}\Big[\frac{1}{(3s-t+1)(2t+1)}\\
&\qquad\qquad \cdot\frac{\xi(2t)}{\xi(2t+1)}\cdot\frac{\xi(3s-t)}{\xi(3s-t+1)}\cdot\frac{\xi(3s+t)}{\xi(3s+t+1)}\Big].
\end{aligned}
$$

Moreover, note that then $t = 3s - 1$ and $2t - 1 = 6s - 3$, $3s + t - 1 = 6s - 2$, $3s - t + 1 = 2$, $3s + t + 1 = 6s$, $2t + 1 = 6s - 1$ (surely we list these trivial relations for the reason to get the next lines straightforward) we conclude that

$$\begin{aligned}
&\mathrm{Res}_{3s-t=1} \int_{\Lambda(G(F)\backslash G(\mathbb{A}))} \Lambda^{(u=1,y=1)} E(\mathbf{1}, \lambda)(g)\, d\mu(g) \\
&= \frac{1}{6s-3} - \frac{1}{6s-2} \cdot \frac{1}{2} \cdot \frac{1}{\zeta(2)} - \frac{1}{6s} \cdot \frac{\xi(6s-2)}{\xi(6s-1)} \cdot \frac{\xi(6s-1)}{\xi(6s)} \\
&\quad - \frac{1}{6s(6s-3)} \cdot \frac{1}{\xi(2)} \cdot \frac{\xi(6s-1)}{\xi(6s)} \\
&\quad + \frac{1}{2\cdot(6s-1)} \cdot \frac{\xi(6s-2)}{\xi(6s-1)} \cdot \frac{1}{\xi(2)} \cdot \frac{\xi(6s-1)}{\xi(6s)} \\
&= \frac{1}{(6s-3)} - \frac{1}{2(6s-2)} \cdot \frac{1}{\zeta(2)} - \frac{1}{6s} \cdot \frac{\xi(6s-2)}{\xi(6s)} \\
&\quad - \frac{1}{6s(6s-3)} \cdot \frac{1}{\xi(2)} \cdot \frac{\xi(6s-1)}{\xi(6s)} \\
&\quad + \frac{1}{2\cdot(6s-1)} \cdot \frac{1}{\xi(2)} \cdot \frac{\xi(6s-2)}{\xi(6s)}.
\end{aligned}$$

This, by Remark 9.2.2, then completes the proof of the following

Theorem 4. *The rank 3 zeta function for the number field $F = \mathbb{Q}$ is given by*

$$\begin{aligned}
\xi_{F,3}(s) = &\xi_F(2) \cdot \frac{1}{3s-3} \cdot \xi_F(3s) - \xi(2) \cdot \frac{1}{3s} \cdot \xi_F(3s-2) \\
&+ \frac{1}{3} \cdot \frac{1}{3s-3} \cdot \xi_F(3s-1) - \frac{1}{3} \cdot \frac{1}{3s} \cdot \xi_F(3s-1) \\
&+ \frac{1}{2} \cdot \frac{1}{3s-1} \cdot \xi_F(3s-2) - \frac{1}{2} \cdot \frac{1}{3s-2} \cdot \xi_F(3s).
\end{aligned}$$

Moreover, we also have the following recent result of M. Suzuki.

Theorem 5 (Suzuki[Su]). *All zeros of $\xi_{\mathbb{Q},3}(s)$ lie on the central line* $\mathrm{Re}\, s = \frac{1}{2}$.

We end this paper with the following generalization of the Riemann Hypothesis:

The RH. *All zeros of zetas, rank one or higher, lie on the central line* $\mathrm{Re}\, s = \frac{1}{2}$.

Acknowledgments: I would like to thank Arthur, Deninger, Elstrodt, Fesenko, Freitag, Hida, Henry Kim, Lagarias, I. Nakamura, Takayuki Oda, Takhtajan, Ueno, Zagier and Zink for their interests; Special thanks also due to M. Suzuki for sharing with me his discovery that rank two and rank three zetas for the field of rationals satisfy the Riemann Hypothesis.

I would like to dedicate this paper to A. Fujiki for his 60 years birthday.

This project is partially supported by JSPS.

REFERENCES

[Ar1] J. Arthur, Eisenstein series and the trace formula, Proc. Sympos. Pure Math., XXXIII, 253-274, AMS, Providence, R.I., 1979.

[Ar2] J. Arthur, A trace formula for reductive groups. I. Terms associated to classes in $G(\mathbb{Q})$. Duke Math. J. **45** (1978), no. 4, 911–952.

[Ar3] J. Arthur, A trace formula for reductive groups. II. Applications of a truncation operator. Compositio Math. **40** (1980), no. 1, 87–121.

[Ar4] J. Arthur, A measure on the unipotent variety, Canad. J. Math **37**, (1985) pp. 1237–1274.

[Ar5] J. Arthur, The trace formula in invariant form, Ann. of Math., 114 (1981), 1-74.

[Bo1] A. Borel, Some finiteness properties of adele groups over number fields, Publ. Math., IHES, **16** (1963) 5-30.

[Bo2] A. Borel, *Introduction aux groupes arithmetictiques*, Hermann, 1969.

[Bor] A. Borisov, Convolution structures and arithmetic cohomology, Compositio Math. 136 (2003), no. 3, 237–254.

[Ca] W. Casselman, *Introduction to the theory of admissible representations of p-adic reductive groups*, unpublished notes.

[Co] A. Connes, Trace formula in noncommutative geometry and the zeros of the Riemann zeta function. Selecta Math. (N.S.) 5 (1999), no. 1, 29–106.

[De1] C. Deninger, On the Γ-factors attached to motives. Invent. Math. **104** (1991), no. 2, 245–261.

[De2] C. Deninger, Local L-factors of motives and regularized determinants. Invent. Math. **107** (1992), no. 1, 135–150.

[De3] C. Deninger, Motivic L-functions and regularized determinants, Motives (Seattle, WA, 1991), 707-743, Proc. Sympos. Pure Math, **55** Part 1, AMS, Providence, RI, 1994.

[Fo] G.B. Folland, *A course in abstract harmonic analysis*, Studies in advanced mathematics, CRC Press, 1995.

[FT] A. Fröhlich & M.J. Taylor, *Algebraic Number Theory*, Cambridge studies in advanced mathematics **27**, Cambridge Univ. Press (1991).

[GS] G. van der Geer & R. Schoof, Effectivity of Arakelov Divisors and the Theta Divisor of a Number Field, Sel. Math., New ser. **6** (2000), 377-398.

[Gr1] D.R. Grayson, Reduction theory using semistability. Comment. Math. Helv. **59** (1984), no. 4, 600–634.

[Gr2] D.R. Grayson, Reduction theory using semistability. II. Comment. Math. Helv. **61** (1986), no. 4, 661–676.

[HN] G. Harder & M.S. Narasimhan, On the cohomology groups of moduli spaces of vector bundles over curves, Math. Ann. **212**, (1975) 215-248.

[Ha] T. Hayashi, Computation of Weng's rank 2 zeta function over an algebraic number field, 2006, submitted.

[Ho] W. Hoffmann, Geometric estimates for the trace formula, preprint, 2003.

[Hu] J. Humphreys, *Introduction to Lie algebras and representations*, Springer-Verlag, 1972.

[Iw1] K. Iwasawa, *Algebraic theory of algebraic functions*, Lecture Notes at Princeton Univ. noted by T. Kimura, 1975.

[Iw2] K. Iwasawa, Letter to Dieudonné, April 8, 1952, in N.Kurokawa and T. Sunuda: *Zeta Functions in Geometry*, Advanced Studies in Pure Math. **21** (1992), 445-450.

[JLR] H. Jacquet, E. Lapid & J. Rogawski, Periods of automorphic forms. J. Amer. Math. Soc. 12 (1999), no. 1, 173–240.

[Ki] H. Ki, All but finitely many non-trivial zeros of the approximations of the Epstein zeta function are simple and on the critical line. Proc. London Math. Soc. (3) 90 (2005), no. 2, 321–344.

[KW] H.H. Kim & L. Weng, Volume of truncated fundamental domains, Proc. AMS, to appear.

[Laf] L. Lafforgue, *Chtoucas de Drinfeld et conjecture de Ramanujan-*

Petersson. Asterisque No. 243 (1997).

[LS] J. Lagarias & M. Suzuki, The Riemann Hypothesis for certain integrals of Eisenstein series, Journal of Number Theory, 118(2006) 98-122.

[L1] S. Lang, *Algebraic Number Theory*, Springer-Verlag, 1986.

[L2] S. Lang, *Fundamentals on Diophantine Geometry*, Springer-Verlag, 1983.

[L3] S. Lang, *Introduction to Arakelov theory*, Springer Verlag, 1988.

[La1] R. Langlands, *On the functional equations satisfied by Eisenstein series*, Springer LNM **544**, 1976.

[La2] R. Langlands, the volume of the fundamental domain for some arithmetical subgroups of Chevalley groups, in *Algebraic Groups and Discontinuous Subgroups,* Proc. Sympos. Pure Math. 9, AMS (1966) pp.143–148.

[La3] R. Langlands, *Euler products*, Yale Math. Monograph, Yale Univ. Press, 1971.

[Li] X. Li, A note on the Riemann-Roch theorem for function fields. *Analytic number theory*, Vol. **2**, 567–570, Progr. Math., **139**, 1996.

[MS] V.B. Mehta & C.S. Seshadri, Moduli of vector bundles on curves with parabolic structures. Math. Ann. 248 (1980), no. 3, 205–239.

[Mi] H. Minkowski, *Geometrie der Zahlen*, Leipzig and Berlin, 1896.

[MW] C. Moeglin & J.-L. Waldspurger, *Spectral decomposition and Eisenstein series.* Cambridge Tracts in Math, **113**. Cambridge University Press, 1995.

[Mo] C. Moreno, *Algebraic curves over finite fields.* Cambridge Tracts in Mathematics, **97**, Cambridge University Press, 1991.

[M] D. Mumford, *Geometric Invariant Theory*, Springer-Verlag, (1965).

[NR] M.S. Narasimhan & S. Ramanan, Moduli of vector bundles on a compact Riemann surfaces, Ann. of Math. **89** 14-51 (1969).

[NS] M.S. Narasimhan & C.S. Seshadri, Stable and unitary vector bundles on a compact Riemann surface. Ann. of Math. (2) 82 1965.

[Ne] J. Neukirch, *Algebraic Number Theory*, Grundlehren der Math. Wissenschaften, Vol. **322**, Springer-Verlag, 1999.

[Pa] A.N. Parshin, On the arithmetic of two-dimensional schemes. I. Distributions and residues. (Russian) Izv. Akad. Nauk SSSR Ser. Mat. 40 (1976), no. 4, 736–773, 949.

[RR] S. Ramanan & A. Ramanathan, Some remarks on the instability flag. Tohoku Math. J. (2) 36 (1984), no. 2, 269–291.

[RZ] M. Rapoport & Th. Zink, *Period spaces for p-divisible groups*. Annals of Mathematics Studies, 141. Princeton University Press, Princeton, NJ, 1996.

[S] J.-P. Serre, *Algebraic Groups and Class Fields*, GTM 117, Springer (1988).

[Se1] C.S. Seshadri, Space of unitary vector bundles on a compact Riemann surface. Ann. of Math. (2) 85 (1967) 253–336.

[Se2] C. S. Seshadri, *Fibrés vectoriels sur les courbes algébriques*, Asterisque **96**, 1982.

[Sie] C.L. Siegel, *Lectures on the geometry of numbers*, notes by B. Friedman, rewritten by K. Chandrasekharan with the assistance of R. Suter, Springer-Verlag, 1989.

[St1] U. Stuhler, Eine Bemerkung zur Reduktionstheorie quadratischer Formen, Arch. Math. (Basel) **27** (1976), no. 6, 604–610.

[St2] U. Stuhler, Zur Reduktionstheorie der positiven quadratischen Formen. II, Arch. Math. (Basel) **28** (1977), no. 6, 611–619.

[Su] M. Suzuki, A proof of the Riemann Hypothesis for for the Weng zeta function of rank 3 for the rationals, this volume.

[T] J. Tate, Fourier analysis in number fields and Hecke's zeta functions, Thesis, Princeton University, 1950.

[Te] A. Terras, *Harmonic analysis on symmetric spaces and applications II*, Springer-Verlag, 1988.

[Ve] A.B. Venkov, On the trace formula for $SL(3,\mathbb{Z})$, J Soviet Math., 12 (1979), 384-424.

[VT] A.I. Vinogradov, & L. Takhtajan, Theory of Eisenstein series for the group $SL(3, \mathbb{R})$ and its application to a binary problem J. Soviet Math., 18 (1982).

[W1] A. Weil, Généralisation des fonctions abéliennes, J. Math Pures et Appl, **17**, (1938) 47-87.

[W2] A. Weil, *Basic Number Theory*, Springer-Verlag, 1973.

[W3] A. Weil, *Adeles and algebraic groups*, with appendices by M. Demazure and Takashi Ono. Progress in Math., 23. 1982.

[We1] L. Weng, Geometric Arithmetic: A Program, *Arithmetic Geometry and Number Theory*, 211-390, World Sci. 2006.

[We2] L. Weng, Stability and New Non-Abelian Zeta Functions, *Number theoretic methods* (Iizuka, 2001), 405–419, Dev. Math., 8, Kluwer Acad. Publ., 2002.

[We3] L. Weng, Refined Brill-Noether Locus and Non-Abelian Zeta Functions for Elliptic Curves, *Algebraic geometry in East Asia* (Kyoto, 2001), 245–262, World Sci. 2002.

[We4] L. Weng, Non-abelian zeta functions for function fields, Amer. J. Math. 127 (2005), no. 5, 973–1017.

[We5] L. Weng, Non-abelian class field theory for Riemann surfaces, Appendix to [We1], *Arithmetic Geometry and Number Theory*, 391-399, World Sci. 2006.

[We6] L. Weng, Rank Two Non-Abelian Zeta and its Zeros, available at http://xxx.lanl.gov/abs/math.NT/0412009.

[We7] L. Weng, Automorphic Forms, Eisenstein Series and Spectral Decompositions, *Arithmetic Geometry and Number Theory*, 123-210, World Sci. 2006.

[We8] L. Weng, A Rank Two Zeta and Its Zeros, J. of Ramanujan Math. Soc. **21**, No. 3 (2006), 205-266.

[Z] Zagier, D. The Rankin-Selberg method for automorphic functions which are not of rapid decay. J. Fac. Sci. Univ. Tokyo Sect. IA Math. 28(3), 415–437 (1982).

Lin WENG
Institute for Fundamental Research,
The *L*-Academy
and
Graduate School of Mathematics,
Kyushu University,
Fukuoka 812-8581, Japan
Email: weng@math.kyushu-u.ac.jp